Lecture Notes in Computer Science　　9272

Commenced Publication in 1973
Founding and Former Series Editors:
Gerhard Goos, Juris Hartmanis, and Jan van Leeuwen

More information about this series at http://www.springer.com/series/7408

Marta Beltrán · William Knottenbelt
Jeremy Bradley (Eds.)

Computer Performance Engineering

12th European Workshop, EPEW 2015
Madrid, Spain, August 31 – September 1, 2015
Proceedings

 Springer

Editors
Marta Beltrán
Universidad Rey Juan Carlos
Mostoles (Madrid)
Spain

Jeremy Bradley
Imperial College London
London
UK

William Knottenbelt
Imperial College London
London
UK

ISSN 0302-9743   ISSN 1611-3349   (electronic)
Lecture Notes in Computer Science
ISBN 978-3-319-23266-9   ISBN 978-3-319-23267-6   (eBook)
DOI 10.1007/978-3-319-23267-6

Library of Congress Control Number: 2015946767

LNCS Sublibrary: SL2 – Programming and Software Engineering

Springer Cham Heidelberg New York Dordrecht London

Printed on acid-free paper

Springer International Publishing AG Switzerland is part of Springer Science+Business Media
(www.springer.com)

Preface

This volume of LNCS contains the proceedings of the 12th European Performance Engineering Workshop, held in Madrid, Spain, from August 31 to September 1, 2015. Following on from the positive experience of Florence 2014, EPEW was once again co-located with a number of related conferences (including QEST and FORMATS) in an experience known as Madrid Meet 2015, and we hope this format provided researchers with the opportunity to engage with a broader range of topics and people than would usually have been possible.

As with previous EPEW workshops, the present event was supported by submissions from all over the world, including Asia, Africa, North America, and Europe. There were 39 submissions of which 19 were selected for publication in the proceedings and presentation at the workshop. We would like to commend the diligent efforts of all those Program Committee members who returned their reviews on time – despite the tight reviewing timetable – and all those who participated enthusiastically in the ensuing discussions.

The papers themselves maintain the tradition of diversity and quality for which the European Performance Engineering Workshop is known. This year several papers featured application studies, while others focused on novel techniques for modelling, simulation and software performance. The domains of the application studies have a diverse and contemporary feel to them, ranging from mobile networks and cloud computing environments to railway control and medical imaging systems. Modelling techniques include fresh perspectives on layered queueing network models and processor sharing systems, innovative applications of model checking, and a novel extension of the Markov Decision Petri Net formalism. Simulation-related advances include improvements to rare-event and stochastic simulations. In the software performance domain, new ways to assess the on-the-fly impact of software code changes are revealed.

As program chairs, we would like to thank everyone involved in making EPEW 2015 a success: Springer for their continued support of the workshop series, the invited speakers (Eva Kalyvianaki and Leandro Soares Indrusiak), the Program Committee and reviewers, and of course the authors of the papers submitted, without whom there could not be a workshop. We would also like to thank Jeremy Bradley, who worked extremely hard as publications chair to bring together the LNCS volume, and Isaac Martin de Díego, who did a great job as publicity chair.

We hope that you, the reader, find the papers in this volume interesting, useful, and inspiring, and we hope to see you at future European Performance Engineering Workshops.

July 2015

Marta Beltrán
William Knottenbelt

Organization

EPEW Co-chairs

Marta Beltrán
William Knottenbelt

Publicity Chair

Isaac Martin de Díego

Proceedings Editor

Jeremy Bradley

Program Committee

Simonetta Balsamo	Università Ca' Foscari di Venezia, Italy
Marco Bernardo	University of Urbino, Italy
Tadeusz Czachórski	Polish Academy of Sciences, Poland
Dieter Fiems	Ghent University, Belgium
Jean-Michel Fourneau	Université de Versailles Saint-Quentin-en-Yvelines, France
José Daniel Garcia	Universidad Carlos III de Madrid, Spain
Stephen Gilmore	University of Edinburgh, UK
András Horváth	Università degli Studi di Torino, Italy
Mauro Iacono	Seconda Università degli Studi di Napoli, Italy
Carlos Juiz	Universitat de les Illes Balears, Spain
Helen Karatza	Aristotle University of Thessaloniki, Greece
Leïla Kloul	Université de Versailles Saint-Quentin-en-Yvelines, France
Samuel Kounev	Universität Würzburg, Germany
Fumio Machida	NEC, Japan
Andrea Marin	Università Ca' Foscari di Venezia, Italy
Marco Paolieri	Università di Firenze, Italy
Roberto Pietrantuono	Università degli Studi di Napoli Federico II, Italy
Philipp Reinecke	Freie Universität Berlin, Germany
Anne Remke	University of Twente, Netherlands
Jacques Resing	Eindhoven University of Technology, Netherlands
Sabina Rossi	Università Ca' Foscari di Venezia, Italy
Markus Siegle	Universität der Bundeswehr München, Germany

Yutaka Takahashi	Kyoto University, Japan
Miklós Telek	Budapest University of Technology and Economics, Hungary
Nigel Thomas	Newcastle University, UK
Max Tschaikowski	University of Southampton, UK
Petr Tůma	Charles University, Czech Republic
Benny Van Houdt	University of Antwerp, Belgium
Maria Grazia Vigliotti	Imperial College London, UK
Jean-Marc Vincent	Laboratoire d'Informatique de Grenoble, France
Katinka Wolter	Freie Universität Berlin, Germany

Additional Reviewers

The chairs would like to thank the reviewers who kindly contributed additional reviews.

Alessio Angius
Shashank Bijwe
Laura Carnevali
Justyna Chromik
Sonja Filiposka
Masaya Fujiwaka
Hamed Ghasemieh
Carlos Guerrero
Vincenzo Gulisano

Boudewijn Haverkort
Nikolas Herbst
Isaac Lera
Mieke Massink
András Mészáros
Simon Spinner
Jóakim von Kistowski
Jürgen Walter

Contents

Applications I

Applicational

A Markovian Queueing System for Modeling a Smart Green Base Station

Ioannis Dimitriou[1], Sara Alouf[2], and Alain Jean-Marie[2,3](✉)

[1] Department of Mathematics, University of Patras, Patras, Greece
idimit@math.upatras.gr
[2] Inria, Sophia Antipolis, France
{Sara.Alouf,Alain.Jean-Marie}@inria.fr
[3] LIRMM, Montpellier, France

Abstract. We investigate a model to assess the performance of a base station (BS) fully powered by renewable energy sources. The BS is modeled as a three-queue system where two of them are coupled. One represents accumulated energy, the second is the data queue and the third one serves as a reserve energy queue. This smart BS is able to dynamically adjust its coverage area (thereby controlling the traffic intensity) and to generate signals to the reserve energy queue that trigger the movement of energy units to the main energy buffer. Given the randomness of renewable energy supply and the internal traffic intensity control, our queueing model is operated in a finite state random environment. Using the matrix analytic formalism we construct a five-dimensional Markovian model to study the performance of the BS. The stationary distribution of the system state is obtained and key performance metrics are calculated. A small numerical example illustrates the model and a simplified product-form approximation is proposed.

Keywords: Coupled queues · QBD processes · Green base station

1 Introduction

The architectural design of cellular networks has evolved in recent years to better satisfy users needs. The traffic load generated by users exhibits a night-day pattern, having a peak of traffic during the day and almost no traffic during the night. A geographical pattern is also observed as offices areas witness a peak traffic during the day while residential areas witness a smaller peak late in the evening. Heterogeneous cellular networks are an attractive deployment solution: large powerful base stations (BSs) are used to ensure coverage and connectivity whereas smaller coverage-limited BSs are used to accommodate the peak load where needed. Those BSs can be analyzed in isolation.

In this paper, we consider a single small BS and study the question of powering it using solely renewable energy. A key factor is that the solar radiation exhibits a night-day pattern that makes solar panels fit to power a small BS. We assume that the BS is "smart" in the sense that it is able to dynamically

© Springer International Publishing Switzerland 2015
M. Beltrán et al. (Eds.): EPEW 2015, LNCS 9272, pp. 3–18, 2015.
DOI: 10.1007/978-3-319-23267-6_1

adjust its coverage area, controlling thereby the number of mobiles with which it communicates, and consequently its offered traffic rate and its energy consumption. The harvested energy is stored in batteries that are used to power the BS. We propose to model this BS as a queueing system that operates in a finite-state Markovian random environment (RE). The behavior of such a system is described by a five-dimensional Markov process, which is a homogeneous finite Quasi Birth-Death (QBD) process.

The literature on QBD processes is abundant and these processes have been used to study many different applications. De Cuypere et al. [4], have studied sensor nodes [3] and kitting processes through a paired queueing system [15]. In [3], a finite energy queue is paired with an infinite data queue, where customers arrive at both queues according to Markovian Arrival Processes. In [4], both queues are finite, but due to the sparsity of the generator matrix of the underlying Markov process, the size of the state space does not cause serious issues. Closely to our work, Takahashi et al. [17] consider a synchronization queue consisting of two buffers with finite capacities, the arrival processes at both buffers are Poisson for one and phase type renewal for the other. In our work we assume that energy is discretized (as in e.g. [3, 8, 9]) but it is also possible to model batteries as fluid queues; see for instance [11, 12].

In the following, we present our model in Section 2 and detail the infinitesimal generator of the five-dimensional continuous-time Markov chain representing the state of the system in Section 3. Section 4 discusses several algorithms that could be used to compute the stationary distribution of the system state and relevant performance metrics, exploiting the QBD structure of the generator. The feasibility of the developed algorithm is demonstrated through a numerical example. First passage times to lower and higher levels in the QBD are discussed in Section 5. In Section 6, we propose an approximate model with product-form solution. We briefly conclude in Section 7.

2   The Model

We are interested in a wireless communications base station (BS), isolated from the electric grid and operating thanks to renewable energy sources. The model supposes a continuous functioning of the station: it does not handles issues like startup, shutdown, malfunction and other transitory phenomena.

The model we develop has a 5-dimensional state space, representing the state of three queues, a service process and an environment process. We describe now these elements and the way they interact. Figure 1 summarizes the model under study. Although we describe here a quite specific situation, we point out that all these elements and their composition can be generalized so as to build models of other device configurations.

Queues. We consider two energy queues (EQ) of finite capacity E_j, $j = 1, 2$ that store energy extracted from the environment, and a data queue (DQ) that keeps track of packets not yet transmitted, having also a buffer of finite capacity N. Energy is assumed to be discretized, as for instance in [8, 9]. Assume that EQ 1,

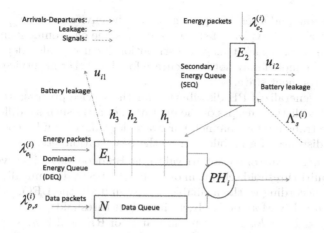

Fig. 1. The model when the RE is in state i.

is coupled with the DQ, whereas EQ 2 is deployed close to the BS and serves as a reserve. Each EQ, is fed by different renewable energy sources, and the choice of the one that will be coupled with the DQ depends on the area where the BS is deployed. We will refer to this EQ, say EQ 1, as the dominant energy queue (DEQ) and the EQ 2 as the secondary energy queue (SEQ).

Environment. The system behavior depends on the state of the RE. This RE is defined by means of an irreducible regular continuous-time Markov chain $Y(t)$ with state space $\{1, ..., M\}$, and infinitesimal generator Q_Y. The state of this environment may represent a variety of factors that make energy arrival or packet arrival processes, energy consumption and service times non time-stationary. For instance, the intermittent nature of wind, the variability of sunlight, can be taken into account with an environment variable. Hourly variations of data traffic can also be modeled that way. Other features can be the variation of transmission power due to global adjustments at the network level. As usual, independent environment features can be combined in a complex environment process with multiple "phases", at the cost of large values for M. Clearly, one can skip the RE and reduce the dimension of the QBD, by assuming that the model migrates quickly from the stationary situation of one environment setting to another.

Packet Service Duration. The service time distribution is of phase type [14,16] of order ν and depends on the state of the RE. For a fixed RE state i, this distribution can be interpreted as a time until some underlying Markov process η_t with finite state space $\{1, ..., \nu\}$ reaches the absorbing state 0 with initial probability vector $(\tau_0^{(i)}, \underline{\tau}^{(i)})$. Transition rates of the process η_t within the set $\{1, ..., \nu\}$ are defined by the sub-generator $T^{(i)}$ and transition rates into the absorbing state (which lead to a service completion) are given by the entities of the column vector $\underline{t}^{(i)} = -T^{(i)}\underline{1}$. Observe that the value ν does not depend on i. A typical situation allowed by this model is to have a packet size with some given distribution, and a service speed given by the environment: $T_{kl}^{(i)} = T_{kl} \times v_i$, v_i begin the "velocity" typical of state i and T_{kl} some environment-independent transition rate. If the packet length distribution itself can also depend on the environment, the model may not be accurate for the packet in service during the change of environment state.

By using a generalized PH distribution for the service process, we can incorporate several realistic concepts of the operation of a BS such as: different levels of noise in the transmission channel, hardware degradation and recovering, variations in the distance of a mobile user to the base station, etc.

Packet Arrival. The rate of packet arrival to the BS depends on its coverage area. We adopt a multi-threshold scheme in order for the BS to dynamically adjust the coverage area according to the available energy units in the DEQ: in this sense, we model a *smart* BS. More precisely, we introduce thresholds, say $h_0 = 0 < h_1 < h_2 < ... < h_K < E_1 = h_{K+1}$. Given the state i of RE, and if $h_s < m_1 \leq h_{s+1}$, $s = 0, ..., K$, the users' arrival rate equals $\lambda_{p,s}^{(i)}$. In practice, it is expected that $\lambda_{p,0}^{(i)} \leq \lambda_{p,1}^{(i)} \leq ... \leq \lambda_{p,K}^{(i)}$ but this is not needed for the definition of the model. Note that the packet traffic potentially includes system control packets, signaling (e.g. for handovers) etc. Superposition of different traffic sources can classically be taken care of.

Energy Arrival and Depletion. For a fixed state i, $i = 1, ..., M$ of the RE, energy units are stored in EQ j at rate $\lambda_{e_j}^{(i)}$. Transmission opportunities occurs only when both the DEQ and the DQ queue are non empty.

The modeling of energy consumption by a packet transmission requires some trick. Indeed, it is unlikely that a phase of the service of some packet will consume exactly an integer amount of energy quanta. Randomness in consumption and/or non-integral energy values can be taken into account using random variables.

The number of energy units required for the transmission of a single packet depends both on the state of the RE and on the phase of the service process. Thus, the transmission of a packet may be interrupted, if the completion of a phase of service consumes the available energy. In this paper, we assume that the transmission of this packet has to restart from scratch (but with a stochastically independent value), whenever there will be available energy. If some mechanism, e.g. error-correcting codes, allows for it, transmission could resume from the same phase. Alternately, the transmission of the packet could be simply canceled in case of energy shortage. Moreover, we can also allow this cancellation to occur only after some timeout, by introducing a timer for each "impatient" packet when energy is 0. None of these alternatives would make the transition matrix much more complex. With the "restart" point of view, given the state i of RE and if there are m_1 available energy units in the DEQ, the completion of phase x, $x = 1, 2, ..., \nu$ requires k energy units with probability $p_{kx}^{(i,m_1)}$, $\sum_{k=0}^{m_1} p_{kx}^{(i,m_1)} = 1$. We allow *a priori* $k = 0$, i.e. that the completion of a phase of service may not consume even a single energy unit. However, in order to initiate a transmission we need at least one energy unit.

Energy leakage from each EQ is unavoidable. More precisely, an energy unit will be lost from EQ j, $j = 1, 2$ at exponential rate u_{ij}, given that the RE is in state i. Note that if the DEQ drains during the transmission of a packet, the packet has to be retransmitted, whenever there will be available energy units.

Energy Transfers. The SEQ serves as backup storage for the DEQ. The control of our smart BS generates signals to the SEQ at a rate $\Lambda_s^{-(i)}$, that trigger the movement of energy units to the DEQ, according to the multi-threshold rule introduced above. Given the state i of the RE, and $h_s < m_1 \leq h_{s+1}$, $s = 0, ..., K$, signals are generated at a rate $\Lambda_s^{-(i)}$, and trigger [7] k energy units from the SEQ towards the DEQ with probability $q_{ks}^{(m_1,m_2,i)}$ ($\sum_{k=1}^{m_2} q_{ks}^{(m_1,m_2,i)} \delta_{m_1+k \leq E_1} = 1$). Define also $\underline{q}_{ks}^{(m_1,m_2)} = diag(q_{ks}^{(m_1,m_2,1)}, ..., q_{ks}^{(m_1,m_2,M)})$. In case the SEQ is empty, the signal will have no effect. As above, it is expected in practice that $\Lambda_0^{-(i)} \geq \Lambda_1^{-(i)} \geq ... \geq \Lambda_K^{-(i)}$: the demand rate for replenishment of DEQ becomes larger as the energy in the DEQ depletes. It is also possible to reflect the urgency of replenishment by modifying the distribution q_{ks}. Moreover, when the DEQ is full, i.e. $m_1 = E_1$, there will be no signal towards the SEQ (i.e., $\Lambda_K^{-(i)} = 0$). We also adopt an overflow operation in the sense that when an EQ buffer is full, its energy traffic is rerouted to the other EQ. Clearly, the signal generation possibly consumes an amount of energy, which is assumed to be negligible in our model.

Thanks to the generality of the definition of the probabilities $p_{kx}^{(i,m_1)}$, $q_{ks}^{(m_1,m_2,i)}$ (which are function of the state of the RE, the phase of service process and the available energy in DEQ, SEQ), the modeler can incorporate additional realistic features regarding the energy consumption, related to the control channel data transmission, or processing and forwarding of packets. More complex packet processing architectures may however require more than one packet queue.

Clearly, recent technological developments are towards smart autonomic wireless networks. The concept of signaling towards the SEQ, that triggers the instantaneous transition of energy units to the DEQ, has become an intelligent modeling tool for communication systems. Queues with signals [7] were introduced to model the behavior of control actions such as the displacement of units from one queue to another using "triggers" resulting in load balancing.

3   Process of the System State

The behavior of the system under consideration can be described in terms of the CTMC $X_t = \{Q_p(t), J(t), Q_{e_1}(t), Q_{e_2}(t), Y(t)\}$, $t \geq 0$, where $Q_p(t)$, $J(t)$, $Q_{e_j}(t)$ and $Y(t)$ are, respectively, the number of data packets, the phase of the service process (present only when $Q_p > 0$ and $Q_{e_1} > 0$), the number of energy units in EQ j and the state of RE at time t. The state space of X_t is $\widehat{H} = \cup_{n=0}^N l(n)$ and the "levels" $l(n)$ are defined as:

$$l(0) = \{(0, m_1, m_2, i); m_j = 0, 1, ..., E_j, j = 1, 2, i = 1, ..., M\}, \qquad (1)$$

and for $1 \leq n \leq N$:

$$l(n) = l(n,0) \cup \bar{l}(n),$$
$$l(n,0) = \{(n,0,m_2,i); m_2 = 0,1,...,E_2, i = 1,...,M\},$$
$$\bar{l}(n) = \{(n,x,m_1,m_2,i); x = 1,\ldots,\nu; m_1 = 1,\ldots,E_1; m_2 = 0,1,...,E_2; \quad (2)$$
$$i = 1,...,M\}.$$

These levels have cardinals $|l(0)| = L_0 := M(E_1 + 1)(E_2 + 1)$ and for $n \geq 1$, $|l(n)| = L := M(E_2+1)(\nu E_1+1)$. For convenience, define also $S := M(E_2+1)$. The state space has then cardinal $|\widehat{H}| = M(E_2 + 1)[N(\nu E_1 + 1) + (E_1 + 1)]$.

Using this decomposition in levels, the infinitesimal generator of X_t has a quasi-birth-death structure with block matrix representation as:

$$Q = \begin{pmatrix} B_0 & \widetilde{C} & 0 & 0... & 0 & 0 \\ A_{10} & A_1 & C & 0... & 0 & 0 \\ 0 & A_{21} & A_1 & C... & 0 & 0 \\ & \ddots & \ddots & \ddots & & \\ 0 & 0 & 0 & ...A_{21} & A_1 & C \\ 0 & 0 & 0 & ...0 & A_{21} & A_2 \end{pmatrix}.$$

We proceed with a detailed description of the blocks. We shall need the definition of the following matrices $\Lambda_{e_j} = diag(\lambda_{e_j}^{(i)})$, $j = 1,2$, $\Lambda_{p,s} = diag(\lambda_{p,s}^{(i)})$, $\Lambda_s^- = diag(\Lambda_s^{-(i)})$, $s = 0,...,K$, $U_j = diag(u_{ij})$, where $i = 1,...,M$, and $t_x = diag(t_x^{(1)},...,t_x^{(M)})$, $p_{lx}^{(m_1)} = diag(p_{lx}^{(1,m_1)},...,p_{lx}^{(M,m_1)})$.

The edge block A_{10} is an $L \times L_0$ matrix that corresponds to packet service and energy consumption events that lead to an empty data queue. It has a block-triangular structure with $A_{10}^{(0,m_1')} = 0_{S \times S}$, for all m_1' (this first row corresponds to the case where there is a packet waiting for service, but the transmission cannot be initiated since there is no available energy in the DEQ), $A_{10}^{(m_1,m_1')} = 0_{\nu S \times S}$ for $1 \leq m_1 < m_1' \leq E_1$ and for $m_1 = 1,...,E_1$, $m_1' = 0,...,m_1$:

$$A_{10}^{(m_1,m_1')} = A_{m_1 m_1'}^{(m_1-m_1')} = \quad (3)$$
$$(I_{(E_2+1)\times(E_2+1)} \otimes [t_1^{(m_1)} p_{m_1-m_1',1}^{(m_1)}], ..., I_{(E_2+1)\times(E_2+1)} \otimes [t_\nu^{(m_1)} p_{m_1-m_1',\nu}^{(m_1)}])'.$$

The sub-diagonal block A_{21} is an $L \times L$ matrix that corresponds to packet service and energy consumption:

$$A_{21} = \begin{pmatrix} 0_{S \times S} & 0_{S \times S} & 0_{S \times S} & ... & 0_{S \times S} & 0_{S \times S} \\ A_{10}^{(1)} & F_{11} & 0_{\nu S \times S} & ... & 0_{\nu S \times S} & 0_{\nu S \times S} \\ A_{20}^{(2)} & F_{21} & F_{22} & ... & 0_{\nu S \times S} & 0_{\nu S \times S} \\ ... & ... & ... & \ddots & & \\ ... & ... & ... & & \ddots & \\ A_{E_10}^{(E_1)} & F_{E_11} & F_{E_12} & ... & F_{E_1 E_1 - 1} & F_{E_1 E_1} \end{pmatrix},$$

where $F_{m_1 m_1'}$ is a $\nu S \times \nu S$ matrix defined from $\tau_x = diag(\tau_x^{(1)}, ..., \tau_x^{(M)})$ as:

$$F_{m_1 m_1'} = I_{(E_2+1) \times (E_2+1)} \otimes (t_x \tau_y p_{m_1 - m_1' x}^{(m_1)}), \quad x, y = 1, ..., \nu. \tag{4}$$

Packet arrivals are represented in block $\widetilde{C} = diag(C_0, C_1, ..., C_K)$, which is an $L_0 \times L$ matrix where C_s is of order $n(s)S \times n(s)S\nu$ for $s = 1, 2, ..., K$, and of order $n(0)S \times [(n(0) - 1)\nu + 1]S$ for $s = 0$, with $n(s) = h_{s+1} - h_s + 1\delta_{s=0}$ and

$$C_0 = diag(C_{00}, \underbrace{G_0, ..., G_0}_{h_1}), \quad C_s = I_{n(s) \times n(s)} \otimes G_s, s = 1, ..., K,$$

$$G_s = (G_s^{(1)}, ..., G_s^{(\nu)}), \quad G_s^{(x)} = I_{(E_2+1) \times (E_2+1)} \otimes (\Lambda_{p,s} \tau_x). \tag{5}$$

Matrix $C = diag(C_0', C_1', ..., C_K')$ is of order $L \times L$, where

$$C_0' = diag(C_{00}, \underbrace{G_0', ..., G_0'}_{h_1}), \quad C_s' = I_{n(s) \times n(s)} \otimes G_s', s = 1, ..., K,$$

$$G_s' = I_{\nu \times \nu} \otimes C_{0s}, \quad C_{0s} = I_{(E_2+1) \times (E_2+1)} \otimes \Lambda_{p,s}, \quad s = 0, ..., K. \tag{6}$$

Next, $B_0 = (B_{m_1, m_1'}^{(0)})$, $0 \le m_1, m_1' \le E_1$, is an $L_0 \times L_0$ matrix formed of square sub-blocks $B_{m_1, m_1'}^{(0)}$ of order S, typical of energy movements in the DEQ that are independent of packet transmissions, such as energy leakage, energy generation and signal-triggered energy unit transition from SEQ to DEQ. Its structure is:

$$B_0 = \begin{pmatrix} B_{00}^{(0)} & B_{01}^{(0)} & B_{02}^{(0)} & \cdots & & B_{0E_1}^{(0)} \\ \widehat{U}_1 & B_{11}^{(0)} & B_{12}^{(0)} & \cdots & & B_{1E_1}^{(0)} \\ \cdots & \cdots & \cdots & \cdots & & \\ 0_{L_0 \times L_0} & 0_{L_0 \times L_0} & 0_{L_0 \times L_0} & \cdots \widehat{U}_1 & B_{E_1-1,E_1-1}^{(0)} & B_{E_1-1,E_1}^{(0)} \\ 0_{L_0 \times L_0} & 0_{L_0 \times L_0} & 0_{L_0 \times L_0} & \cdots 0_{L_0 \times L_0} & \widehat{U}_1 & B_{E_1 E_1}^{(0)} \end{pmatrix},$$

where $\widehat{U}_1 = I_{(E_2+1) \times (E_2+1)} \otimes U_1$ represents energy leakage from DEQ, and for $0 \le m_1 < E_1$, $h_s < m_1 \le h_{s+1}$,

$$B_{m_1, m_1+1}^{(0)} = \begin{pmatrix} \Lambda_{e_1} & & \cdots & \\ \Lambda_s^- \underline{q}_{1s}^{(m_1,1)} & \Lambda_{e_1} & \cdots & \\ \cdots & \cdots & \cdots & \\ & & \cdots \Lambda_s^- \underline{q}_{1s}^{(m_1,E_2-1)} & \Lambda_{e_1} \\ & & \cdots & \Lambda_s^- \underline{q}_{1s}^{(m_1,E_2)} & \Lambda_{e_1} + \Lambda_{e_2} \end{pmatrix}.$$

The remaining sub-blocks concern the energy replenishment due to signals. Assuming $E_1 \le E_2$, for $k = m_1 + 2, ..., E_1$, $h_s < m_1 \le h_{s+1}$,

$$B_{m_1k}^{(0)} = \begin{pmatrix} \cdots & & \cdots & \cdots & \cdots & \cdots \\ \Lambda_s^- \underline{q}_{k-m_1s}^{(m_1,k-m_1)} & & & & & \cdots \\ & & \Lambda_s^- \underline{q}_{k-m_1s}^{(m_1,k-m_1+1)} & & & \cdots \\ \ddots & & \ddots & \ddots & \ddots & \\ & & & & \cdots \Lambda_s^- \underline{q}_{k-m_1s}^{(m_1,E_2)} & \cdots \end{pmatrix}.$$

Remark: In this work we assume without loss of generality that $E_1 \leq E_2$. For $E_1 > E_2$, $k = m_1+2, ..., m_1+E_2$, $B_{m_1k}^{(0)}$ has also the same form but $B_{m_1k}^{(0)} = 0_{S \times S}$, for $k = m_1 + E_2 + 1, ..., E_1$. □

Furthermore, for $0 \leq m_1 \leq E_1$ and $h_s < m_1 \leq h_{s+1}$, set $F_s = Q_Y - \sum_{j=1}^2 (\Lambda_{e_j} \delta_{m_j < E_j} + U_j \delta_{m_j > 0}) - \Lambda_{p,s} - \Lambda_s^- \delta_{m_1 < E_1}$. Then for $m_2, m_2' = 0, ..., E_2$, $B_{m_1m_1}^{(0)} = (B_{m_1m_1}^{(0;m_2,m_2')})$ where, $B_{m_1m_1}^{(0;m_2,m_2-1)} = U_2$ and,

$$B_{m_1m_1}^{(0;m_2,m_2+1)} = \Lambda_{e_2} + \Lambda_{e_1} \delta_{m_1 \neq E_1}, \quad B_{m_1m_1}^{(0;m_2,m_2)} = F_s, \quad h_s < m_1 \leq h_{s+1}. \quad (7)$$

Finally, the main diagonal blocks A_1 and A_2 are square matrices of order L, also composed of sub-blocks $A_l = (A_{m_1m_1'}^{(l)})$, $m_1, m_1' = 0, 1, ..., E_1$, for $l = 1, 2$. Here, $A_{00}^{(l)}$ is of order $S \times S$, $A_{0m_1'}^{(l)}$, $m_1' = 1, ..., E_1$ are of order $S \times \nu S$, $A_{m_10}^{(l)}$, $m_1 = 1, ..., E_1$, of order $\nu S \times S$, and $A_{m_1m_1'}^{(l)}$, $m_1, m_1' = 1, ..., E_1$ of order $\nu S \times \nu S$.

They are given by, $A_{00}^{(1)} = B_{00}^{(0)}$ and $A_{0m_1}^{(1)} = (Z_1^{(0,m_1)}, ..., Z_\nu^{(0,m_1)})$ where $Z_x^{(0,m_1)} = (Z_{x;m_2,m_2'}^{(0,m_1)})$, $x = 1, ..., \nu$, are $S \times S$ matrices where, $Z_{x;m_2,m_2}^{(0,1)} = (\Lambda_{e_1} + \Lambda_{e_2} \delta_{m_2=E_2}) \tau_x$, $m_2 = 0, 1, ..., E_2$, $Z_{x;m_2,m_2-1}^{(0,1)} = \Lambda_1^- \underline{q}_{11}^{(0,m_2)} \tau_x$, $Z_{x;m_2,m_2'}^{(0,1)} = 0_{M \times M}$, elsewhere. For $m_1 = 2, ..., E_1$, $h_s < k \leq h_{s+1}$,

$$Z_{x;m_2,m_2'}^{(0,m_1)} = \Lambda_s^- \underline{q}_{m_1s}^{(0,m_2)} \tau_x, \quad m_2 = m_1, ..., E_2, \quad m_2' = 0, 1, ..., m_2 - m_1, \quad (8)$$

and $Z_{x;m_2,m_2'}^{(0,m_1)} = 0_{M \times M}$ elsewhere. Furthermore, for $l = 1, 2$, $m_1 = 1, ..., E_1$, $A_{m_10}^{(l)} = (\widehat{S}_{m_11}, ..., \widehat{S}_{m_1\nu})'$, where

$$\widehat{S}_{m_1x} = I_{(E_2+1) \times (E_2+1)} \otimes [U_1 \delta_{m_1=1}$$
$$+ diag(\sum_{c=1, c \neq x}^\nu T_{xc}^{(1)} p_{m_1x}^{(1,m_1)}, ..., + \sum_{c=1, c \neq x}^\nu T_{xc}^{(M)} p_{m_1x}^{(M,m_1)})], \quad (9)$$
$$A_{m_1k}^{(l)} = I_{\nu \times \nu} \otimes B_{m_1k}^{(0)}, \quad k = m_1 + 1, ..., E_1, \quad m_1 \geq 1.$$

Moreover, $A_{m_1,m_1-1}^{(l)} = (A_{m_1,m_1-1}^{(l;x,y)})$, $k = 1, 2, ..., \nu$, and $A_{m_1,m_1-1}^{(l;x,x)} = \widehat{U}_1$, $x = 1, 2, ..., \nu$, $A_{m_1,m_1-1}^{(l;x,y)} = I_{(E_2+1) \times (E_2+1)} \otimes [T_{xy} p_{1x}^{(m_1)}]$, $x, y = 1, 2, ..., \nu$, $x \neq y$.

For $k = m_1 - 2, ..., 1$, $A_{m_1k}^{(l)} = (A_{m_1k}^{(l;x,y)})$, $x, y = 1, ..., \nu$, where, $A_{m_1k}^{(l;x,x)} = 0_{S \times S}$, $x = 1, ..., \nu$, and $A_{m_1k}^{(l;x,y)} = I_{(E_2+1) \times (E_2+1)} \otimes T_{xy} p_{m_1-kx}^{(m_1)}$, $x, y = 1, ..., \nu$, $x \neq y$ and $A_{m_1k}^{(2)} = A_{m_1k}^{(1)}$, $k = m_1 + 1, ..., E_1$.

Moreover, $A_{m_1 m_1}^{(l)} = (\Phi_{xy;m_2,m_2'}^{(l,m_1,m_1)})$, $l = 1, 2$, $x, y = 1, ..., \nu$, $m_l = 1, ..., E_l$, where for $T_{xy} = diag(T_{xy}^{(1)}, ..., T_{xy}^{(M)})$,

$$\Phi_{xy;m_2,m_2'}^{(l,m_1,m_1)} = I_{(E_2+1)\times(E_2+1)} \otimes [T_{xy} p_{0x}^{(m_1)}], m_1 = 0, 1, ..., E_1, x \neq y,$$
$$\Phi_{xx;m_2,m_2'}^{(l,m_1,m_1)} = I_{(E_2+1)\times(E_2+1)} \otimes (T_{xx} + \Lambda_{p,s}\delta_{m_1<E_1}) + B_{m_1 m_1}^{(0)} + W_s\delta_{m_1<E_1},$$
$$W_s = diag(0_{M\times M}, I_{(ME_2\times ME_2)} \otimes \Lambda_s^-).$$

$$(10)$$

Sparsity of the Generator. The number of non-zero transition rates in the generator Q can be roughly evaluated as follows. Neglecting the "border" cases, transitions from a typical state (n, x, m_1, m_2, i) occur as: a) packet arrival, energy arrival, energy leakage: one transition each; b) service phase change: $\nu - 1$ possible transitions with service continuation, plus service completion, which triggers energy consumption with up to m_1 possible transitions; c) environment phase change: $M - 1$ possible transitions (this could be much less if the phase process is itself a superposition, see Section 2); d) energy transfers: up to m_2 transitions. Summing over m_1 and m_2, we conclude that storing the elementary blocks of Q requires $O\left((\nu + M)ME_1^2E_2^2\right)$ space. Storing all Q in sparse form would require $O\left(N(\nu + M)ME_1^2E_2^2\right)$ space. The ratio of this quantity with the total number of entries in Q: $O\left((NM\nu E_1 E_2)^2\right)$, is of order $\alpha = O((\nu + M)/N\nu M)$. It may even be smaller if the RE transition matrix Q_Y is itself sparse, or if the distributions $p_{kx}^{(i,m_1)}$ or $q_{ks}^{(m_1,m_2,i)}$ have restricted supports. Note that some algorithms work with storing only the different blocks.

4   Stationary Solution

4.1   Algorithms for Stationary Probabilities

Under reasonable assumptions on packet arrival rates and the representation of service distributions, the chain X_t is irreducible. Since it is finite, it then admits a unique stationary distribution given by (referring to the notation in (1) and (2)):

$$p(n, x, m_1, m_2, i)$$
$$= \lim_{t\to\infty} P(Q_p(t) = n, J(t) = x, Q_{e_1}(t) = m_1, Q_{e_2}(t) = m_2, Y(t) = i)$$
$$p(n, m_1, m_2, i) = \lim_{t\to\infty} P(Q_p(t) = n, Q_{e_1}(t) = m_1, Q_{e_2}(t) = m_2, Y(t) = i).$$

In order to compute this stationary probability vector, several methods are available. On the one hand, the QBD structure of Q offers the possibility to use direct numerical methods, such as the Matrix Analytic ones. Those are "exact" (up to round off errors) and in general they are efficient especially for systems where their generator matrix has a special form, which is the case here. On the other hand, the sparsity of Q points at the use of iterative and other advanced techniques [18].

One way or the other, a closer investigation of the block structure of Q is fruitful. The stationary distribution vector $\underline{p}$ is represented using the level structure as $\underline{p} = (\underline{p}_0, \underline{p}_1, ..., \underline{p}_N)$. Global balance equations $\underline{p}Q = 0$ can then be written as

$$\underline{p}_0 B_0 + \underline{p}_1 A_{10} = 0, \tag{11}$$

$$\underline{p}_0 \widetilde{C} + \underline{p}_1 A_1 + \underline{p}_2 A_{21} = 0, \tag{12}$$

$$\underline{p}_{i-1} C + \underline{p}_i A_1 + \underline{p}_{i+1} A_{21} = 0, \quad i = 2, ..., N-1, \tag{13}$$

$$\underline{p}_{N-1} C + \underline{p}_N A_2 = 0. \tag{14}$$

The matrix analytic formalism gives the opportunity to obtain the stationary probability vector $\underline{p}$ using several algorithms (see [14] and references within, and later contributions among which [1,5,19]). The methods of [6,19] are highlighted in [14, Chap. 10]. The key step in [6] resides especially in the inversion of a matrix of order L_0 and N matrices of order L. Given that $L_0 \leq L$, the overall complexity is $O(NL^3)$. The "folding method" in [19] is less expensive, reducing the "N" to $\log_2 N$, but at a higher implementation cost. Moreover, [6] also provides an algorithm for the computation of the expected first passage times between the neighbour levels. This result is interesting for further capacity planning investigation, see Section 5 and our conclusion. In our situation, the limiting factor is clearly the value of $L \approx M\nu E_1 E_2$.

The form of the block generator matrix will help us to improve the efficiency in obtaining the stationary probabilities, since the matrix C is a diagonal non-singular matrix. Solving recursively system (11)–(14) starting from (14) we get that, $\underline{p}_n = \underline{p}_N R_n$ for $n = 0, ..., N$, with: $R_N = I$, $R_{N-1} = -A_2 C^{-1}$, $R_0 = (R_2 A_1 + R_3 A_{21}) A_{10} B_0^{-1}$, and

$$R_n = -(R_{n+1} A_1 + R_{n+2} A_{21}) C^{-1}, \, n = 1, ..., N-2. \tag{15}$$

To obtain $\underline{p}_N$, we solve equation (12) along with the normalizing condition:

$$\underline{p}_N (R_0 \widetilde{C} + R_1 A_1 + R_2 A_{21}) = 0, \qquad \underline{p}_N (R_0 \underline{1}_0 + \sum_{n=1}^{N} R_n \underline{1}) = 1, \tag{16}$$

where $\underline{1}$, $\underline{1}_0$ are column vectors of order L and L_0 respectively.

The computation of the R_n's involves only matrix additions and multiplications. Since C is diagonal and non singular, multiplying by its inverse is not costly. The computation of the p_n's [5] requires two matrix inversions (one to compute B_0^{-1}, one to solve the system (16)), $4N + 5$ matrix multiplications and $2N + 1$ matrix additions. The resulting complexity is $O(L^3)$. Using an iterative method to solve (16) and exploiting the inner structure of blocks (see Section 2) may help overcome difficulties about the computational complexity.

4.2   Performance Metrics

By calculating the stationary distribution vector of the underlying Markov process we can obtain some important performance metrics, such as the depletion probability (DP) (i.e., the probability of an empty DEQ), and the expected number of data packets and energy packets in each EQ:

$$
\begin{aligned}
DP &= \sum_{n=0}^{N} \sum_{m_2=0}^{E_2} \sum_{i=1}^{M} p(n, 0, m_2, i), \\
E(Q_p) &= \sum_{n=1}^{N} \sum_{m_2=0}^{E_2} \sum_{i=1}^{M} n \left[p(n, 0, m_2, i) \right. \\
&\qquad\qquad \left. + \sum_{m_1=1}^{E_1} \sum_{x=1}^{\nu} p(n, x, m_1, m_2, i) \right], \\
E(Q_{e_1}) &= \sum_{m_1=1}^{E_1} \sum_{m_2=0}^{E_2} \sum_{i=1}^{M} m_1 \left[p(0, m_1, m_2, i), \right. \\
&\qquad\qquad \left. + \sum_{n=1}^{N} \sum_{x=1}^{\nu} p(n, x, m_1, m_2, i) \right],
\end{aligned}
\tag{17}
$$

$$
E(Q_{e_2}) = \sum_{m_2=1}^{E_2} \sum_{i=1}^{M} m_2 \left[\sum_{m_1=0}^{E_1} p(0, m_1, m_2, i) \right. \\
\left. + \sum_{n=1}^{N} \left(p(n, 0, m_2, i) + \sum_{m_1=1}^{E_1} \sum_{x=1}^{\nu} p(n, x, m_1, m_2, i) \right) \right].
\tag{18}
$$

Using Little's law we can obtain the mean waiting time for a data packet to be served. Moreover, various optimization problems can be formulated, such as finding optimal values of $\lambda_{p,s}^{(i)}$ that minimizes $E(Q_p)$, asking $DP \le a$, or finding an optimal value for E_1 which minimizes $DP = DP(E_1)$.

4.3 Numerical Example

To demonstrate the feasibility of the developed algorithm and numerically show some features of the model, we present the results of a preliminary numerical experiment, inspired from values taken from [13]. Assume that the RE is defined by the 2-state infinitesimal generator with rates $(Q_Y)_{12} = 0.01$ and $(Q_Y)_{21} = 0.1$. We may interpret that the first state of the RE corresponds to a mode when the system is overloaded by the data units (peak time) and the second state to a normal mode of the system. Under the first and second state of the RE the service time distribution is characterized by the vectors $\underline{\tau}^{(1)} = (0.2, 0.8)$, $\underline{\tau}^{(2)} = (0.7, 0.3)$ and the matrices

$$
T^{(1)} = \begin{pmatrix} -0.4368 & 0.40768 \\ 0.42608 & -1.71809 \end{pmatrix}, \quad T^{(2)} = \begin{pmatrix} -0.6402 & 0.56812 \\ 0.22308 & -1.34325 \end{pmatrix},
$$

respectively. We assumed for $m_1 + m_2 \le E_1$, $q_{ks}^{(m_1, m_2, i)} = 2^{ki} / \sum_{j=1}^{m_2} 2^{ji}$, $k = 1, ..., m_2$ and if $m_1 + m_2 > E_1$, $q_{ks}^{(m_1, m_2, i)} = 2^{ki} / \sum_{j=1}^{E_1 - m_1} 2^{ji}$, $k = 1, ..., E_1 - m_1$. Furthermore, $p_{kx}^{(i, m_1)} = \left(\sum_{j=1}^{m_1} (ix)^{k-j} \right)^{-1}$, $k = 1, ..., m_1$. Table 1 contains the values of the parameters of the system.

Table 1. Overview of system's parameters

$E_1 = 10, \ E_2 = 12$	$M = 2, \ \nu = 2$	$N = 15,$	$h_1 = 8, \ h_2 = 12$
$\Lambda_{p,1} = diag(1.5, 1)$	$\Lambda_{p,2} = diag(2, 1.5)$	$U_1 = diag(0.1, 0.3)$	$U_2 = diag(0.3, 0.4)$
$\Lambda_1^- = diag(0.3, 0.2)$	$\Lambda_2^- = diag(0.2, 0.1)$	$L_{e_1} = diag(1.5, 1)$	$L_{e_2} = diag(1.8, 1.2)$

We focus on the depletion probability. Figure 2 (left) describes the way the depletion probability is affected by the RE for increasing values of the data arrival rate when $m_1 \in [0,8]$ ($s = 0$). Clearly, the depletion probability will increase, but this increase will become more apparent during the overloaded period, as expected. In Figure 2 (right) we can observe the benefits of the smart operation of the base station due to the presence of signals. By increasing the signal generation rate, the depletion probability is strongly reduced. As a result, the quality of service is thoroughly increased.

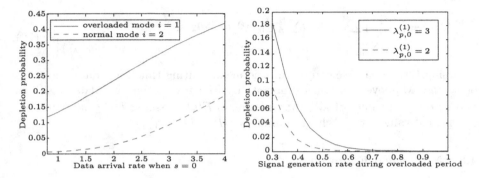

Fig. 2. Sensibility of DP w.r.t. data rate and RE for $\Lambda_0^- = diag(0.35, 0.25)$ (left) and w.r.t. signal rate when the RE is in overload

5   First Passage Times

Gaver et al. [6] analyzed the first passage times for general finite QBD processes. They obtained two systems of recurrence equations for the Laplace-Stieltjes transforms of passage times to higher and lower levels.

In the following we turn our attention to the computation of the *average* first passage times. Let the expected values of the first passage time from the states of $l(n-1)$ to the states of $l(n)$ be $c^{(n)}$ and that from the states of $l(n+1)$ to $l(n)$ be $v^{(n)}$. Define the following matrices:

$$
\begin{aligned}
H_0 &= B_0 & H_1 &= A_1 + A_{10}(-H_0^{-1})\widetilde{C}, \\
H_j &= A_1 + A_{21}(-H_{j-1}^{-1})C, & 2 &\leq j \leq N-1, \\
H_N &= A_2 + A_{21}(-H_{N-1}^{-1})C. &
\end{aligned}
\tag{19}
$$

Then we have the following recurrence relations for the expected values of first passage times to higher levels:

$$
c^{(n)} = \begin{cases}
-H_0^{-1}\underline{1}, & n = 1, \\
-H_1^{-1}(\underline{1} + A_{10}c^{(1)}), & n = 2, \\
-H_{n-1}^{-1}(\underline{1} + A_{21}c^{(m_1-1)}), & 3 \leq n \leq N.
\end{cases}
$$

For the determination of the expected values of the first passage time to lower levels we have first to denote by $\widehat{H}_n$, $n = 0, 1, ..., N$ the infinitesimal generator of the restriction of the process X_t observed during those intervals of time spent at $l(n)$ before the original process moves to $l(n-1)$ for the first time. Then, all $\widehat{H}_n$-processes, $1 \leq n \leq N$ are transient while $\widehat{H}_0$ is positive recurrent. The matrices $\widehat{H}_n$ are recursively computed as follows

$$\widehat{H}_n = \begin{cases} B_0 + \widetilde{C}(-\widehat{H}_1^{-1})A_{10}, & n = 0, \\ A_1 + C(-\widehat{H}_{n+1}^{-1})A_{21}, & 1 \leq n \leq N-1, \\ H_N, & n = N. \end{cases}$$

Then,

$$v^{(n)} = \begin{cases} -\widehat{H}_1^{-1}(\underline{1} + \widetilde{C}v^{(1)}), & n = 0, \\ -\widehat{H}_{n+1}^{-1}(\underline{1} + Cv^{(n+1)}), & 1 \leq n \leq N-2, \\ -\widehat{H}_N^{-1}\underline{1}, & n = N-1. \end{cases}$$

These computations can be exploited, for instance, for controlling the (expected) time at which the packet queue overflows for the first time. Based on them, values for the control thresholds h_s and packet arrival rates $\lambda_{p,s}^{(i)}$, $s = 0, ..., K$ can be determined numerically offline. If the computation is fast enough, online adaptation becomes possible. Another use of first passage times is to compute the average length of busy periods. From this metric, analyses about the opportunity to put the BS in sleep mode during idle periods can be envisioned.

Computing first passage times for energy levels is also of great interest, in order to determine and optimize the autonomy of the BS. However, the model we have presented is not directly a QBD for energy levels, and a more elaborate analysis is the topic of a work in progress.

6   An Approximated Product-form Model

In this section, we describe a simplification of the model proposed in Section 2, strongly motivated by recent works in [8,9]. This new model is a G-network and therefore has product-form solution. The simplification applies to instances of our model where: a) there is no threshold control for the DEQ, therefore just environment-dependent packet as well as signal arrival rates $\lambda_p^{(i)}$, $\Lambda^{-(i)}$, $i = 1, ..., M$; b) exactly one energy unit is required to transmit any data packet.

Consider the following simplifications: 1) infinite capacity buffers for the DEQ, SEQ, and the data queue; 2) as soon as the BS has both a data packet and enough energy to transmit that packet, the transmission is instantaneous; 3) environment-dependent parameters are averaged out using the stationary distribution $\underline{q}_Y$ of the generator Q_Y. Thus, defining the vector $\underline{\lambda}_p = (\lambda_p^{(1)}, ..., \lambda_p^{(M)})$

and $\underline{A}^- = (A^{-(1)}, ..., A^{-(M)})$, then the arrival rates for packets, DEQ and SEQ energy units, and signals in the simplified model are as:

$$\widehat{\lambda}_p = \underline{\lambda}_p \underline{q}'_Y, \quad \widehat{\lambda}^- = \underline{A}^- \underline{q}'_Y, \quad \widehat{\lambda}_{e_j} = \underline{1} A_{e_j} \underline{q}'_Y, \; j = 1, 2. \tag{20}$$

Similarly, we will take into account the average leakage rate for each energy queue, i.e., if $\underline{u}_j = \underline{1} U_j \underline{q}'_Y$, $j = 1, 2$, where $\underline{1}$ is an $1 \times M$ vector of 1.

The state of the approximation model can be represented by the pair (n, m), where $n = 0$, means that the BS has neither energy units in the DEQ, nor data packets to transmit, while $n > 0$, means that it currently stores n data packets but no energy units in DEQ, while $n < 0$, means that it stores $-n$ energy units in DEQ but no data packets. The element $m \geq 0$, counts the number of stored energy units in SEQ. Note that our system is a special type of G-networks. The underlying Markov process $\widetilde{X} = \{(n, m); n \in Z, m \in Z^+ \cup \{0\}\}$ is ergodic provided that:

$$\widehat{\lambda}_p < \widehat{\lambda}_{e_1} + \widehat{\lambda}^- q_2, \quad \widehat{\lambda}_{e_2} < \widehat{u}_2 + \widehat{\lambda}^-, \quad \widehat{\lambda}_{e_1} + \widehat{\lambda}^- q_2 < \widehat{\lambda}_p + \widehat{u}_1, \tag{S}$$

where $q_2 = \widehat{\lambda}_{e_2}/(\widehat{u}_2 + \widehat{\lambda}^-)$, is the probability that the SEQ is not empty. These conditions come from the fact that the system must be stable with respect to the data packets ($n > 0$, left hand condition), with respect to the energy units in SEQ, (the condition at the middle), and with respect to the energy units in DEQ ($n < 0$, right hand condition). To conclude we state the following proposition.

Proposition 1. *Under conditions (S) and with $f(m) = q_2^m$, $m \geq 0$, the joint stationary distribution of $\widetilde{X}$ has the product form:*

$$p(n, m) = C g(n) f(m), \tag{21}$$

$$g(0) = 1, \qquad g(n) = q_1^n, n > 0, \qquad g(n) = (\tilde{q}_1)^{-n}, n < 0, \tag{22}$$

$$q_1 = \frac{\widehat{\lambda}_p}{\widehat{\lambda}_{e_1} + \widehat{\lambda}^- q_2}, \quad \tilde{q}_1 = \frac{\widehat{\lambda}_{e_1} + \widehat{\lambda}^- q_2}{\widehat{\lambda}_p + \widehat{u}_1}, \quad C = \frac{(1 - q_1)(1 - \tilde{q}_1)(1 - q_2)}{1 - q_1 \tilde{q}_1}. \tag{23}$$

The proof proceeds with the substitution of (21) into the global balance equations, and is omitted due to page constraint. For alternative approaches see [2,10]. C is obtained using the normalizing condition $\sum_{n=-\infty}^{+\infty} \sum_{m=0}^{+\infty} p(n, m) = 1$.

Extension to a Network of SEQs. The approximation model is flexible enough to incorporate an arbitrary number of SEQs that are deployed close to the target BS while preserving the product-form solution. Indeed, assume that there are K reserve energy queues deployed close to our BS. In such a case our smart BS will generate signals at SEQ k, at a Poisson rate $\widehat{\lambda}_k^-$, while energy units arrive at SEQ k at a Poisson rate $\widehat{\lambda}_{e_k}^r$, and energy leakage at a Poisson rate $\widehat{u}_{2k}$ $k = 1, ..., K$. The system's state is now $(n, \underline{m}) = (n, m_1, ..., m_K)$, $n \in Z$, $m_k = 0, 1, ...$, where now m_k, counts the number of stored energy units in SEQ k.

Then, provided that $\check{q}_1 = \frac{\widehat{\lambda}_p}{\widehat{\lambda}_{e_1} + \sum_{k=1}^{K} \widehat{\lambda}_k^- q_{2k}} < 1$, $\check{\tilde{q}}_1 = \frac{\widehat{\lambda}_{e_1} + \sum_{k=1}^{K} \widehat{\lambda}_k^- q_{2k}}{\widehat{\lambda}_p + \widehat{u}_1} < 1$,

$q_{2k} = \frac{\widehat{\lambda}_{e_k}^r}{\widehat{\lambda}_k^- + \widehat{u}_{2k}} < 1$, $k = 1, ..., K$, a modification of Proposition 1 can be proved, and the joint stationary distribution of $(n, \underline{m})$ is for $f_k(m_k) = q_{2k}^{m_k}$, $k = 1, ..., K$, $m_k \geq 0$,

$$p(n, \underline{m}) = C' v(n) \prod_{k=1}^{K} f_k(m_k), \tag{24}$$

$$v(0) = 1, \qquad v(n) = \breve{q}_1^n, n > 0, \qquad v(n) = (\tilde{\breve{q}}_1)^{-n}, n < 0, \tag{25}$$

and the normalization constant is $C' = (1 - \breve{q}_1)(1 - \tilde{\breve{q}}_1) \prod_{k=1}^{K}(1 - q_{2k})/(1 - q_1 \tilde{q}_1)$.

7 Conclusion

We have given the detailed specification of a versatile model of energy supply for base stations or similar devices and presented preliminary numerical results. Our future investigations will concentrate on the resolution of models with a challenging size, the comparison with the product-form approximation, and also on ways to compute efficiently transient measures such as the mean time to depletion. With that objective, we will also investigate whether an approximate analysis through time decomposition is accurate. In a future work, we intent to specify the component interactions of our model by means of high level formalism such as a Markovian process algebra.

References

1. Akar, N., Oguz, N.C., Sohraby, K.: A novel computational method for solving finite QBD processes. Stoch. Models **16**(2), 273–311 (2000)
2. Chao, X., Miyazawa, M., Pinedo, M.: Queueing Networks: Customers, Signals and Product Form Solutions. Wiley (1999)
3. De Cuypere, E., De Turck, K., Fiems, D.: Stochastic modelling of energy harvesting for low power sensor nodes. In: Proc. of QTNA 2012 (2012)
4. De Cuypere, E., De Turck, K., Fiems, D.: Performance analysis of a kitting process as a paired queue. Mathematical Problems in Engineering, 10 (2013)
5. Elhafsi, E.H., Molle, M.: On the solution to QBD processes with finite state space. Stoch. Anal. and Appl. **25**(4), 763–779 (2007)
6. Gaver, D.P., Jacobs, P.A., Latouche, G.: Finite birth-and-death models in randomly changing environments. Adv. Appl. Prob. **16**(4), 715–731 (1984)
7. Gelenbe, E.: G-networks with triggered customer movement. J. Appl. Prob. **30**(3), 742–748 (1993)
8. Gelenbe, E.: A sensor node with energy harvesting. ACM Sigmetrics Perform. Evaluation Review **42**(2), 37–39 (2014)
9. Gelenbe, E., Marin, A.: Interconnected wireless sensors with energy harvesting. In: Remke, A., Manini, D., Gribaudo, M. (eds.) ASMTA 2015. LNCS, vol. 9081, pp. 87–99. Springer, Heidelberg (2015)
10. Harrison, P., Marin, A.: Product-forms in multi-way synchronizations. Comp. Journal **57**(11), 1693–1710 (2014)
11. Jones, G.L., Harrison, P.G., Harder, U., Field, A.J.: Fluid queue models of battery life. In: Proc. of MASCOTS 2011, pp. 278–285 (2011)

12. Jones, G.L., Harrison, P.G., Harder, U., Field, A.J.: Fluid queue models of renewable energy storage. In: Proc. of VALUETOOLS 2012, pp. 224–225 (2012)
13. Kim, C., Dudin, A., Dudin, S., Dudina, O.: Analysis of an $MMAP/PH_1$, $PH_2/N/\infty$ queueing system operating in a random environment. Int. J. Appl. Math. Comput. Sci. **24**(3), 485–501 (2014)
14. Latouche, G., Ramaswami, V.: Introduction to Matrix Analytic Methods in Stochastic Modeling. ASA-SIAM (1999)
15. Latouche, G.: Queues with paired customers. J. Appl. Prob. **18**(3), 684–696 (1981)
16. Neuts, M.F.: Matrix-geometric solutions in stochastic models: An Algorithmic Approach. The John Hopkins University Press (1981)
17. Takahashi, M., Osawa, H., Fujisawa, T.: On a synchronization queue with two finite buffers. Queueing Systems **36**(1–3), 107–123 (2000)
18. Wimmer, R., Derisavi, S., Hermanns, H.: Symbolic partition refinement with automatic balancing of time and space. Perform. Evaluation **67**(9), 816–836 (2010)
19. Ye, J., Li, S.Q.: Folding algorithm: A computational method for finite QBD processes with level-dependent transitions. IEEE Trans. Comm. **42**(2/3/4), 625–639 (1994)

Static and Dynamic Hosting of Cloud Servers

Paul Ezhilchelvan[✉] and Isi Mitrani

School of Computing Science,
Newcastle University, Newcastle upon Tyne NE1 7RU, UK
{paul.ezhilchelvan,isi.mitrani}@ncl.ac.uk

Abstract. The problem of maximizing the profit achieved by hiring servers from a Cloud and offering virtual machines to paying customers is examined. A number of VMs, each running a user job, can share a server. Hiring a server incurs an initial set-up cost, as well as running costs proportional to the duration of hire. New jobs that cannot start immediately may be lost, or they may be queued. It may or may not be possible to move running VMs from server to server. The effect of these different conditions on several hiring policies, both static and dynamic, is analyzed and evaluated.

1 Introduction

This paper addresses a problem that arises in the market for computer services. A host gains income by running user jobs on servers that it hires from a Cloud provider. To run a job, a Virtual Machine (VM) is instantiated on one of the servers. However, there is a limit on the number of VMs, and hence jobs, that can run in parallel on one server without unduly degrading each other's performance. When that limit is reached for all currently hired servers, the host may

(a) reject newly arriving jobs, thus losing revenue;
(b) queue newly arriving jobs, possibly having to pay penalties mandated by a Service Level Agreement (SLA);
(c) hire more servers, incurring more costs.

The cost of hiring a server may include a fixed initial set-up component, plus a cost proportional to the duration of hire. In the case of queued jobs, the SLA may guarantee a bound on waiting, with a penalty payable when that bound is exceeded. In all cases, the problem is to decide what actions to take so as to maximize the long-term average profit (revenues minus costs) obtained per unit time.

We analyze, evaluate and compare several server hiring policies. Some of these are static, choosing a fixed number of servers and keeping them for as long as the input parameters remain the same. Others are dynamic, hiring and releasing servers in response to changes in the number of jobs in the system. The majority of policies reject incoming jobs which cannot start immediately. However, the possibility of queueing such jobs subject to waiting time guarantees is also considered.

© Springer International Publishing Switzerland 2015
M. Beltrán et al. (Eds.): EPEW 2015, LNCS 9272, pp. 19–31, 2015.
DOI: 10.1007/978-3-319-23267-6_2

In a practical application, these policies would have to be combined with some monitoring and parameter estimation technique that would detect when the loading parameters change. We do not dwell on that aspect because it has already been covered quite extensively in the literature (see below). Our assumption is that the system reaches steady state during a period where the parameters stay the same.

It should be pointed out that the behaviour of a dynamic policy depends on whether a running VM can be moved from one server to another or not. In the former case, jobs can be packed into the smallest number of servers required, whereas in the latter one may need to keep an unnecessarily large number of partially filled servers.

A special queueing model is analyzed and solved for the case of a dynamic hiring policy with non-movable VMs.

The general conclusion reached after a number of numerical experiments comparing the different hiring policies is that a static policy can perform really well, provided that it is chosen optimally (this proviso is important!). Dynamic policies do tend to produce higher profits, but the improvements rarely exceed 10%.

There has been quite a lot of work on server allocation, often in the context of the trade-off between performance and energy consumption. In most cases the focus has been on static policies, with an emphasis on estimating the traffic and reacting to changes in the parameters. Such studies were carried out by Mazzucco et al. [7,8], using models and empirical observations. Bodík et al. [2] use statistical machine learning to estimate the workload during the next period.

Chaisiri et al. [3] attempt to exploit the lower costs of future reservations in order to minimize the overall cost of hiring Cloud servers. They use stochastic and deterministic programming techniques, coupled with approximations. This study has some dynamic features. However, the actual demand process is not modelled and therefore neither losses nor waiting can be taken into account.

A dynamic optimization using Markov decision theory was carried out by McGough and Mitrani [9] for a model with batch arrivals and also when hiring decisions are made at fixed intervals. Gandhi et al [5], and Mitrani [12] analyzed certain dynamic server allocation policies with set-up costs. In these studies jobs are queued but there are no SLAs, and the possibility of rejections is not considered.

More distantly related work concerns the maximization of throughput and the minimization of waiting or response time in different scheduling contexts, e.g. Urgaonkar et al. [13], Chandra et al. [4] and Bennani and Menascé [1]. A deterministic example of job scheduling with migration in order to minimize the number of servers was considered by Ghribi et al [6].

In all of the above papers, servers are assumed to serve one job at a time (VMs are mentioned in [8] for the purpose of parameter estimation, but are not included in the analysis). Where a dynamic policy has been compared to a static one (e.g. in [9]), the latter has been chosen in an ad-hoc manner, rather than optimally. The effect of not being able to move VMs between servers has not been examined.

Section 2 introduces a number of static and dynamic policies and evaluates the profit they achieve. The models involve job losses and also queueing. The dynamic policies assume that VMs can move instantaneously from one server to another. The model of a dynamic policy that does not move VMs is analyzed in section 3. Section 4 summarizes the conclusions and outlines some directions for further research.

2   Static and Dynamic Policies

A host hires servers from a cloud provider in order to offer services to paying customers, Servers can be hired and released instantaneously and at any time. Similarly, VMs can be initiated and terminated instantaneously and at any time. In this section we also assume that VMs can be moved from server to server without delay and without incurring costs.

The service provided by a VM during its lifetime is referred to as a 'job'. A server can run efficiently up to m parallel VMs, so if there are n active servers at a given moment, there is room for a maximum of nm jobs.

The cost of a server which is used for a period of length t is $c_1 + c_2 t$. The first term, c_1, if non-zero, may be considered as a 'set-up' cost, or it may be introduced by the provider in order to discourage short-term hire. The coefficient c_2 reflects the cost of operating a server per unit time.

Jobs arrive into the system in a Poisson stream at rate λ. Their lifetimes may have arbitrary distribution with mean $1/\mu$. The offered load is thus $\rho = \lambda/\mu$. The values of these parameters are assumed to remain constant long enough so that the system can be treated as being in steady state.

The assumption of easily movable VMs implies that jobs can be 'packed' efficiently. Suppose that the servers currently hired are numbered $1, 2, \ldots$. When accepting an incoming job, allocate it to the server with the lowest index that has room for it. When a job is completed and its VM is terminated, move a job from the non-empty server with the highest index (if different) to the vacated place. This ensures that if there are j jobs present, they can be run on $\lceil j/m \rceil$ servers, where $\lceil x \rceil$ is the smallest integer exceeding or equal to x.

The problem that needs to be addressed in this context is: When, and how many, servers should be hired or released? One possibility is to employ a static policy whereby a fixed number of servers, n, is hired and kept for as long as the parameters λ and μ retain their values. An incoming job that finds all servers full, i.e. nm jobs present, is rejected. If the policy is static, the question of moving jobs between servers does not arise.

In such a system, the number of jobs present behaves like an $M/M/K/K$ queue, i.e. an Erlang loss model where $K = nm$ is the maximum number of jobs that can be accepted. The steady-state probability, q_j, that there are j jobs present is equal to (e.g., see [11])

$$q_j = \frac{\rho^j}{j!} \left[\sum_{k=0}^{nm} \frac{\rho^k}{k!} \right]^{-1} \; ; \; j = 0, 1, \ldots, nm \, . \tag{1}$$

The decision on what value of n to choose depends on the objective function to be optimized. If, for example, the aim is simply to avoid job losses, one could fix a desirable value, ε, and hire the smallest number of servers which ensures that the probability of rejection does not exceed ε:

$$n^* = \min\{n \ : \ q_{(nm)} \leq \varepsilon\} \, . \tag{2}$$

This will be referred to as the 'fix-ε' policy.

Alternatively, one could attempt to maximize profit. Suppose that every accepted job brings in a revenue of r. Then the average long-run profit produced by n servers per unit time, $R(n)$, is given by

$$R(n) = r\lambda(1 - q_{(nm)}) - c_2 n \, , \tag{3}$$

The long-run set-up costs incurred per unit time are zero, because after the initial moment there are no new hiring events.

It is known that the Erlang loss probability, $q_{(nm)}$, is convex in n (see [10]). Hence, the profit function $R(n)$ is concave in n and has a single maximum. The optimal number of servers, and the corresponding maximum achievable profit, can therefore be computed quite easily by evaluating $R(n)$ for $n = 1, 2, \ldots$, and stopping as soon as $R(n+1) < R(n)$. The resulting hiring policy will be referred to as 'fix-opt'.

Now consider the possibility of hiring and releasing servers dynamically, in response to changes in the system state. A rather general policy of this type would work as follows: Hire a block of k_1 servers; if there are $k_1 m$ jobs present and a new job arrives, hire a new block of k_2 servers. This goes on up to a maximum of b blocks with a total of $n = k_1 + k_2 + \ldots + k_b$ servers. When a job completes, a job from a block with a higher index (if any) is moved into its place so as to maintain optimal packing. If, as a result of this completion, a block empties, all the servers in it are released. One may also decide to keep block 1 permanently hired. This policy will be referred to as 'blocks-b', with a bound of n.

The number of jobs in the system under the blocks-b policy with bound n has the same distribution, given by (1), as under the static policy with n servers. In particular, the probability that an incoming job is accepted is the same. However, the dynamic policy incurs set-up costs, while reducing the operating costs.

Let K_i be the total number of servers in the first i blocks: $K_i = k_1 + k_2 + \ldots + k_i$; $i = 1, 2, \ldots, b$; $K_b = n$ and, by definition, $K_0 = 0$. Since block $i + 1$ is hired whenever an incoming job finds exactly $K_i m$ jobs present, the average number of hiring events per unit time is

$$S = \lambda \sum_{i=0}^{b-1} q_{(K_i m)} k_{i+1} \, . \tag{4}$$

For a given number, j, of jobs present $(1 \leq j \leq nm)$, denote by $K(j)$ the number of servers currently hired. That is the smallest K_i such that $K_i \geq \lceil j/m \rceil$.

With that notation, the average number of servers hired, L, can be written as

$$L = \sum_{j=1}^{nm} q_j K(j) \ . \tag{5}$$

Hence, the average long-term profit obtained per unit time under the blocks-b policy is equal to

$$R(n) = r\lambda(1 - q_{(nm)}) - c_1 S - c_2 L \ . \tag{6}$$

To determine the best blocks-b policy with a bound n, one would have to search not only with respect to n, but also with respect to b and k_i. That search may be quite expensive. However, two special cases can be handled quite easily. At one extreme is the policy which we shall call 'one-by-one': it hires and releases servers one at a time ($b = n$ and $k_i = 1$ for all i). The best one-by-one policy can be found by a simple search with respect to n. At the other extreme is blocks-2, where only two blocks are used; $b = 2$, $k_1 + k_2 = n$. The search in this case is with respect to n and k_1.

Figure 1 illustrates the performance of the above policies in the context of a system where each server runs up to 5 VMs in parallel. The average residence of a VM is taken as the unit of time, $\mu = 1$. The revenue per job is $r = 1$, while the server costs are $c_1 = 0.1$ and $c_2 = 3$. Thus, a server can make a profit by running 5 jobs, but the margin is not large.

For the policies fix-opt, one-by-one and blocks-2, the profit produced by the best server bound n (and, in the case of blocks-2, the best block sizes k_1 and k_2, all determined by a search), is plotted against the offered load by increasing the arrival rate. For the fix-eps policy, the value of n is chosen so that no more than one job in a thousand is lost, $\varepsilon = 0.001$.

We observe that the fix-ε policy has the worst performance, actually losing money unless the arrival rate exceeds 30. This is not surprising, since the value of ε is quite small, and the costs of servers are disregarded when choosing n. Of the dynamic policies, one-by-one is better than blocks-2, but not by much. Both are better than the static fix-opt policy, but again not by much.

The server allocations for each value of λ are shown in the table below (the values of λ now go up to 100). n^* is the best number of servers found for the fix-opt policy; $n1^*$ is the best upper bound for the one-by-one policy; k_1 and k_2 are the best block sizes for the blocks-2 policy. Note that $n1^*$ is always larger than n^*, while k_1 (at least in this example) is the same as n^*. The improvement in profit produced by the blocks-2 policy is due to the small second block which is brought into play when the first block is full.

The above behaviour is observed for other parameter values, as long as the set-up costs are quite small compared to the operating costs. When that is not the case, the comparison is less clear-cut. This is illustrated in Figure 2, where the performance of the fix-opt, one-by-one and blocks-2 policies is plotted against the set-up cost c_1. The job arrival rate is $\lambda = 40$ and the other parameters are as in figure 1.

Since the fix-opt policy is not affected by the set-up cost, its plot is a horizontal line. The one-by-one policy starts off as the best of the three, but eventually

.

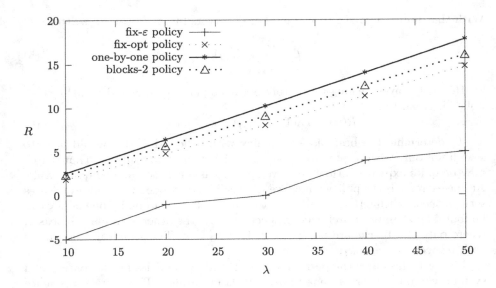

Fig. 1. Comparison of hiring policies $m = 5$, $\mu = 1$, $r = 1$, $c_1 = 0.1$ $c_2 = 3$

Table 1. Server allocations

λ	n^*	$n1^*$	k_1	k_2
20	4	14	4	1
40	8	20	8	1
60	12	27	12	1
80	16	33	16	2
100	20	39	20	2

becomes the worst. This is because it keeps hiring and releasing servers even when that is not warranted by the high set-up costs. The blocks-2 policy is more conservative. It yields slightly lower profits than one-by-one when c_1 is low, but on the other hand it never becomes worse than fix-opt as c_1 increases. What happens is that for high values of c_1, the best blocks-2 policy is of the form $k_1 = n$, $k_2 = 0$. In other words, it becomes identical to fix-opt.

The conclusions that can be drawn from these results, as well as from others derived with different parameter values and cost coefficients, can be summarized as follows:

The optimally chosen static policy fix-opt, which does not incur repeated set-up costs and does not require moving VMs from server to server, performs very well. The dynamic policies from the blocks-b family can achieve $10\% - 15\%$ higher profits when the set-up costs are low. The best blocks-2 policy is always at least as good as the fix-opt policy and is never much worse than one-by-one.

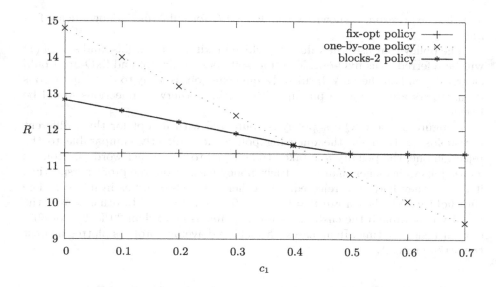

Fig. 2. Comparison of hiring policies; different set-up costs $m = 5$, $\mu = 1$, $r = 1$, $\lambda = 40$ $c_2 = 3$

2.1 Queued Jobs

Instead of rejecting the jobs that find all available VMs occupied, it may be possible to queue them. That would avoid the revenue loss due to rejections, but would raise the question of quality of service. Since customers do not like waiting, the host would normally offer some *Service Level Agreement* (SLA), e.g. promising to pay a penalty u for any job whose waiting time exceeds a given threshold, v.

Would that be worth doing? To evaluate the trade-off, assume that the service times are distributed exponentially. Consider the static policy that hires a fixed number of servers, n, and queues jobs. The long-run average profit produced per unit time is now

$$R(n) = \lambda[r - uP(w > v)] - c_2 n \,, \tag{7}$$

where $P(w > v)$ is the steady-state probability that the waiting time in the $M/M/(nm)$ queue exceeds v. For a stable queue ($\rho < nm$), that probability is given by (e.g., see [11])

$$P(w > v) = qe^{-\mu(nm-\rho)v} \,, \tag{8}$$

where q is the steady-state probability that an incoming job would have to wait:

$$q = \frac{\rho^{nm}}{(nm-1)!(nm-\rho)} \left[\sum_{j=0}^{nm-1} \frac{\rho^j}{j!} + \frac{\rho^{nm}}{(nm-1)!(nm-\rho)} \right]^{-1} \,. \tag{9}$$

This last expression is known as the 'Erlang-C formula', or 'Erlang's delay formula'.

The policy that uses the value of n which maximizes the right-hand side of (7) will be referred to as fixQ-opt. The trade-off between fix-opt and fixQ-opt would clearly depend on the SLA. Intuitively, queueing jobs is likely to be advantageous if customers are willing to put up with waiting, otherwise rejections would be better.

In figure 3, the fixQ-opt policy is compared with fix-opt for three different thresholds v. To make the queueing policy more directly comparable to the rejection one, the penalty u is taken to be equal to r. In other words, customers whose jobs wait longer than v get their money back. From the profit perspective, it is as if they had been rejected. The other parameters are as in figure 1. The threshold values chosen are $v = 0.2$, $v = 0.4$ and $v = 0.6$. In other words, the penalty is payable if the customer's waiting time is more than 20%, 40% or 60% of their residence time. In all cases, the achieved average profit is plotted against the arrival rate, λ.

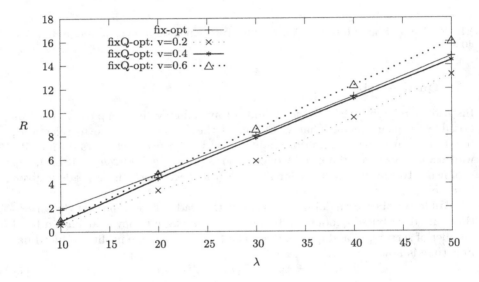

Fig. 3. Rejection vs queueing; different thresholds v $m = 5$, $\mu = 1$, $r = u = 1$, $c_2 = 3$

We observe that when customers are impatient ($v = 0.2$), it is better to reject jobs than to queue them. The situation is reversed when customers are quite tolerant of waiting ($v = 0.6$) and the load exceeds 20. However, the differences are not large in either case. For the intermediate threshold ($v = 0.4$), there is even less to choose between queueing and rejecting. Moreover, increasing v beyond 0.6 does not improve the profits achieved by fixQ-opt significantly.

In summary, one can say again that, as long as the number of servers is chosen optimally, the management of the jobs is of secondary importance.

N.B. The Markovian assumptions of Poisson arrivals and exponential service times can be relaxed, at the price of replacing the exact results with approximations. There are approximate results for the $G/GI/n/n$ loss model, as well as for the $G/GI/n$ queue. To use those expressions one would need to estimate not only the average interarrival and service times, but also their second moments.

3   Virtual Machines Do Not Move

The dynamic policies examined in the last section relied for their operation on the ability to move virtual machines from server to server instantaneously and without cost. It would be interesting to evaluate the extent to which profits are affected when that flexibility is removed. In order to do that, we shall assume that the job service times are distributed exponentially, with mean $1/\mu$.

Consider the dynamic policy blocks-2, where n servers are divided into two blocks, 1 and 2, of sizes k_1 and k_2 respectively ($k_1 + k_2 = n$). A VM, once started in a server, remains there until its job completes. Let us assume that the servers in block 1 are permanently hired. Incoming jobs are allocated VMs in block 1 whenever there are fewer than $k_1 m$ jobs present there; otherwise they go to block 2, if there are fewer than $k_2 m$ jobs there; when all nm VMs are busy, jobs are rejected. The servers in block 2 are released when a departure leaves all of them empty; they are re-hired at the next arrival instant which finds block 1 full.

A single integer - the total number of jobs present - is no longer enough to describe the state of the system. One needs to specify the number of jobs, I, present in block 1, and the number of jobs, J, present in block 2. Those two random variables are not independent: J can increase only when $I = k_1 m$. Let $p_{i,j}$ be the joint steady-state probability that $I = i$ and $J = j$ ($i = 0, 1, \ldots, k_1 m$; $j = 0, 1, \ldots, k_2 m$).

The servers in block 2 are being used whenever $J > 0$. Hence, the average number of servers hired, L, is given by

$$L = k_1 + k_2 \left[1 - \sum_{i=0}^{k_1 m} p_{i,0} \right] . \tag{10}$$

The hiring of block 2 occurs whenever an incoming job finds the system in state $I = k_1 m$, $J = 0$. Hence, the average number of server hiring events per unit time, S, is

$$S = \lambda k_2 p_{k_1 m, 0} . \tag{11}$$

Jobs are rejected when both blocks are full, which occurs with probability $p_{k_1 m, k_2 m}$. The average profit obtained per unit time is thus given by an expression similar to (6):

$$R(n) = r\lambda(1 - p_{k_1 m, k_2 m}) - c_1 S - c_2 L . \tag{12}$$

It now remains to determine the joint distribution $p_{i,j}$. The instantaneous transition diagram for the Markov process (I, J) is illustrated in figure 4.

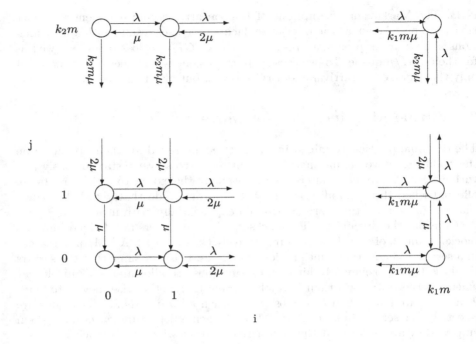

Fig. 4. Transition diagram

For $i = 0, 1, \ldots, k_1m - 1$ and $j = 0, 1, \ldots, k_2m$, the probabilities $p_{i,j}$ satisfy the following balance equations:

$$(\lambda + i\mu + j\mu)p_{i,j} = \lambda p_{i-1,j} + (i + 1)\mu p_{i+1,j} + (j + 1)\mu p_{i,j+1} , \qquad (13)$$

where $p_{-1,j} = 0$ and $p_{i,k_2m+1} = 0$ by definition. When $i = k_1m$, the equations become:

$$(\lambda + k_1m\mu + j\mu)p_{k_1m,j} = \lambda p_{k_1m-1,j} + \lambda p_{k_1m,j-1} + (j + 1)\mu p_{k_1m,j+1} , \qquad (14)$$

where $p_{k_1m,k_2m+1} = 0$ by definition.

The numerical complexity of solving this set of simultaneous equations, plus the normalizing equation, by Gaussian elimination, can be high. It is on the order of $O[(k_1m + 1)^3(k_2m + 1)^3]$. Fortunately, we can exploit the special structure of this Markov process in order to reduce that complexity considerably.

Note first that the total number of jobs present, $I+J$, behaves like the number of calls in an Erlang loss system $M/M/nm/nm$ with offered traffic $\rho = \lambda/\mu$. In particular, the rejection probability, p_{k_1m,k_2m}, is equal to the probability q_{nm}, given by expression (1) for $j = nm$.

Examining equations (13) for $j = k_2m$ and $i = 0, 1, \ldots, k_1m - 1$ in turn, we see that they can be transformed into recurrence relations. Denoting k_2m by K, these can be written as

$$p_{i,K} = a_{i+1}p_{i+1,K} \;\; ; \;\; i = 0, 1, \ldots, k_1 m - 1 \,, \tag{15}$$

where $a_0 = 0$ and

$$a_{i+1} = \frac{i+1}{(1 - a_i)\rho + i + K} \;\; ; \;\; i = 0, 1, \ldots, k_1 m - 1 \,. \tag{16}$$

After evaluating $p_{k_1 m, k_2 m}$ and the coefficients a_i, the recurrences (15) enable the probabilities, $p_{k_1 m - 1, K}$, $p_{k_1 m - 2, K}$, $\ldots$, $p_{0,K}$ to be computed by successive substitution. Denote the sum of those probabilities, i.e. the marginal probability that block 2 is full, by $p_{.,K}$. Equating the rate at which the number of jobs in block 2 decreases from K to $K - 1$ with that at which it increases from $K - 1$ to K, we obtain

$$K\mu p_{.,K} = \lambda p_{k_1 m, K-1} \,. \tag{17}$$

This equation determines $p_{k_1 m, K-1}$. In general, if the marginal probability, $p_{.,j+1}$, that there are $j + 1$ jobs in block 2 is known, then the probability $p_{k_1 m, j}$ is determined from

$$(j + 1)\mu p_{.,j+1} = \lambda p_{k_1 m, j} \;\; ; \;\; j = 0, 1, \ldots, K - 1 \,. \tag{18}$$

Now consider the balance equations corresponding to row j in the diagram, for $j = 0, 1, \ldots, K - 1$. They can be written in the form

$$p_{i,j} = a_{i+1,j}p_{i+1,j} + b_{i+1,j} \;\; ; \;\; i = 0, 1, \ldots, k_1 m - 1 \,, \tag{19}$$

with $a_{0,j} = 0$,

$$a_{i+1,j} = \frac{i+1}{(1 - a_{i,j})\rho + i + j} \;\; ; \;\; i = 0, 1, \ldots, k_1 m - 1 \,, \tag{20}$$

and $b_{0,j} = 0$,

$$b_{i+1,j} = \frac{\rho b_{i,j} + (j + 1)p_{i,j+1}}{(1 - a_{i,j})\rho + i + j} \;\; ; \;\; i = 0, 1, \ldots, k_1 m - 1 \,. \tag{21}$$

Having determined the probabilities in row $j + 1$, and consequently $p_{k_1 m, j}$ from (18), these relations determine all the other probabilities in row j.

Proceeding in this manner through rows $K - 1$, $K - 2$, $\ldots$, 0, one can compute all unknown probabilities. The numerical complexity of that procedure is on the order of $O[(k_1 m + 1)(k_2 m + 1)]$, i.e. it is linear in the number of unknowns.

It may be expected that the inability to move VMs and pack them into the smallest number of servers would reduce the effectiveness, and hence the profits, of a dynamic hiring policy. The extent of that reduction is illustrated in figure 5. The blocks-2 policy of this section is compared with the fix-opt and blocks-2 policies of section 2, where VMs could move instantaneously from server to server. For both versions of the dynamic policy, the block sizes k_1 and k_2 are

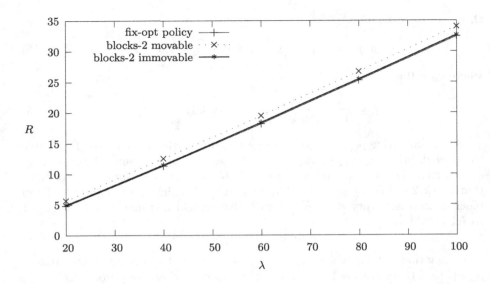

Fig. 5. Static and dynamic policies with movable and immovable VMs $m = 5$, $\mu = 1$, $r = 1$, $c_1 = 0.1$ $c_2 = 3$

chosen optimally, by performing a search. In all cases, the achieved profit is plotted against the arrival rate λ, while the other parameters are as in figure 1.

The figure shows that when VMs cannot be moved, the advantage of dynamic hiring is almost eliminated. The blocks-2 policy with immovable VMs still produces slightly higher profits than the fix-opt policy, but the improvements are on the order of 1%.

4   Conclusion

We have evaluated and compared the performance of several static and dynamic server hiring policies. The results of a number of numerical experiments lead to the rather general conclusion that a well-chosen static policy can be nearly as good as a dynamic one. Moreover, either an increase in set-up costs, or a restriction in the movement of VMs, tends to reduce the gains achieved by a dynamic policy. That will be the case, for example, if VMs can be moved, but not instantaneously. The costs and durations of moving VMs were examined in more detail in Voorsluys et al [14].

Among the dynamic hiring policies, the simple blocks-2 policy may be recommended. The best block sizes are quite easily determined, and it is robust with respect to rising set-up costs. However, its advantages are significantly reduced if VMs cannot move from server to server.

If the incoming jobs belong to different classes, with different parameters, and if servers are hired and dedicated to separate classes, then our results would apply to each individual class. The situation would be more complicated if VMs of different classes were allocated to the same server. That would require further work.

What if jobs do not arrive in a Poisson stream? Some answers may be obtained by applying approximate results, but those would probably need to be validated by simulations.

References

1. Bennani, M.N., Menascé, D.: Resource allocation for autonomic data centers using analytic performance methods. In: Procs., 2nd IEEE Conf. on Autonomic Computing (ICAC 2005), pp. 229–240 (2005)
2. Bodík, P., Griffith, R., Sutton, C., Fox, A., Jordan, M., Patterson, D.: Statistical machine learning makes automatic control practical for internet datacenters. In: Conf. on Hot Topics in Cloud Computing (HotCloud 2009), Berkeley, CA, USA (2009)
3. Chaisiri, S., Lee, B.S., Niyato, D.: Optimization of resource provisioning cost in cloud computing. IEEE Transactions on Services Computing 5(2), 164–177 (2012)
4. Chandra, A., Gong, W., Shenoy, P.D.: Dynamic resource allocation for shared data centers using online measurements. In: Jeffay, K., Stoica, I., Wehrle, K. (eds.) IWQoS 2003. LNCS, vol. 2707, pp. 381–400. Springer, Heidelberg (2003)
5. Gandhi, A., Harchol-Balter, M., Adan, I.: Server farms with setup costs. Performance Evaluation 67(11), 1123–1138 (2010)
6. Ghribi, C., Hadji, M., Zeghlache, D.: Energy efficient VM scheduling for cloud data centers: exact allocation and migration algorithms. In: 13th IEEE/ACM International Symposium on Cluster, Cloud and Grid Computingpp, 671–678 (2013)
7. Mazzucco, M., Dyachuk, D., Dikaiakos, M.: Profit-aware server allocation for green internet services. In: IEEE Int. Symp. on Modeling, Analysis and Simulation of Computer and Telecommunication Systems (MASCOTS), pp. 277–284 (2010)
8. Mazzucco, M., Vasar, M., Dumas, M.: Squeezing out the cloud via profit-maximizing resource allocation policies. In: IEEE Int. Symp. on Modeling, Analysis and Simulation of Computer and Telecommunication Systems (MASCOTS), pp. 19–28 (2012)
9. McGough, A.S., Mitrani, I.: Optimal hiring of cloud servers. In: Horváth, A., Wolter, K. (eds.) EPEW 2014. LNCS, vol. 8721, pp. 1–15. Springer, Heidelberg (2014)
10. Messerli, E.J.: Proof of a convexity property of the Erlang B formula. Bell System Technical Journal 51, 951–953 (1972)
11. Mitrani, I.: Probabilistic Modelling. Cambridge University Press (1998)
12. Mitrani, I.: Trading power consumption against performance by reserving blocks of servers. In: Tribastone, M., Gilmore, S. (eds.) UKPEW 2012 and EPEW 2012. LNCS, vol. 7587, pp. 1–15. Springer, Heidelberg (2013)
13. Urgaonkar, R., Kozat, U.C., Igarashi, K., Neely, M.J.: Dynamic resource allocation and power management in virtualized data centers. In: IEEE/IFIP NOMS 2010, Osaka, Japan (2010)
14. Voorsluys, W., Broberg, J., Venugopal, S., Buyya, R.: Cost of virtual machine live migration in clouds: a performance evaluation. In: Jaatun, M.G., Zhao, G., Rong, C. (eds.) Cloud Computing. LNCS, vol. 5931, pp. 254–265. Springer, Heidelberg (2009)

Security and Performance Tradeoff Analysis of Mobile Offloading Systems Under Timing Attacks

Tianhui Meng[✉], Katinka Wolter, and Qiushi Wang

Department of Mathematics and Computer Science, Freie Universität Berlin,
Takustr. 9, 14195 Berlin, Germany
{tianhui.meng,katinka.wolter,qiushi.wang}@fu-berlin.de

Abstract. Mobile offloading systems have been proposed to migrate complex computations from mobile devices to powerful servers. While this may be beneficial from the performance and energy perspective, it certainly exhibits new challenges in terms of security due to increased data transmission over networks with potentially unknown threats. Among possible security issues are timing attacks which are not prevented by traditional cryptographic security. Metrics on which offloading decisions are based must include security aspects in addition to performance and energy-efficiency. This paper aims at quantifying the security attributes and their impact on the performance of mobile offloading systems. The offloading system is modeled as a hybrid CTMC and queueing model. The proposed model focuses on state transition and state-based control. The quantification analysis is carried out for steady-state behavior of the CTMC model as to optimize the weighted-sum cost measure. By transforming the security model to a model with absorbing state, we compute the "mean time to security failure" (MTTSF) measure. Finally, a security and performance tradeoff measure is computed based on the system model and optimum parameter set is found for the system.

Keywords: Mobile offloading · Security attributes · Quantitative analysis · Phase-type distribution

1 Introduction

Mobile phones are no longer used only for voice communication and short message service (SMS); instead, they are used for watching videos, gaming, web surfing, and many other applications. While the last decades witness great advances in hardware technology, mobile devices still face the restriction of resources, such as battery life, network bandwidth, storage capacity, and processor performance. Offloading is a solution to augment these mobile systems' capabilities by migrating computation to more resourceful computers (i.e., servers). This is different from the traditional client-server architecture, where a thin client always migrates computation to a server [1]. In many scenarios, the limited computing speeds of mobile systems can be enhanced by offloading. One example is

© Springer International Publishing Switzerland 2015
M. Beltrán et al. (Eds.): EPEW 2015, LNCS 9272, pp. 32–46, 2015.
DOI: 10.1007/978-3-319-23267-6_3

context-aware computing infrastructure [2] –where multiple streams of data from different sources like GPS, maps, accelerometers and temperature sensors need to be analyzed together in order to obtain real-time information about a user's context. Even though battery technology has been steadily improving, it has not been able to keep up with the rapid growth of power consumption of mobile systems. Offloading may extend battery life by migrating the energy-intensive parts of the computation to servers [3].

The smooth offloading of computation depends on a fast and stable network connection, which guarantees seamless communication. While in unreliable network condition, task completion can be delayed or interrupted by congestion or packet loss, when offloading may not always benefit [4]. This involves making a decision regarding whether and what computation to migrate, which usually depends on many parameters such as the network bandwidths and the amounts of data exchanged through the networks. A vast body of research exists on offloading decisions for improving system performance and saving energy [5][6][7][8].

While offloading becomes an attractive solution for mobile systems from the performance and energy perspective as applications become increasingly complex, it certainly exhibits new challenges in terms of security due to increased data transmission over networks with potentially unknown threats. Protecting user privacy and data/application secrecy from an adversary is a key to establish and maintain consumers' trust in the mobile platform, especially in mobile cloud computing. Metrics on which offloading decisions are based must include security aspects in addition to performance and energy-efficiency. Indeed, security is such a big area covering large numbers of issues. In this paper, we deal with the specific threat of timing attacks whose remote feasibility has been proved [9]. Timing attacks based on information gained from the service response time are so effective that they pose a real threat to mobile offloading systems.

Quantitative analyses of system dependability and reliability have received great attention for several decades. However quantification of security has only recently attracted more attention, and some initial conceptual work has been published already decades ago, serious model-based evaluation of security mechanisms has been published only recently. The authors in [10] have shown how a key distribution centre can be modelled and analysed, and how to find an optimal key refresh rate for such a system. Previous work on the security of computing and information systems has been mostly assessed from a level point of view. A system is assigned a given security level with respect to the presence or absence of certain functional characteristics and the use of certain development techniques. In 2013, Zhang [11] proposed an approach to evaluate the network security situation objectively using Network Security Index System (NSIS). Only a few studies have considered the quantitative evaluation of security. The authors in [12] make an effort to examine the security vulnerabilities of operating systems of routers within the cloud carrier by assessing the risk based on the National Vulnerability Database (NVD) and gives a quantifiable security metrics for cloud carrier, which is very useful in the Service Level Agreement (SLA) negotiation between a cloud consumer and a cloud provider.

In order to proceed to a quantitative treatment of the performance-security tradeoff of offloading system we propose a hybrid Continuous-time Markov chain (CTMC) and queueing model which treats the security and performance attributes respectively. In this paper, we show how to formulate measures that include both, performance and security aspects and that optimize the tradeoff between the two. Our model is aimed to deal with a general offloading system with a master secret stored on the server side, where the attacking client can get normal offloading service. Of course many security problems are relevant to the mobile offloading scenario, but in this paper we only address timing attacks. In a timing attack the attacker deduces information about a secret key from runtime measurements of successive requests. This process can be interrupted by frequently changing the key [13]. By solving the proposed model, we propose different metrics on which offloading decisions can be based. The system cost metric is based on a weighted sum of security and system effort made for rekeying. The tradeoff metric is the product of a security metric and a performance metric. Also, optimal rekeying rates are found for the two measures.

The remainder of this paper is structured as follows. In Section 2, we overview the system and attackers' behavior and propose a hybrid model for a offloading system under timing attacks. The system metrics on which the evaluation based are addressed in Section 3. Section 4 shows the model analysis by solving the model. Section 5 gives numeral results of the analysis performed on the model for a sample. And finally, the paper is concluded and future work are presented in Section 6.

2   System Overview and The Model

A mobile offloading system is a common solution to enhance the capabilities of the mobile system by migrating computation to more resourceful computers (i.e., servers). To quantitatively analyze the performance and security attributes of such a system under the threat of timing attacks, we have to incorporate the actions of an attacker who is trying to capture sensitive information in conjunction with the protective actions taken by the system. Therefore, we have to develop a hybrid CTMC and queueing model that takes into account the behavior of both actors.

The state transition model represents the system behavior for a specific attack and given system configuration that depends on the actual security requirements. In our scenario, the system is assumed to be vulnerable to timing attacks in which the attacker in the worst case will eventually decrypt the system private key saved in the server. We assume that the server is configured as to renew its key regularly to prevent or handle these attacks. At the same time, the queueing model presents the offloading decision and jobs processing operation. A job is either processed locally by the mobile device or offloaded and served by cloud servers.

2.1   Behavior of System and Attackers

In the offloading system we consider, a master key stored in the server is used for the encryption and decryption operations of all user data. In order to improve security, the server regularly or irregularly changes the master key, which is called the rekeying process. The system has to process all user-files with both the new and the old master key. In this process, the system does not accept any other user commands. When user data is very large, this process will take long. Therefore, it is reasonable to recommend an optimal interval time for the master key replacement cycle, and select a suitable time, when there is a low amount of user access(e.g. at night).

Implementations of cryptographic algorithms often perform computations in non-constant time, due to performance optimization. If such operations involve secret parameters, these timing variations can leak some information and a careful statistical analysis could even lead to the total recovery of these secret keys. Because of timing attacks gain secret information from the server response time, they are a real threat to mobile offloading systems. However this threat is not covered by traditional notions of cryptographic security [14]. It was commonly believed that timing attacks can be directed only towards smart cards or affect inter-process locally, but more recent research reveals that remote timing attacks are also possible and should be taken into consideration [15][9]. Mobile offloading requires access to resourceful servers for short duration through wireless networks. These servers may use virtualization techniques to provide services so that they can isolate and protect different programs and their data. However, the author in [16] shows that using a cache timing attack, an attacker can bypass the isolated environment provided by virtualization characteristics, where sensitive code is executed in isolation from untrustworthy applications. It is worth mentioning that a timing attack also poses a threat to other types of systems. Timing attacks can be detrimental in the mix-zone construction and usage model over road networks [17].

In timing attacks to our offloading system, an attacker continues to send requests to the server and the obtained offloading service will be properly performed by the server. In addition the attacker records each response time for a certain service and tries to find clues to the master secret of the server by comparing time differences from several request queues. If the attacker successfully breaks the secret information from the timing results, he may hack into the system, read and even modify other users' information without authorization.

2.2   The System Model

As compared to our previous work [18], in which the model only considered the security attributes of offloading systems, the proposed hybrid CTMC and queueingngnmodel in this work takes the performance properties of a generic offloading system into account (Fig. 1). When jobs are generated by a mobile device, they are either offloaded to the cloud or executed locally, expressed by the two queues, respectively. The parameters λ and λ' indicate the arrival rates for the two queues. A job dispatched to offload comes to the upper queue and is served by the server with service rate μ, which also includes the data transmission time. For jobs dispatched to execute locally, the service rate is μ' which is assumed to be lower than μ.

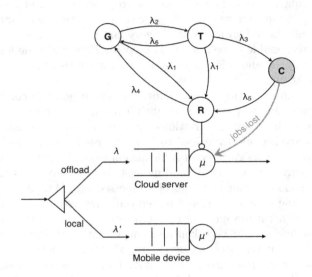

Fig. 1. State transition diagram for a generic offloading system

The states and parameters of the CTMC model are summarized here:

- G Good state in which the offloading system works properly
- T Timing attack happening state
- C Compromised state after the attacker knows the secret of the system
- R Rekeying state in which system renews its master secret
- λ_1 rate at which the system launches the rekeying process in state G and state T
- λ_2 rate at which an attacker triggers a timing attack to the system
- λ_3 rate at which a timing attack succeeds to break the system secret
- λ_4 rate at which the system is brought back to the good state by the rekeying process
- λ_5 rate at which the system launches the rekeying process in state C
- λ_6 rate at which the attacker successfully breaks the key, while fails at accessing the data or he just fails to conduct a successful timing attack

The upper part of Fig. 1 shows a CTMC model representing the states of the system. After initialization, the system starts to operate properly in the good state G. The system is under the specific threat of timing attacks conducted by random attackers. We describe the events that trigger transitions among states in terms of transition rates. It is assumed that there is only one attacker in the system at one time. If an attack happens, the system is brought to the timing attack state T at rate λ_2. In this state the attacker tries to break the server encryption key by making time observations. So while the system is in state T, the attacker is not yet able to access confidential information.

It takes a certain time to perform the timing attack after which the attacker may know the encryption key and the system moves to the compromised state C at rate λ_3. So λ_3^{-1} is the mean time a timing attack takes. There is a possibility indicated by the arc λ_6 that the attacker successfully breaks the key, while fails at accessing the data or he just fails to conduct a successful timing attack. If the attacker succeeds to determine the encryption key through time measurements, confidential data will be disclosed which is assumed to incur a high cost. This can only happen if the system is in the compromised state C and we call the incident of entering the compromised state a security failure. In this state, all jobs dispatched to offload are not secure any more, therefore they must be repeated and do not contribute to the throughput. The jobs lost is represented by the red arc in Fig. 1.

Renewing the server encryption key can prevent or interrupt a timing attack. The arcs from other states to state R represent these operations in the server. The rekeying rate is the parameter one can tune as a system administrator. It indicates how often the system launches the rekeying process. The rate λ_1 is the rekeying rate when the system is in good state G or in the timing attack state T. We assume the offloading system has intrusion detection mechanisms running on it, that can find clues of compromised behavior, in which case the system will trigger the rekeying process more frequently. So in the compromised state C, we assume the rekeying process is triggered at a different rate, $\lambda_5 = n\lambda_1$. The parameter n is called the coefficient of rekeying in the compromised state because it represent the relationship between the rekeying rate (or rekeying frequency) in good state and the rekeying rate in compromised state. All these three paths transfer the system to the rekeying state R from which it will finally return to the initial state G. The challenge is to find an optimal value for the rekeying interval. The rekeying should in the optimal case happen before or soon after the system enters the compromised state. We consider 4 rekeying options in this paper, that is $n = 0.5, 1, 2, 3$ respectively.

In the rekeying state the system refuses all user requests. So we put a inhibitor arc on the cloud server. All the jobs are dispatched to the local queue and some jobs will be lost in this state. As a result, the system throughput is degraded. The rekeying process will bring the system back to the initial state G at rate λ_4. Consequently, the mean time to perform the rekeying process is λ_4^{-1} and during this time the server refuses user requests.

For the system parameters we use experimental data from an offloading engine and OCR (Optical Character Recognition) implementation [4]. The mean local execution time for an OCR job on the mobile device was 2377 ms. We set $\mu' = 1/2.377 \approx 0.42$. The mean offloading time including the data transition time is 1191 ms. Then $\mu = 1/1.191 \approx 0.84$. For the queues to be stable, we assume $\lambda = 0.8$ and $\lambda' = 0.4$.

In this work, we use the CTMC model for the brevity of analysis which means we assume all the state transition times are exponential distributed. While in practice, depending on the nature of an attack, attacking time may follow one of the several distribution functions. After getting the realistic time distribution of timing attacks, we will do phase-type fitting using Hyperstar [19] to get more accurate parameters for our model.

3   Metrics

After defining the model and its parameters, we must now establish the measures we want to investigate. We present security and performance metrics, respectively.

3.1   Security Metrics

As in our previous work [18], the security measures are defined as confidentiality and system (security) cost that are functions of the steady-state probabilities of the CTMC model. The steady-state probabilities π_i may be interpreted as the proportion of time that the CTMC spends in state i, where $i \in \{R, G, T, C\}$.

If a timing attack to the offloading system is successful, the attacker obtains the master key and can browse unauthorized files thereafter. The entered states denote the loss of confidentiality. Therefore, the steady-state confidentiality measure can be computed as

$$Confid = 1 - \pi_C . \tag{1}$$

We also define a system cost metric. In our scenario, the offloading system suffers from cost in two states, the compromised state C and the rekeying state R. The system loses sensitive information in the compromised state, and cost is also incurred when the system deploys a rekeying process. The rekeying cost and the data disclosure cost are both interpreted as the proportion of system life time, that is, the steady-state probability of the CTMC. We define a weight w and its complement $1 - w$ for the two kinds of cost. We use normalised weights for simplicity. So the system cost is defined as:

$$Cost = w\pi_R + (1 - w)\pi_C , \tag{2}$$

where $\pi_i, i \in \{R, C\}$ denotes the steady-state probability that the continuous-time Markov process is in state i. $0 \leq w \leq 1$ is the weighting parameter used to share relative importance between the loss of sensitive information and the effort needed to rekey regularly.

3.2   Performance Metrics

The performance metrics we are interested in describe the system in terms of its throughput, completion times, or response times, as defined e.g. in queueing theory or networking. In this paper we use the throughput as the performance metric for the offloading system. By Little's Law, the throughput (denoted X) is defined as:

$$X = \frac{E[N]}{E[R]} \, . \tag{3}$$

For each queue, the throughput equals the average number of jobs in the queueing station ($E[N]$) divided by the average time a job spends in the queueing station ($E[R]$). The system throughput equals the sum of the two queues.

3.3 Tradeoff Metric

In order to investigate how system security will interact with performance, we also define a tradeoff metric. An objective function formed from the product of the security attribute confidentiality and system throughput is created to demonstrate the tradeoff situation. As a system designer, one may look forward to maintaining the confidentiality of sensitive information with higher throughput, as for the tradeoff measure, the larger the better.

$$Tradeoff = Confid \times X \, . \tag{4}$$

The security and performance metrics defined here will be used to evaluate the system attributes in the rest of this paper. In the next Section, we will evaluate these measures by computing the steady-state probability of the CTMC model and solving the queueing model.

4 Model Analysis

In this section, we derive and evaluate the security and performance attributes of the offloading system using methods for quantitative assessment of dependability, known as the dependability attributes, e.g. reliability, availability, and safety which have been well established quantitatively.

4.1 CTMC Steady-State Probability Computation

For the system security attributes, we have described the system's dynamic behavior by a CTMC model with the state space $X_s = \{R, G, T, C\}$ and the transitions between these states. In order to carry out the security quantification analysis, we need to determine the stationary distribution of the CTMC model.

The steady-state probabilities $\{\pi_i, i \in X_s\}$ of the CTMC can be computed by solving the system of linear equations [20]

$$\pi \mathbf{Q} = 0, \tag{5}$$

where $\pi = [\pi_R, \pi_G, \pi_T, \pi_C]$ and $\mathbf{Q}$ is the infinitesimal generator (or transition-rate matrix) which can be written as:

$$\mathbf{Q} = \begin{array}{c} \\ R \\ G \\ T \\ C \end{array} \begin{pmatrix} \overset{R}{-\lambda_4} & \overset{G}{\lambda_4} & \overset{T}{0} & \overset{C}{0} \\ \lambda_1 & -\lambda_1 - \lambda_2 & \lambda_2 & 0 \\ \lambda_1 & \lambda_6 & -\lambda_1 - \lambda_3 - \lambda_6 & \lambda_3 \\ \lambda_5 & 0 & 0 & -\lambda_5 \end{pmatrix} \tag{6}$$

In addition, we have the total probability relationship:

$$\sum_i \pi_i = 1 \quad i \in X_s. \tag{7}$$

The transition-rate matrix $\mathbf{Q}$ describes the dynamic behavior of the security model as shown in Fig. 1. The first step towards quantitatively evaluating security attributes is to find the steady-state probability vector π of the CTMC states by solving Eqs. 5 and 7. We can get solutions:

$$\pi_R = \frac{[(\lambda_1 + \lambda_2)(\lambda_1 + \lambda_3) + \lambda_1\lambda_6]\lambda_5}{\phi}, \tag{8}$$

$$\pi_G = \frac{(\lambda_1 + \lambda_3 + \lambda_6)\lambda_4\lambda_5}{\phi}, \ \pi_T = \frac{\lambda_2\lambda_4\lambda_5}{\phi}, \ \pi_C = \frac{\lambda_2\lambda_3\lambda_4}{\phi}.$$

For the sake of brevity, we assume:
$\phi = (\lambda_1 + \lambda_4)(\lambda_1 + \lambda_3 + \lambda_6)\lambda_5 + [(\lambda_1 + \lambda_4)\lambda_5 + (\lambda_4 + \lambda_5)\lambda_3]\lambda_2.$

Given the steady-state probabilities of CTMC model, various measures, such as, confidentiality and cost can be computed via Eqs. 1 to 4.

4.2   CTMC with Absorbing State - MTTSF Analysis

For quantifying the reliability of a software system, *mean time to failure* (MTTF) is a widely used reliability measure. MTTF provides the mean time it takes for the system to reach one of the designated failure states, given that the system starts in a good state. In reliability analysis, the failed states are made absorbing states. Once the system reaches one of the absorbing states, the probability of moving out of this state is 0, i.e., there are no outgoing arcs from such states. In this section, we use *mean time to security failure* (MTTSF) as the measure for quantifying the security of our offloading system. MTTF or MTTSF can be evaluated by making the compromised state of the CTMC an absorbing state, as shown in Fig. 2.

Given a Continuous-Time Markov Chain (CTMC) with one absorbing state, we may enter this chain at some state i with probability α_i. For each state that is visited, the time before going to the next state follows an exponential distribution. Thus, the time required to reach the absorbing state from an initial state i is a sum of samples from exponential distributions. For a given CTMC, a phase-type distribution is defined as the distribution of the time to absorption that can be observed along the paths in a CTMC with one absorbing state [21].

In our scenario, the MTTSF is the mean time it takes for the system to reach the security failure state C. The first moment of a PH-distribution exactly expresses the mean time to absorption in an absorbing CTMC. So for our model

$$MTTSF = E[X] = -\underline{\alpha}\mathbf{T}^{-1}\underline{1}. \tag{9}$$

The parameters are

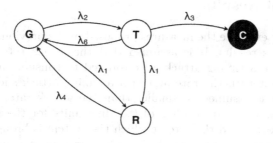

Fig. 2. CTMC model with an absorbing state

$$\mathbf{T} = \begin{pmatrix} -\lambda_4 & \lambda_4 & 0 \\ \lambda_1 & -\lambda_1 - \lambda_2 & \lambda_2 \\ \lambda_1 & \lambda_6 & -\lambda_1 - \lambda_3 - \lambda_6 \end{pmatrix}, \tag{10}$$

and the initial probability vector is

$$\underline{\alpha} = \begin{pmatrix} 0 & 1 & 0 \end{pmatrix}. \tag{11}$$

Substituting into Eq. 9, we get

$$MTTSF = \frac{(\lambda_1 + \lambda_4)(\lambda_1 + \lambda_2 + \lambda_3 + \lambda_6)}{\lambda_2 \lambda_3 \lambda_4}. \tag{12}$$

4.3 Throughput Analysis

The steady-state probabilities π_i may be interpreted as the proportion of time that the CTMC spends in state i. We have defined the throughput metric in Eqe. 3. We assume the total system life time is T. In the good state G and timing attack state T, the number of jobs served by the system should be $\lambda(\pi_G + \pi_T)T + \lambda'(\pi_G + \pi_T)T$, given the queues are stable. While in the rekeying state R, the server refuses all the users' requests and all jobs must be executed locally. Assuming $\mu' < \lambda + \lambda'$, the number of jobs served then is $\mu'\pi_R T$. In the compromised state C, all the jobs dispatched to offload are not secure, so they do not contribute to the throughput. In this state, the system throughput only covers the jobs executed locally $\lambda'\pi_C T$ Therefore, we get the system throughput as

$$X = \frac{\lambda(\pi_G + \pi_T)T + \lambda'(\pi_G + \pi_T)T + \mu'\pi_R T + \lambda'\pi_C T}{T} \tag{13}$$
$$= (\pi_G + \pi_T)\lambda + (1 - \pi_R)\lambda' + \pi_R \mu'.$$

After determining the security and performance measures for the model, we conduct tradeoff analysis in the following Section.

5   Numerical Results

In this section, we evaluate the measures proposed in the previous sections using the model analysis results. It is assumed that the attack rate to the system is $\lambda_2 = 1$. Because a timing attack is considered to consume more time than attacker coming interval, the rate at which a timing attack succeeds to break the system secret is assumed to smaller than the attack rate, i.e. $\lambda_3 = 0.3$. We also assume it takes an average of 0.5 time units for the server to carry out a rekeying process and the rate at which the system is brought back to the good state by the rekeying process is $\lambda_4 = 2$. The rate at which the attacker successfully breaks the key, while fails at accessing the data or he just fails to conduct a successful timing attack is assumed to be $\lambda_6 = 1$.

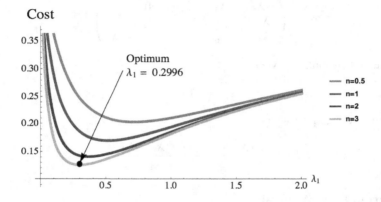

Fig. 3. System measure $Cost$ as a function of the rekeying rate λ_1

Fig. 3 shows the system measure cost, which is defined in Eq. 2, changing with the rekeying rate λ_1. We set the weighting parameter $w = 0.5$ to put equal importance to the loss of sensitive information cost and the effort needed to rekey regularly. The parameter n in this figure is the coefficient of rekeying in the compromised state, i.e. $\lambda_5 = n\lambda_1$. In this work, we consider 4 rekeying options, that is $n = 0.5, 1, 2, 3$ respectively. The rekeying rate λ_1 indicates how often the system launches the rekeying process. When the rekeying rate is low, the system cost is very large due to the high probability of an insecure state. We find the optimum rekeying rate $\lambda_1 = 0.2996$ for the lowest system cost when $n = 3$. After the lowest value, because of the increasing effort to perform rekeying process, the cost is also getting larger at high rekeying rate. We further see that the system cost decreases with increasing coefficient n. This is because for all rekeying durations, the mean time in the compromised state decreases as we rekey more frequently in this situation. On practical bound, it is to trigger more rekeying process when the intrusion detection mechanism finds some clues.

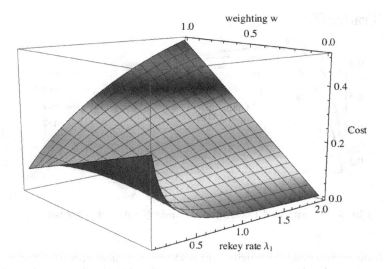

Fig. 4. *Cost* over rekeying rate λ_1 and weighting parameter w

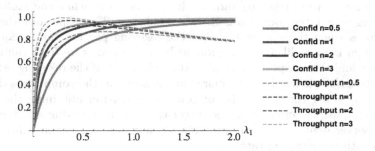

Fig. 5. *Confidentiality* and *Throughput* over the rekeying rate λ_1

We study the effect of the weighting parameter w on the system cost in Fig. 4. Also 4 rekeying options are considered here, i.e. $n = 0.5, 1, 2, 3$ respectively. We look at the marginal values first. It can be seen from the figure that the cost decreases monotonically with the rekeying rate λ_1 when $w = 0$, where we only consider the costs of losing sensitive information in the compromised system. Intuitively, in this case when we trigger the rekeying process more often, the security cost will decrease. When we put all weight on the rekeying effort ($w = 1$), the cost increases with the rekeying rate. The light color in the middle of the figure shows the optimum rekeying rate. For the middle values of the weighting parameter w, the optimum rekeying rate for the lowest cost decreases when we put more weight on rekeying effort cost. For each specific rekeying rate λ_1 the system cost is a straight line weighting the two kinds of cost. In this figure, we get the largest rekeying effort cost and lowest security cost at rate $\lambda_1 = 2.0$.

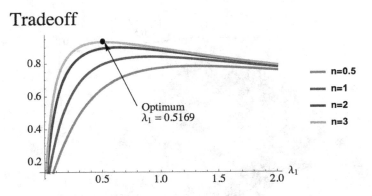

Fig. 6. Security and Performance *Tradeoff* with rekeying rate λ_1

Fig. 5 shows the system security and performance measures, i.e. system confidentiality (defined in Eq. 1) and throughput (defined in Eq. 3), changing with the rekeying rate λ_1. It can be seen that the confidentiality measure monotonically increases with growing rekeying rate λ_1. It also increases when the multiple of the rekeying rate in the compromised state n is larger. This is because the security improves when the system launches the rekeying process more frequently, as the system is more likely to be brought back to good state from the timing attack state and the compromised state. At small values of the rekeying rate, the system throughput is low because more time is spent in the compromised state when the offloading throughput is not contributing. After obtaining the maximum throughput, the more often the server triggers the rekeying act, the more often the server denies offloading requests. As a result, the system throughput decreases with the rekeying rate.

At last, we present the security and performance tradeoff analysis for the offloading system in Fig. 6. The tradeoff measure increases rapidly with the rekeying rate λ_1 at its low values, as the system security improves quickly. We find the optimum rekeying rate for the best security and performance tradeoff at $\lambda_1 = 0.5169$, when $n = 3$ (n is the multiple of rekeying rate in compromised state). This optimum rekeying rate is different from the one for the lowest cost measure since they look at different aspects of evaluating the system. However, after reaching the optimum value, the tradeoff measure decreases much slower as the rekeying rate is getting larger. When the rekeying rate has a large value, the multiple parameter n does not affect the tradeoff measure much as the rekeying act is triggered frequently enough. The system tradeoff metric decreases because of the degrading system throughput at large rekeying rates.

We show a method that can be used by the system administrator to find out how to tune the security mechanism (rekeying) in the system. The system cost measure can act as a criterion for service providers to charge their users. If users need higher security level, they have to pay more for offloading services.

The system administrator can also use the result to obtain the minimum cost or maximum security performance tradeoff for the system.

6   Conclusion and Future Work

In order to proceed to a quantitative treatment of the security-performance tradeoff of offloading systems we have proposed a hybrid CTMC queueing model for an offloading system under the specific threat of timing attacks. We have shown how to formulate measures that include both security and performance attributes and that optimize the tradeoff between the two. System metrics are also proposed which take into account both the rekeying effort a system makes and the sensitive information loss. An optimum rekeying rate is found for the tradeoff measure depending on n, the parameter for the rekeying rate in compromised state. On the metrics discussed in this paper an offloading decision can be based.

Extending the analysis to include a key refresh protocol and validating against implementation will be the future work. At the same time, the model will be extended including retries. Another objective of our future work is to conduct experiments to get the precise input parameters for our model. We have implemented a simple timing attack demonstrator which we will develop further and use to determine the time distribution for timing attacks.

References

1. Kumar, K., Liu, J., Lu, Y.-H., Bhargava, B.: A survey of computation offloading for mobile systems. Mobile Networks and Applications 18(1), 129–140 (2013)
2. Hong, J.I., Landay, J.A.: An infrastructure approach to context-aware computing. Human-Computer Interaction 16(2), 287–303 (2001)
3. Kumar, K., Lu, Y.-H.: Cloud computing for mobile users: Can offloading computation save energy? Computer (4), 51–56 (2010)
4. Wang, Q., Wolter, K.: Reducing task completion time in mobile offloading systems through online adaptive local restart. In: Proceedings of the 6th ACM/SPEC International Conference on Performance Engineering, ICPE 2015, pp. 3–13. ACM, New York (2015)
5. Wolski, R., Gurun, S., Krintz, C., Nurmi, D.: Using bandwidth data to make computation offloading decisions. In: IEEE International Symposium on Parallel and Distributed Processing, IPDPS 2008, pp. 1–8. IEEE (2008)
6. Hong, Y.-J., Kumar, K., Lu, Y.-H.: Energy efficient content-based image retrieval for mobile systems. In: IEEE International Symposium on Circuits and Systems, ISCAS 2009, pp. 1673–1676. IEEE (2009)
7. Kosta, S., Aucinas, A., Hui, P., Mortier, R., Zhang, X.: Thinkair: dynamic resource allocation and parallel execution in the cloud for mobile code offloading. In: 2012 Proceedings IEEE INFOCOM, pp. 945–953. IEEE (2012)
8. Ding, A.Y., Han, B., Xiao, Y., Hui, P., Srinivasan, A., Kojo, M., Tarkoma, S.: Enabling energy-aware collaborative mobile data offloading for smartphones. In: 2013 10th Annual IEEE Communications Society Conference on Sensor, Mesh and Ad Hoc Communications and Networks (SECON), pp. 487–495. IEEE (2013)

9. Brumley, D., Boneh, D.: Remote timing attacks are practical. Computer Networks **48**(5), 701–716 (2005)
10. Zhao, Y., Thomas, N.: Efficient solutions of a PEPA model of a key distribution centre. Performance Evaluation **67**(8), 740–756 (2010)
11. Zhang, J.-F., Liu, F., Zheng, L.-M., Jia, Y., Zou, P.: Using network security index system to evaluate network security. In: Qi, E., Shen, J., Dou, R. (eds.) The 19th International Conference on Industrial Engineering and Engineering Management, pp. 989–1000. Springer, Heidelberg (2013)
12. Lenkala, S.R., Shetty, S., Xiong, K.: Security risk assessment of cloud carrier. In: 2013 13th IEEE/ACM International Symposium on Cluster, Cloud and Grid Computing (CCGrid), pp. 442–449. IEEE (2013)
13. Rebeiro, C., Mukhopadhyay, D., Bhattacharya, S.: An introduction to timing attacks. In: Timing Channels in Cryptography, pp. 1–11. Springer (2015)
14. Köpf, B., Basin, D.: Automatically deriving information-theoretic bounds for adaptive side-channel attacks. Journal of Computer Security **19**(1), 1–31 (2011)
15. Brumley, B.B., Tuveri, N.: Remote timing attacks are still practical. In: Atluri, V., Diaz, C. (eds.) ESORICS 2011. LNCS, vol. 6879, pp. 355–371. Springer, Heidelberg (2011)
16. Weiss, M., Heinz, B., Stumpf, F.: A cache timing attack on AES in virtualization environments. In: Financial Cryptography and Data Security, pp. 314–328. Springer (2012)
17. Palanisamy, B., Liu, L.: Mobimix: protecting location privacy with mix-zones over road networks. In: 2011 IEEE 27th International Conference on Data Engineering (ICDE), pp. 494–505. IEEE (2011)
18. Meng, T., Wang, Q., Wolter, K.: Model-based quantitative security analysis of mobile offloading systems under timing attacks. In: Remke, A., Manini, D., Gribaudo, M. (eds.) ASMTA 2015. LNCS, vol. 9081, pp. 143–157. Springer, Heidelberg (2015)
19. Reinecke, P., Krauß, T., Wolter, K.: Phase-type fitting using hyperstar. In: Balsamo, M.S., Knottenbelt, W.J., Marin, A. (eds.) EPEW 2013. LNCS, vol. 8168, pp. 164–175. Springer, Heidelberg (2013)
20. Stewart, W.J.: Probability, Markov chains, queues, and simulation: The mathematical basis of performance modeling. Princeton University Press (2009)
21. Neuts, M.F.: Matrix-geometric solutions in stochastic models: an algorithmic approach. Courier Corporation (1981)

Non-Markovian Performability Evaluation of ERTMS/ETCS Level 3

Laura Carnevali[1], Francesco Flammini[2], Marco Paolieri[1(✉)], and Enrico Vicario[1]

[1] Department of Information Engineering, University of Florence, Florence, Italy
{laura.carnevali,marco.paolieri,enrico.vicario}@unifi.it
[2] Ansaldo STS, Naples, Italy
francesco.flammini@ansaldo-sts.com

Abstract. The European Rail Traffic Management System/European Train Control System (ERTMS/ETCS) is an innovative standard introduced to enhance reliability, safety, performance, and interoperability of trans-European railways. In Level 3, the standard replaces fixed-block safety mechanisms, in which only one train at a time is allowed to be in each railway block, with *moving blocks*: a train proceeds as long as it receives radio messages ensuring that the track ahead is clear of other trains. This mechanism increases line capacity, but relies crucially on the communication link: if messages are lost, the train must stop within a safe deadline even if the track ahead is clear. We develop upon results of the literature to propose an approach for the evaluation of *transient* availability of the communication channel and probability of train stops due to lost messages. We formulate a non-Markovian model of communication availability and system operation, and leverage solution techniques of the ORIS Tool to provide experimental results in the presence of multiple concurrent activities with non-exponential durations.

Keywords: European Rail Traffic Management System (ERTMS) · European Train Control System (ETCS) · Real-time systems design · Markov Regenerative Process (MRP) · Transient analysis · Stochastic state classes

1 Introduction

Quantitative evaluation of models with stochastic timers often provides crucial support in the engineering of dependability requirements. Both analytic and simulative approaches can serve the objective, with different limitations and advantages. In particular, when applicable, analytic approaches facilitate the exploration of the space of preliminary design, especially in the presence of rare events. The limits for applicability are determined by various factors, and notably by the class of the underlying stochastic process of the model [10].

If all activity durations are distributed according to (memoryless) exponential distributions, the underlying stochastic process is a continuous-time Markov

© Springer International Publishing Switzerland 2015
M. Beltrán et al. (Eds.): EPEW 2015, LNCS 9272, pp. 47–62, 2015.
DOI: 10.1007/978-3-319-23267-6_4

chain and evaluation can resort to consolidated and efficient analytic approaches [11,23,26,27]. However, the system under analysis is sometimes strongly characterized by activity durations that are deterministic (e.g., timeouts) or distributed according to general (i.e., nonexponential) distributions, imposing hard constraints on the minimum or maximum value. In this case, the underlying stochastic process is non-Markovian, but it can still satisfy the Markov property (conditional independence of future evolution from past history, given the current state) at selected time instants called *regeneration points*. In Markov Regenerative Processes (MRPs) [20], a new regeneration is eventually reached with probability 1, and the analysis can be formulated in terms of a local and a global kernel that characterize the behavior of the process between subsequent regeneration points. Solutions for the evaluation of kernels have been consolidated only for models satisfying the so-called *enabling restriction*, which requires that at most one generally distributed transition is enabled in each state [5,9,10,21]. Recent results based on the method of stochastic state classes have overcome the limit [18,32], enabling the numerical solution of models beyond the enabling restriction, and in particular MRPs reaching regenerations in a bounded number of state transitions [18]. The ORIS Tool provides an implementation of the approach [8], opening the way to the analysis of a large class of problems.

Level 3 is the most promising operation level of the European Rail Traffic Management System/European Train Control System (ERTMS/ETCS) [15,16] in terms of capacity gains and trackside installation savings, and it represents a challenging case study in the engineering of future train control systems. As in Communication Based Train Control (CBTC) for metro-railways, the ERTMS/ETCS Level 3 standard adopts a radio-based moving-block technology, where the maximum distance before a virtual stop is computed in real-time from locations and speeds of all trains, requiring continuous two-way communication between each train and the control center.

The case study has been widely addressed in the literature of quantitative evaluation of dependability [1,2,13,14,17,22,30,35]. Notably, in [30,35], probabilistic parameters were derived from the analysis of the standard specification and cast into a hierarchical modeling and evaluation approach, based on rare events simulation and analysis of non-Markovian models under enabling restriction, both supported by the TimeNET Tool [33,34]. Reliability analysis is addressed in [17] by leveraging the MODEST language [6] and the Möbius Tool [12]. Reliability aspects are also assessed in [14] by means of a compositional approach that integrates analysis of fault trees and evaluation of Bayesian networks. Dependability and safety metrics are evaluated in [2] focusing on the parameters that affect the probability of an emergency train stop. A multi-formalism model is used in [13] to evaluate the influence of basic design parameters on the probability of system-level failure modes, showing that system availability is lower than the threshold prescribed by the specification. In [22], the ERTMS/ETCS Level 2 railway signaling system (using radio communication but not moving blocks) is modeled as a system of systems and its dependability parameters are evaluated using statecharts, taking into account human factors, network failures, and imprecise failure

Acronym	Meaning
BTS	Base Transceiver Station
DET	Transition with deterministic duration
ERTMS	European Rail Traffic Management System
ETCS	European Train Control System
EVC	European Vital Computer
EXP	Exponentially distributed transition
GEN	Transition with general (i.e., nonexponential) distribution
IMM	Immediate transition
MA	Movement Authority
MRP	Markov Regenerative Process
PR	Position Report
RBC	Radio Block Center
sTPN	Stochastic Time Petri Net

Fig. 1. Summary of acronyms used in the rest of the paper.

and repair rates. Reverse engineering is used in [1] to perform static analysis of the software of a complex safety-critical subsystem of the ERTMS/ETCS, supporting both correctness verification of software and its refactoring.

In this paper, we develop upon the results of [30,35] to propose a model of communication availability including multiple concurrent activities with generally distributed durations. The model accounts for the concurrent nature of communication failures due to handovers between neighboring radio stations, and random burst noise or connection losses. We provide a safe approximation of the *transient* availability of the communication layer, and leverage this measure in a higher-level operational model of moving-block signaling, in which a train proceeds as long as it receives messages ensuring that the track ahead is clear of other trains. Through a first-passage analysis of this model, we compute the *transient* probability that the train has stopped due to lost messages, as opposed to previous work focusing on steady state analysis [30,35]. Since the "arrive and depart" mechanism of trains is inherently transient, the results provide a further step in the analysis of the effects of communication failures on moving-block signaling. The evaluation leverages the analysis of MRPs based on stochastic state classes [18] and its implementation within the ORIS Tool [8].

The rest of the paper is organized in four sections. In Sect. 2, we examine the ERTMS/ETCS case study, with specific focus on the Level 3 implementation. In Sect. 3, we recall syntax and semantics of a non-Markovian variant of stochastic Petri nets [32] and the salient traits of regenerative analysis [18]. In Sect. 4, we present a non-Markovian model of communication availability and derive a safe approximation that is used, in turn, to compute transient performability measures on the operational model based on moving blocks. Conclusions are drawn in Sect. 5.

2   The ERTMS/ETCS L3 Case-Study

The ERTMS/ETCS is a recent standard that has been developed to enhance performance, reliability and safety of trans-European railway networks. In fact, the standard has been an intercontinental success, so that ERTMS/ETCS compliant railways have been or are being engineered in several installations even outside Europe (e.g., China, United Arab Emirates). Though actual systems can be very complex, heterogeneous and highly distributed, the working principles of ERTMS/ETCS are rather straightforward. Trains are equipped with on-board automatic train control devices, which are embedded real-time computers known as European Vital Computers (EVCs). EVCs are connected with train-borne apparels (e.g., odometer, brakes) to allow automatic braking in case the speed is over the allowed limit. To compute the maximum allowed speed (known as the *braking curve* or *dynamic speed profile*), the EVC needs to receive the following information from the trackside subsystems: *i*) Movement Authority (MA), i.e., the maximum distance before a virtual stop signal; *ii*) Static speed profile, i.e., the maximum speed depending on track morphology; *iii*) Possible temporary speed restrictions or conditional/unconditional emergency stops. Such information can be provided to the EVCs using different communication means in the three levels of operation defined by the standard. At Level 2 and 3, the so-called Radio Block Centers (RBCs) are employed, enabling continuous radio-signalling using GSM-R (similar to the well-known mobile phone standard) and the safe Euroradio protocol. In turn, the RBC needs to know the Position Reports (PRs), that is the precise position of all the trains on the track. The EVC obtains this information by reading "telegrams" sent by the so-called *balises*, which are devices installed between the track lines and acting like milestones. PRs are sent by the EVC to the RBC either periodically, at each newly encountered balise, or upon specific RBC requests.

Most of the lines that are currently operational, starting from the first Rome-Naples Italian high-speed railway, are based on the ERTMS/ETCS Level 2, which employs fixed-block signaling. That means the MA is computed by summing track circuits and routes that are neither occupied by any train nor in out-of-service/exclusion conditions. Such an implementation needs an interface between the RBC and the underlying (possibly legacy) interlocking systems. The ERTMS/ETCS Level 2 is generally considered safer at the expense of longer headways due to the obviously less fine-grained spacing.

To increase line capacity, Level 3 introduces *moving-block signaling*: the MA of the chasing train is computed considering the minimum safe rear-end of the foregoing train, and not the status of track-circuits. In those conditions, it is rather intuitive that system safety is highly dependent on train integrity checks, hence the EVC has to provide this additional information to the RBC. Moving-block signaling has received higher attention in mass-transit (e.g., subways), due to the required high-frequency of trains (few minutes waiting times), and it is adopted in Communication Based Train Control (CBTC) for metro-railways. To justify the adoption of Level 3 for new high-capacity railways, where the braking distances and data latencies are essential factors to take into account, it is very important to preliminarily evaluate the real expected performance by model-based analysis. Such an analysis can also assess which are the variables having a higher impact on system performance. Actually, since both performance (computing latencies, communication delays) and reliability (data transmission errors, connection faults) aspects need to be addressed, this kind of assessment comprises a classical problem of performability evaluation.

In [35,36], the maximum delay d (i.e., deadline) after which automatic braking is activated by the on-board system is derived for the condition of a train that runs at speed $v = 300\,\mathrm{km\,h^{-1}}$, as usual in most real installations, and it is expressed as a function of the following factors: i) the train headway s, i.e., the distance between the maximum safe front-end of the train and the minimum safe rear-end of its predecessor, with those positions corrected taking into account the estimated odometric measurement errors; ii) the braking distance s_{brake}, assumed to be approximately 3 km including the aforementioned position errors; iii) the packet age p_{age}, i.e., the maximum staleness of a received packet (p_{age} is assumed to be 12 s in the worst case). More specifically, in [35,36] it is shown that $d = (s - s_{braking})/v - p_{age}$. Based on these assumptions, it is evinced that headways cannot be shorter than 4 km, that is the theoretical minimum. In such a scenario, model-based analysis is essential to evaluate the train stop probability as a function of the required headways, or, conversely, the minimum headways allowing acceptable system availability measures. Also, sensitivity to other parameters can be evaluated in order to support system design choices in industrial and operational settings.

3 Non-Markovian Modeling and Analysis

The system is modeled using a variant of non-Markovian stochastic Petri nets called *stochastic Time Petri Nets* (sTPN) [32], enriched with enabling functions, flush functions, and priorities, augmenting the modeling convenience without impacting on the nature and complexity of the analysis, as in [25,28].

3.1 Stochastic Time Petri Nets

Syntax. An sTPN is a tuple $\langle P; T; A^-; A^+; A^\bullet; m_0; EFT^s; LFT^s; \mathcal{F}; \mathcal{C}; E; L; R \rangle$, where: P is a set of places; T is a set of transitions; $A^- \subseteq P \times T$, $A^+ \subseteq T \times P$, and $A^\bullet \subseteq P \times T$ are the sets of precondition, postcondition, and inhibitor arcs, respectively; $m_0 : P \to \mathbb{N}$ is the initial marking associating each place with a non-negative number of tokens; $EFT^s : T \to \mathbb{Q}_0^+$ and $LFT^s : T \to \mathbb{Q}_0^+ \cup \{\infty\}$ associate each transition with a *static Earliest Firing Time* and a static *Latest Firing Time*, respectively, such that $EFT^s(t) \leq LFT^s(t) \ \forall \ t \in T$; $\mathcal{F} : T \to F_t^s$ associates each transition with a static Cumulative Distribution Function (CDF) such that $x < EFT^s(t) \Rightarrow F_t^s(x) = 0$ and $x > LFT^s(t) \Rightarrow F_t^s(x) = 1$; $\mathcal{C} : T \to \mathbb{R}^+$ associates each transition with a weight; $E : T \to \{true, false\}^{\mathbb{N}^P}$ associates each transition t with an *enabling function* $E(t) : \mathbb{N}^P \to \{true, false\}$ that, in turn, associates each marking with a boolean value; $L : T \to \mathcal{P}(P)$

is a *flush function* associating each transition with a subset of P; $R : T \to \mathbb{N}$ associates each transition with a priority. A place p is called an *input*, an *output*, or an *inhibitor* place for a transition t if $\langle p, t \rangle \in A^-$, $\langle t, p \rangle \in A^+$, or $\langle p, t \rangle \in A^\bullet$, respectively. A transition t is called *immediate* (IMM) if $[EFT^s(t), LFT^s(t)] = [0, 0]$ and *timed* otherwise. A timed transition t is called *exponential* (EXP) if $F_t^s(x) = 1 - e^{\lambda x}$ over $[0, \infty]$ for some $\lambda \in \mathbb{R}_0^+$ and *general* (GEN) otherwise. A GEN transition t is called *deterministic* (DET) if $EFT^s(t) = LFT^s(t) > 0$ and *distributed* otherwise (i.e., if $EFT^s(t) \neq LFT^s(t)$). For each distributed transition t, we assume that F_t^s is absolutely continuous over $[EFT^s(t), LFT^s(t)]$ and, thus, that there exists a Probability Density Function (PDF) f_t^s such that $F_t^s(x) = \int_0^x f_t^s(y)dy$.

Semantics. The *state* of an sTPN is a pair $\langle m, \tau \rangle$, where $m : P \to \mathbb{N}$ is a marking that associates each place with a non-negative number of tokens and $\tau : T \to \mathbb{R}_0^+$ associates each transition with a (dynamic) real-valued time-to-fire. A transition is *enabled* by a marking if each of its input places contains at least one token, none of its inhibitor places contains any token, and its enabling function evaluates to true. An enabled transition t is *firable* if its time-to-fire is not higher than that of any other enabled transition and, in case t is IMM or DET, if its priority is not lower than that of any other enabled IMM/DET transition. When multiple transitions are firable, one of them is selected as the firing transition with probability $Prob\{t \text{ is selected}\} = \mathcal{C}(t)/\sum_{t_i \in T^f(s)} \mathcal{C}(t_i)$, where $T^f(s)$ is the set of transitions that are firable in s. When a transition t fires, the state $s = \langle m, \tau \rangle$ is replaced by a new state $s' = \langle m', \tau' \rangle$. Marking m' is derived from m by: *i)* removing a token from each input place of t and removing all tokens from the places in $L(t) \subseteq P$, which yields an intermediate marking m_{tmp}, *ii)* adding a token to each output place of t. Transitions that are enabled both by m_{tmp} and by m' are called *persistent*, while those that are enabled by m' but not by m_{tmp} or m are called *newly-enabled*. If the fired transition t is still enabled after its own firing, it is always regarded as newly enabled [4,31]. While the time-to-fire of persistent transitions is reduced by the time elapsed in s, that of newly-enabled transitions takes a random value sampled according to the static CDF.

3.2   Regenerative Transient Analysis Through Stochastic State Classes

The method of stochastic state classes [7,32] faces the analysis of the underlying stochastic process of models with multiple concurrent GEN transitions. To this end, the marking and the vector of times to fire of GEN transitions are characterized after each firing. This yields an embedded discrete time Markov chain encoded in a so-called *stochastic graph*, whose states are called *stochastic state classes*. Each class is made of a marking plus the joint support and PDF of the times-to-fire of enabled GEN transitions. To support transient evaluation, in [18] a fresh clock named τ_{age} is added to each class to account for the absolute elapsed time. The marginal PDF of τ_{age} permits to derive the PDF of the absolute time at which a class can be entered, enabling the evaluation of continuous-time transient probabilities of reachable markings within a given time horizon, provided that the number of classes that can be reached within that time interval is either bounded or can be truncated under the assumption of some approximation threshold on the total unallocated probability.

In general, the approach of [18] supports transient analysis of models with underlying Generalized Semi-Markov Process (GSMP) with equal-speed timers [10, 24]. Nevertheless, the complexity of the solution technique can be significantly reduced when applied to models with underlying Markov Regenerative Process (MRP) that always reaches a regeneration point within a bounded number of steps, i.e., a state where the future behavior is independent from the past. In fact, transient analysis can be restrained within the first regeneration epoch from each regenerative point, and finalized to the derivation of the local and global kernels that characterize the behavior of the MRP [5, 9, 10]. Transient probabilities of reachable markings at any time can then be derived by numerical integration of generalized Markov renewal equations [20].

4   Performability Evaluation of ERTMS/ETCS Level 3

We consider a model of communication availability that features a non-Markovian representation of failures due to handovers (Sect. 4.1), and we derive a safe estimation of the transient evolution of communication availability through a 3-step function (Sect. 4.2). Such approximation is cast within a non-Markovian model of communication beyond the limits of the enabling restriction (Sect. 4.3), evaluating the transient probability that a timeout occurs within time t (Sect. 4.4).

4.1   A Non-Markovian Model of Communication Availability

At the ERTMS/ETCS Level 3, the GSM-R communication channel appears the most relevant source of unreliability, due to almost unavoidable data transmission errors, connection losses, and Base Transceiver Station (BTS) handovers. In [35, 36], stochastic parameters characterizing communication failures are derived from specification documents and guidelines, evaluating the probability of stops through a combined use of analytic evaluation under enabling restriction and rare-event simulation, both supported within the TimeNET Tool [33].

Here we present a model of communication failure that develops upon the results of [35, 36], leveraging the same stochastic parameters while extending the model structure to encompass a non-Markovian representation of handovers. The model is shown in Fig. 2. As in [35, 36]: the arrival and duration of "bursts" of noise are modeled by the EXP transitions startBurst and endBurst with rate $0.007\,33\,\mathrm{s}^{-1}$ and $3\,\mathrm{s}^{-1}$, respectively, derived by fitting the specification that the mean time between two bursts is at least $7\,\mathrm{s}$, with each burst shorter than $1\,\mathrm{s}$ in 95% of the cases; the occurrence of a connection loss is represented by the EXP transition loss with rate $2.77 \times 10^{-8}\,\mathrm{s}^{-1}$, derived from the specification that the probability to have a connection loss per hour is less than 10^{-4}; the

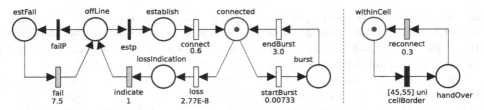

Fig. 2. The sTPN model of communication availability (times expressed in s), combining the sub-models of failures due to handovers (right) and transmission errors or connection losses (left). IMM, EXP, DET, and GEN transitions are represented by thin bars, thick empty bars, thick gray bars, and thick black bars, respectively.

time needed to detect a connection loss is required to be not greater than 1 s, thus it is accounted by the DET transition `indicate`; the reconnection attempt is required to be successful with a probability higher than 99.9%, which is represented as a switch between the IMM transitions `estP` and `failP`, having weight 0.999 and 0.001, respectively; in case of reconnection success, the establishment time must be less than 5 s in 95% of the cases, which is modeled by the EXP transition `connect` with rate $0.6\,\text{s}^{-1}$; in case of reconnection failure, a reconnection is retried after 7.5 s, which is modeled by the DET transition `fail`.

In [35, 36], connection loss due to handovers is modeled by an EXP transition conflicting with `startBurst` and `loss`, whose rate is derived as the inverse of the time spent by a train that runs at the maximum speed $v = 500\,\text{km}\,\text{h}^{-1}$ to cover the 7 km distance between BTS, i.e, $1/0.0198 = 50.4\,\text{s}^{-1}$. As opposed to [35, 36], we model failures due to hand-overs and failures due to transmission errors or connection losses as concurrent events. Actually, this reflects the nature of the phenomenon, as handovers indeed occur in parallel to transmission errors and connection losses. According to this, in the model of Fig. 2, the sub-model that accounts for failures due to handovers (the right part) is concurrent to the sub-model that represents failures due to transmission errors and connection losses (the left part). The time between subsequent communication failures due to handovers is modeled by a GEN distribution with bounded support rather than with an EXP distribution over $[0, \infty)$. This captures the fact that the distance between subsequent BTS is nearly constant and the speed of trains ranges within a min-max interval. In the present experimentation, a uniform distribution supported over [45, 55] s is associated with the GEN transition `cellBorder` in the model of Fig. 2. The mean value of such transition (i.e., 50 s) is a conservative approximation of the mean value of the namesake EXP transition in the model of [35, 36] (i.e., 50.4 s). As in [35, 36], the time to reconnection is modeled by the DET transition `reconnect`, whose duration equal to 0.3 s is the maximum time allowed by the requirements specification.

4.2 Evaluation of the Communication Availability Model

To reduce the stiffness of the problem due to failures that occur with different time-scales, we separate the analysis of independent events. According to this,

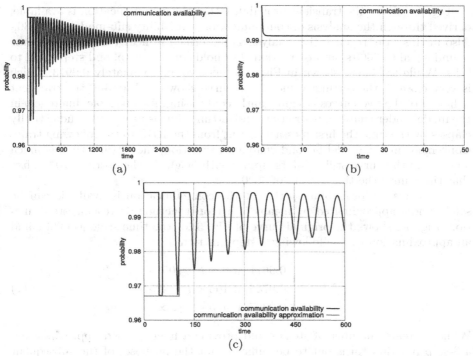

Fig. 3. Transient probability that the communication is available, derived through regenerative analysis of: (a) the model of Fig. 2 and (b) the communication availability sub-model of [35,36]. (c) A conservative approximation of the curve of Fig. 3-a for the time scale $[0, 600]$ s through a 3-step function. Times are expressed in s.

the two sub-models shown in Fig. 2 are separately analyzed using the ORIS Tool [8]. Overall, regenerative analysis of both sub-models is performed in nearly 15 min on a machine equipped with an Intel Xeon 2.67 GHz and 32 GB RAM, assuming time bound equal to 3600 s, approximation threshold equal to zero (i.e., exact analysis), and discretization step in the integration of renewal equations equal to 0.1 s. The analysis yields the transient probability that the communication is not available due to a transmission error or a connection loss, i.e., $p_c(t) = Prob\{\texttt{connected} == 1 \text{ at time } t\}$, and the transient probability that the train is not crossing the border between the communication areas of two neighboring BTS, i.e., $p_w(t) = Prob\{\texttt{withinCell} == 1 \text{ at time } t\}$. By multiplying the obtained numerical results, we derive the transient probability that the communication is available, i.e., $p(t) = p_c(t) \cdot p_w(t)$, whose plot is shown in Fig. 3-a. The plot shows an oscillating pattern, with ripples of decreasing heights, converging to a neighborhood of 0.9912 after a settling time of about 3000 s. This is mainly due to the floating trend of $p_w(t)$, which in fact has a settling time around 3000 s. Conversely, $p_c(t)$ exhibits an exponential trend with a much shorter settling time around 5 s.

Fig. 3-b shows the transient probability that the communication is available, derived through the analysis of communication availability sub-model of [35,36]. Also in this case, regenerative analysis is performed in nearly 20 min with time bound equal to 3600 s, approximation threshold equal to zero, and step equal to 0.1 s. While the curve shown in Fig. 3-b tends to approximately 0.9913, which is very close to the settling value of the curve shown in Fig. 3-a, the transient behavior of the two curves is significantly contrasting. As a notable difference, in the model under enabling restriction, the settling time is nearly 10 s and actually elapses by the time the first message is sent from the RBC to the following train. Conversely, in the model beyond enabling restriction, the settling time is much longer and the curve still exhibits ripples with height in the order of 10^{-4} after that time, until the time bound of 3600 s.

The transient probability $p(t)$ that the communication is available can be safely under-approximated by means of a monotone non-decreasing step function. Fig. 3-c shows the original curve of Fig. 3 for the time scale $[0, 600]$ s and an approximation by the following 3-step function:

$$f(t) = \begin{cases} 0.9671 & \text{if } 0 \leq t \leq 105, \\ 0.9746 & \text{if } 105 < t \leq 405, \\ 0.9827 & \text{if } 405 < t < \infty. \end{cases} \quad (1)$$

While a greater number of steps could provide a finer grained approximation, a 3-step function turns out to be sufficient for the purposes of the subsequent treatment. Note that the complexity of the subsequent analysis is substantially insensitive to the number of steps used in the approximation, and it only depends on the time at which the last jump of the step function is positioned, i.e., 405 s in the present experimentation. In fact, beyond that time instant, the estimate on communication availability is constant and does not carry memory over time, reducing by 1 the number of GEN transitions that are concurrently enabled.

4.3   A Non-Markovian Model of ERTMS/ETCS Level 3

Following the results of [35,36], the proposed approach resorts to the hierarchical composition of models, by relying on the assumption that the availability of communication is independent of the exchange of PR between track-side equipments and on-board devices. In so doing, the method also takes advantage of some approximations of model variables that are guaranteed to be stochastically ordered. As opposed to [35,36], the approach leverages a solution technique that goes beyond the limits of the enabling restriction. In the methodological perspective, this largely relaxes modeling restrictions, as the requirement that the underlying stochastic process is a Markov regenerative process poses less constraints on the model expressivity than the limitation on the number of concurrent GEN timers. In the applicative perspective, this permits to refine the models presented in [35,36] and the way they are composed through a more accurate representation of communication failures due to hand-overs. As a major result, solution can be attained through a fully analytic treatment without facing complexities and limits of simulation in the presence of rare events.

The overall model of communication failure is shown in Fig. 4. As discussed in [35,36], at ERTMS/ETCS Level 3, train integrity checks are performed in 5 s in order to maximize track throughput, and the results are sent together with the PR (transition genMsg). RBC processing time for PR is assumed to be 0.5 s at most, while message transmission delays up-link and down-link are required to be: between 0.4 s and 0.5 s on average, less than 0.5 s in 95% of the cases, less than 1.2 s in 99% of the cases, and less than 2.4 s in 99.99% of the cases. RBC processing and up-link transmission delays are accounted by the GEN transition transmitUp, while down-link transmission delays are represented by the GEN transition transmitDown; in the present experimentation, transmitUp and transmitDown have a uniform distribution over their respective support. Whenever the deadline on the time between subsequent messages is violated (transition timeout), the train starts to brake until it comes to a complete stop: in this case, the resetup/restart delay (not considered in the model of Fig. 4) is assumed to be 15 min long, with all MA dropped during this time.

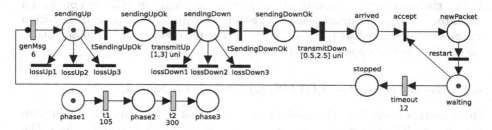

Fig. 4. The sTPN model of the ERTMS/ETCS Level 3 case study beyond the limits of the enabling restriction (times expressed in s). IMM, DET, and GEN transitions are represented by thin bars, thick gray bars, and thick black bars, respectively.

To manage the complexity of the analysis, in [35,36], the sub-model of communication availability is analyzed in isolation and its results are used to derive the rates of a condensed birth-death process made of 2-states, which is then recast in the overall model of communication failure. Yet, such model is not amenable to analysis with methods operating within the limits of the enabling restriction, and the evaluation is thus performed through rare-event simulation [19] supported by the TimeNET Tool, deriving the probability that the train is stopped for different values of the packet age and the head-to-head distance between trains. Conversely, in the proposed approach, the approximation of communication availability is recast within the overall system by means of a phased sub-model with phases of deterministic duration, so that, in each phase, failures of the communication up-link and down-link are accounted by a switch between IMM transitions, whose weights reflect the value of the approximating step function in the corresponding time interval. More specifically, in the model of the overall system shown in Fig. 4: the IMM transitions `lossUp1`, `lossUp2`, and `lossUp3` represent failures of the communication up-link, while the IMM transition `tSendingUpOk` accounts for its availability; `lossUp1`, `lossUp2`, and `lossUp3` have an enabling condition that evaluates to true only during the corresponding phase (i.e., `phase1 == 1`, `phase2 == 1`, and `phase3 == 1`, respectively); their weights are set equal to 0.03402, 0.02607, and 0.01761, respectively, while the weight of `tSendingUpOk` is maintained equal to 1, so that the probability that the communication is available turns out to be equal to 0.9671, 0.9746, and 0.9827 in phase 1, phase 2, and phase 3, respectively, as defined in Eq. 1. Similarly, the IMM transitions `lossDown1`, `lossDown2` and `lossDown3` model failures of the communication down-link, while the IMM transition `tSendingDownOk` represents its availability; they have the same enabling condition and weight of `lossUp1`, `lossUp2`, `lossUp3`, and `tSendingDownOk`, respectively.

Transition `restart` is associated with an enabling function that evaluates to true if `sendingUp` contains a token, and it has higher priority than `lossUp1`, `lossUp2`, `lossUp3`, and `tSendingUpOk`. This guarantees that, whenever the timeout fires, the last received packet has an age equal to (12+timeout) s. Moreover, since we evaluate the transient probability that a timeout occurs within time t (i.e., the transient probability of the first token arrival in place `stopped`), an inhibitor arc is added from place `stopped` to transition `genMsg`, and transition `timeout` is associated with a flush function that removes any token in any place, except for place `stopped`, upon its firing.

4.4   Evaluation of the ERTMS/ETCS Level 3 Model

Regenerative analysis of the model of Fig. 4 is performed in nearly 1 min with time bound 3600 s, time step 1 s, and approximation threshold equal to zero. The analysis is repeated for different values of the DET transitions `genMsg` (i.e., 6 s, 8 s, and 10 s) and `timeout` (i.e., 12 s, 15 s, and 18 s), and the obtained results are shown in Fig. 5. Such values of `genMsg` are selected based on the requirement that the time between two subsequent PR is ≥ 5 s. The values of `timeout` are thereby chosen with the purpose of showing the variability of the studied reward.

For an assigned value of `genMsg`, the probability that the train is stopped within time t decreases as the timeout increases. In fact, in Figs. 5-a, 5-b, and 5-c, the black curve dominates the gray curve which, in turn, dominates the light gray curve. Nevertheless, the gap between the curves may significantly vary among cases. For instance, in Fig. 5-a (i.e., `genMsg` = 6 s), the stop probability is nearly equal to 0.2018, 0.1514, and 0.0114 at time 600 s, and nearly equal to 0.5486, 0.4396, and 0.0305 at time 3600 s, for *timeout* equal to 12 s, 15 s, and 18 s, respectively. Conversely, in Fig. 5-b (i.e., `genMsg` = 8 s), the stop probability is approximately equal to 0.6654 at time 600 s and reaches nearly 0.9930 at time 3600 s for *timeout* equal to 12 s, while it has substantially the same trend for *timeout* equal to 15 s and 18 s (with a difference in the order of 10^{-2} at time 3600 s). Overall, this motivates the opportunity of a sensitivity analysis to assess the considered reward depending on the system parameters.

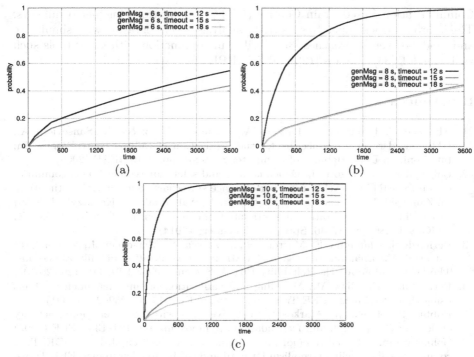

Fig. 5. Transient probability that a timeout occurs within time t (expressed in s).

5   Conclusions

We address performability modeling and evaluation of the ERTMS/ETCS Level 3 case study, supporting exploration of the space of model parameters through regenerative transient analysis [8,18,32]. As in [35,36], the approach relies on the hierarchical composition of sub-models to manage the complexity and stiffness of the problem. Yet, in this paper, we evaluate a model beyond the limits of the enabling restriction. While the approach of [35,36] is concerned with the evaluation of the steady state probability that a timeout occurs, we focus on the transient behavior and derive the probability that a timeout expires within an assigned time. This comprises a measure that is not of less interest than the one studied in [35,36]. In fact, whether the steady state value of the timeout probability is greater than the required threshold or not, it is valuable to study its trend over time and the factors that mainly affect it.

Overall, the approach provides insight in the problem characterization, showing that working beyond the limits of a Markovian setting poses major complexities, but it also provides an advantage in composing results. While this paper specifically addressed the ERTMS/ETCS Level 3 case-study, the model used for system performability evaluation is general enough to be easily adapted to suit similar systems featuring radio-signalling and moving block, like the modern

Communication Based Train Control (CBTC) applications for metro railways. The approach is also open to sensitivity analysis and integration with simulative and model-driven approaches [3], possibly in conjunction with other tools such as TimeNET [33], Möbius [12], and SHARPE [29].

References

1. Abbaneo, C., Flammini, F., Lazzaro, A., Marmo, P., Mazzocca, N., Sanseviero, A.: UML based reverse engineering for the verification of railway control logics. In: Int. Conf. on Dependability of Computer Systems, pp. 3–10. IEEE (2006)
2. Babczyński, T., Magott, J.: Dependability and safety analysis of ETCS communication for ERTMS level 3 using performance statecharts and analytic estimation. In: Zamojski, W., Mazurkiewicz, J., Sugier, J., Walkowiak, T., Kacprzyk, J. (eds.) Proceedings of the Ninth International Conference on DepCoS-RELCOMEX. AISC, vol. 286, pp. 37–46. Springer, Heidelberg (2014)
3. Bernardi, S., Flammini, F., Marrone, S., Mazzocca, N., Merseguer, J., Nardone, R., Vittorini, V.: Enabling the usage of UML in the verification of railway systems: The DAM-rail approach. Reliability Eng. & System Safety 120, 112–126 (2013)
4. Berthomieu, B., Diaz, M.: Modeling and verification of time dependent systems using time Petri nets. IEEE Trans. on Software Eng. 17(3), 259–273 (1991)
5. Bobbio, A., Telek, M.: Markov regenerative SPN with non-overlapping activity cycles. In: Comp. Perf. and Dependability Symposium, pp. 124–133. IEEE (1995)
6. Bohnenkamp, H.C., D'Argenio, P.R., Hermanns, H., Katoen, J.-P.: MODEST: A compositional modeling formalism for hard and softly timed systems. IEEE Trans. on Software Eng. 32(10), 812–830 (2006)
7. Carnevali, L., Grassi, L., Vicario, E.: State-density functions over DBM domains in the analysis of non-Markovian models. IEEE Trans. on Software Eng. 35(2), 178–194 (2009)
8. Carnevali, L., Ridi, L., Vicario, E.: A framework for simulation and symbolic state space analysis of non-Markovian models. In: Flammini, F., Bologna, S., Vittorini, V. (eds.) SAFECOMP 2011. LNCS, vol. 6894, pp. 409–422. Springer, Heidelberg (2011)
9. Choi, H., Kulkarni, V.G., Trivedi, K.S.: Markov regenerative stochastic Petri nets. Performance Evaluation 20(1–3), 337–357 (1994)
10. Ciardo, G., German, R., Lindemann, C.: A characterization of the stochastic process underlying a stochastic Petri net. IEEE Trans. on Software Engineering 20(7), 506–515 (1994)
11. Ciardo, G., Trivedi, K.: SPNP: stochastic Petri net package. In: Int. Workshop on Petri Nets and Performance Models, pp. 142–151. IEEE (1989)
12. Courtney, T., Gaonkar, S., Keefe, K., Rozier, E., Sanders, W.H.: Möbius 2.3: an extensible tool for dependability, security, and performance evaluation of large and complex system models. In: IEEE/IFIP Int. Conf. on Dependable Systems and Networks (DSN), pp. 353–358 (2009)
13. Flammini, F., Marrone, S., Iacono, M., Mazzocca, N., Vittorini, V.: A multi-formalism modular approach to ERTMS/ETCS failure modeling. International Journal of Reliability, Quality and Safety Engineering 21(1) (2014)

14. Flammini, F., Marrone, S., Mazzocca, N., Vittorini, V.: Modelling structural reliability aspects of ERTMS/ETCS by fault trees and Bayesian networks. In: Proc. of the European Safety & Reliability Conference, ESREL, vol. 6 (2006)
15. EEIG Ertms User Group. ERTMS/ETCS RAMS System Requirements Specification, UIC, Brussels (1999)
16. EEIG Ertms User Group. ERTMS/ETCS Systems Requirements Specification, UIC, Brussels (1999)
17. Hermanns, H., Jansen, D.N., Usenko, Y.S.: From StoCharts to MoDeST: a comparative reliability analysis of train radio communications. In: Int. Workshop on Software and performance, pp. 13–23. ACM (2005)
18. Horváth, A., Paolieri, M., Ridi, L., Vicario, E.: Transient analysis of non-Markovian models using stochastic state classes. Perf. Eval. **69**(7–8), 315–335 (2012)
19. Kelling, C.: A framework for rare event simulation of stochastic Petri nets using RESTART. In: Conf. on Winter Simulation, pp. 317–324. IEEE (1996)
20. Kulkarni, V.G.: Modeling and analysis of stochastic systems. CRC Press (1996)
21. Lindemann, C., Thümmler, A.: Transient analysis of Deterministic and Stochastic Petri Nets with concurrent deterministic transitions. Performance Evaluation **36**, 35–54 (1999)
22. Qiu, S., Sallak, M., Schon, W.: Modeling of ERTMS level 2 as an SoS and evaluation of its dependability parameters using statecharts. IEEE Systems Journal **8**(4), 1169–1181 (2014)
23. Miner, A.S., Parker, D.: Symbolic representations and analysis of large probabilistic systems. In: Baier, C., Haverkort, B.R., Hermanns, H., Katoen, J.-P., Siegle, M. (eds.) Validation of Stochastic Systems. LNCS, vol. 2925, pp. 296–338. Springer, Heidelberg (2004)
24. Glynn, P.W.: A GSMP formalism for discrete-event systems. Proceedings of the IEEE **77**, 14–23 (1989)
25. Sanders, W.H., Meyer, J.F.: Stochastic activity networks: formal definitions and concepts. In: Brinksma, E., Hermanns, H., Katoen, J.-P. (eds.) EEF School 2000 and FMPA 2000. LNCS, vol. 2090, pp. 315–343. Springer, Heidelberg (2001)
26. Stewart, W.J.: Introduction to the numerical solution of Markov chains, vol. 41. Princeton University Press, Princeton (1994)
27. Telek, M., Rácz, S.: Numerical analysis of large Markov reward models. Performance Evaluation **36**, 95–114 (1999)
28. Trivedi, K.S.: Probability and statistics with reliability, queuing, and computer science applications. John Wiley and Sons, New York (2001)
29. Trivedi, K.S., Sahner, R.A.: SHARPE at the age of twenty two. ACM SIGMETRICS Perf. Eval. Review **36**(4), 52–57 (2009)
30. Trowitzsch, J., Zimmermann, A.: Using UML state machines and Petri nets for the quantitative investigation of ETCS. In: Int. Conf. on Performance evaluation methodologies and tools, pp. 34. ACM (2006)
31. Vicario, E.: Static analysis and dynamic steering of time dependent systems using time Petri nets. IEEE Trans. on SW Eng. **27**(1), 728–748 (2001)
32. Vicario, E., Sassoli, L., Carnevali, L.: Using stochastic state classes in quantitative evaluation of dense-time reactive systems. IEEE Trans. on Software Eng. **35**(5), 703–719 (2009)

33. Zimmermann, A.: Dependability evaluation of complex systems with TimeNET. In: Int. Workshop on Dynamic Aspects in Dependability Models for Fault-Tolerant Systems, (DYADEM-FTS 2010) (2010)

34. Zimmermann, A., Freiheit, J., German, R., Hommel, G.: Petri net modelling and performability evaluation with TimeNET 3.0. In: Haverkort, B.R., Bohnenkamp, H.C., Smith, C.U. (eds.) TOOLS 2000. LNCS, vol. 1786, pp. 188–202. Springer, Heidelberg (2000)

35. Zimmermann, A., Hommel, G.: A train control system case study in model-based real time system design. In: Int. Parallel and Distributed Processing Symposium, pp. 118–126. IEEE (2003)

36. Zimmermann, A., Hommel, G.: Towards modeling and evaluation of ETCS real-time communication and operation. Journal of Sys. and Soft. **77**(1), 47–54 (2005)

Modelling Techniques I

Simplifying Layered Queuing Network Models

Farhana Islam[✉], Dorina Petriu, and Murray Woodside

Department of Systems and Computer Engineering, Carleton University, Ottawa, Canada
{fislam,petriu,cmw}@sce.carleton.ca

Abstract. The amount of detail to include in a performance model is usually regarded as a judgment to be made by an expert modeler and the question "how much detail is necessary?" is seldom asked and is difficult to answer. However, if a simpler model gives essentially the same performance predictions, it may be more useful than a detailed model. It may solve more quickly, for instance, and may be easier to understand. Or a model for a complex sub-system such as a database server may be usefully simplified so it can be included in larger system models. This paper describes an aggregation process for layered queuing models that reduces the number of queues (called tasks and processors, in layered models) while preserving the total execution demand and the bottleneck characteristics of the detailed model. It demonstrates that this process can greatly reduce the number of tasks and processors with a very small relative error.

Keywords: Performance models · Layered queuing networks · Model simplification

1 Introduction

A performance model may include a very large amount of detail about resources and operations, which makes it difficult to create, maintain and understand, and expensive to solve. This is often true of models created from a system design, for example, because the model includes every operation and component. Frequently many of the model entities have little impact on the performance, and can be aggregated or ignored.

This paper considers layered queuing (LQ) models of service systems with a single class of users, and with distributed and layered operations and resources. It examines a process for aggregating operations and entities in the model, and its impact on performance predictions. The ultimate goal is a process for automatically simplifying a model to an essential core level of detail governed by an accuracy requirement over a range of cases. The first step is to find operations that successfully simplify some details, and this is what is reported here.

The paper examines model-simplification operations that aggregate sub-operations (in LQ terms, activities), operations (in LQ terms, entries), software processes (in LQ terms, tasks), and physical resources (processors). Aggregation may be vertical (along a calling path) or horizontal (across multiple calling paths and classes of operation). Restrictions on simplification that preserve the bottleneck characteristics of the model (which in turn determine its capacity) are investigated. More sophisticated methods

© Springer International Publishing Switzerland 2015
M. Beltrán et al. (Eds.): EPEW 2015, LNCS 9272, pp. 65–79, 2015.
DOI: 10.1007/978-3-319-23267-6_5

may be found, but there is value in simplicity, and furthermore this simple process is remarkably good.

The end result is a simplified model containing one non-bottleneck task and processor, and one or more bottleneck tasks or processors. The process is demonstrated on two small models and one large complex model taken from the literature, and the results show remarkably small errors in most cases. The simplification operations are evaluated by their effect on the system response time or, equivalently, the system throughput with a finite user population.

Section 2 discusses related work, Section 3 describes layered queuing network models, and Section 4 presents heuristic principles for simplification using two example LQN models. Section 5 shows application of simplification principles on a case study that presents and compares performance results among different levels of simplifications. Conclusions, limitations and future works are discussed in Section 6.

2   Related Work

In performance models which are product-form queuing networks, there is a powerful and much-used simplification result in the Norton Theorem for Queues [1], by which any subnetwork of queues can be replaced by a single server with a state-dependent service rate. The replacement is exact in the sense that the throughput and delay at the subnetwork interface is the same for the single server. The original result was for a single class of customers, and it was extended to multiple classes in [2]. The exact simplification for product form networks, and approximations that use the same construction technique for other models, can be referred to as flow-equivalent server (FES) methods [3]. When a submodel is replaced by a FES centre, the entire model is smaller and easier to solve, and parameter changes outside the submodel can be studied efficiently. However the FES construction method requires solving the subnetwork many times, once for every user population that it may experience, which does not scale well to large systems with thousands of customers.

Surrogate delay methods (e.g. [3]) replace a subsystem by a delay which is found by solving an auxiliary model. A surrogate delay is somewhat like a FES with a fixed delay rather than a state-dependent rate, but the construction method is different and requires an iterative solution which includes the auxiliary model. Surrogate delays are most useful to address problems of simultaneous resource possession, rather than particularly for model simplification.

When performance models are fitted by regression methods as in [4], a choice must be made for the level of detail in the model and the modeler can select a simple structure to fit (and test the goodness of fit afterwards). Regression thus automatically raises the question of detail, and can answer it through tests of goodness of fit as discussed in [4] (the reference describes fitting ordinary queuing models but it applies equally to layered models). However this approach cannot be applied to models constructed from a design, before a system is built.

There does not appear to be any prior work on deriving a simplified layered queuing model directly from a detailed one. In particular there is a lack of simplification techniques that avoid the scalability problems of calibrating an FES. This work approximates the system by a model with ordinary multiservers with parameters derived as part of the aggregation.

3   The Layered Queuing Network (LQN) Model

Layered queuing networks (LQNs) are an elegant way to express simultaneous resource possession and are particularly intended to model layered software systems, in which a software server depends not just on its processor, but on other software servers as well [8]. An LQN model basically presents software processes as tasks, one or more operations (or service classes) of a process as entries, interactions among different entries as calls or requests for service, and the host processors at which tasks are deployed. Tasks and processors are servers with queues. Fig. 1 shows an example LQN model of a three tiered (three layered) architecture. Each task is represented as a parallelogram (labeled by the task's name and thread multiplicity m), containing parallelograms for its entries (each labeled by entry name and host demand s_e for one invocation of the entry e). Every task is deployed on a host drawn as a circle. Icons are stacked to represent tasks or processors with multiplicity. A call from one entry to another is represented as an arrow labeled with the mean number of calls y_{de} from entry d to entry e. A task is a multiserver (the threads are the servers) with a single queue, usually FIFO, to hold all the calls to its entries; thus the calls are targeted to the entries but actually go first to the task queue.

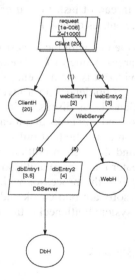

Fig. 1. LQN model of a three-tier architecture

In Fig. 1, the LQN model has three tasks - *Client*, *WebServer* and *DBServer*, each of which is deployed on its own host - *ClientH*, *WebH* and *DbH*, respectively. The 20 users each takes 1000 ms think time (Z) between requests. They are modeled as 20 tasks each running on its own processor *ClientH*. Both *WebServer* and *DBServer* are single threaded tasks and they each have two entries with host service demands indicated in braces (i.e. *webEntry1* has service demand *2 ms*). A single client operation includes one request to *webEntry1* and two to *webEntry2*. Storage devices are not shown but they can be modeled by a task representing the storage logic (read, write operations for example) running on a host representing the device. An entry may include an activity subgraph, as shown in Fig. 2.

LQN models of real systems can be very large, if they describe systems with many servers, replicated servers, and storage devices. Models with a dozen layers and dozens of tasks are common, and hundreds of tasks may arise in complex cases or with large scale-out by replication. These large models are cumbersome and most of the detail does not impact the performance.

Some asymptotic (bottleneck) properties of the model can be deduced from its parameters and will be used to guide the simplification. Let:

- Y_e = the number of calls to entry e, per user Request. $Y_e = \Sigma_d Y_d y_{de}$, where the sum is over all the entries, with $Y_{request} = 1$;
- X_e = the service time of one request to entry e, including its own execution s_e, waiting for its host, and waiting for replies to calls it makes to other entries;
- U_h = utilization of each core in host h, per user response = $(1/ m_h) \Sigma_{e(h)} Y_e s_e$, where the sum is over entries of tasks deployed on h;
- U_t = utilization of each thread of task t per user response = $(1/ m_t) \Sigma_{e(t)} Y_e X_e$, where the sum is over the entries of task t.

Then the most-saturated host is the one with the largest value of U_h and the most saturated task is the one with the largest value of U_t. The system bottleneck is the entity with the largest utilization, provided it is not a client of an entity that also has a large utilization. To identify the system bottleneck in a layered system we must consider the possibility of software bottlenecks as discussed in [9]. Considering any task, we say its "servers" are its processor and any tasks that it calls. The bottleneck strength of a task is the ratio of its utilization to the highest utilization among its servers. Then a task is a software bottleneck (and the system bottleneck) if it has the largest bottleneck strength (considerably greater than unity) and also a high utilization (say greater than 0.9). If no task qualifies, then the processor with the highest utilization is the system bottleneck. If there is a software bottleneck and a saturated processor or processors, then there are multiple system bottlenecks; this is uncommon but possible.

4   The Simplification Process

An LQN model is simplified by aggregating the activities, entries, tasks and processors, using the following four operations. The goal is to reduce the number of tasks and processors in the model while retaining the externally visible performance measures, in this case the mean throughput and response time seen by the users.

1. Substitute the activities of an entry by a total entry demand equal to the sum of the demands caused by executing the activities. Substitute the calls from these activities by calls from the entry, so for each destination entry the number of calls equals the sum of the calls from the activities.
2. Merge the entries of a task. Thus all calls to these entries are redirected to the merged entry, and all calls from these entries now originate from the merged entry. If this gives multiple call arcs between one pair of entries, they are merged also.
3. Merge a set of tasks deployed on a common processor into one task. The entries of each task are first merged separately, and then the merged entries are merged. The merged call rates are calculated based on the relative throughputs of the merged entries, as weights. The merged task's multiplicity is the summation of multiplicities of all the tasks that are being merged.
4. Merge a set of processors and all their tasks. The set of processors is replaced by a single processor whose multiplicity is the sum of the multiplicities in the set, and the merged task is assigned to the merged processor.

Simplification rules using these operations are applied with the goal of retaining the externally visible performance measures, in this case the mean throughput and response time seen by the users. The rules sequence the operations partly as indicated within the operation descriptions (activities, then entries, then tasks, then processors), and partly guided by the location of the system bottleneck.

The first principle of the simplification rules is to preserve the bottleneck task or processor, since the capacity limit of a system is a key property. Thus operations 1 and 2 are applied to all tasks, but operations 3 and 4 are not applied to a task or processor identified as a bottleneck.

A second principle is to preserve the total workload, so that the total throughput and host demand of a merged entry or task, per user request, is the same as for the entities that were merged. The third principle is to preserve concurrency, by which the total multiplicity of a merged task or processor is equal to the sum of multiplicities of the entities that were merged. These three principles are respected in the description of the operations, given above.

4.1   Details of the Operations: Example 1

The detailed application of the operations, including the parameter calculations, will be described with a running example defined by the LQN model from [6] presented in Fig. 2(a). Each of a number of users ($N = 20$) make one visit to the *Server* task, which has one entry *server* with a number of activities. Some requests from different activities are delegated to the pseudo-task *BigLoop* and some are requested from the task *Disk* for *diskread* and *diskwrite* operations. *Server* and *BigLoop* are deployed on the same processor *ServerP* which has a processor-shared queuing discipline. Task *Disk* is deployed on *DiskP* with FIFO queuing discipline. From the initial experiments, *Disk* and its processor are found to be the bottleneck in this model. Thus, *Disk* and its processor are to be preserved in the simplification process.

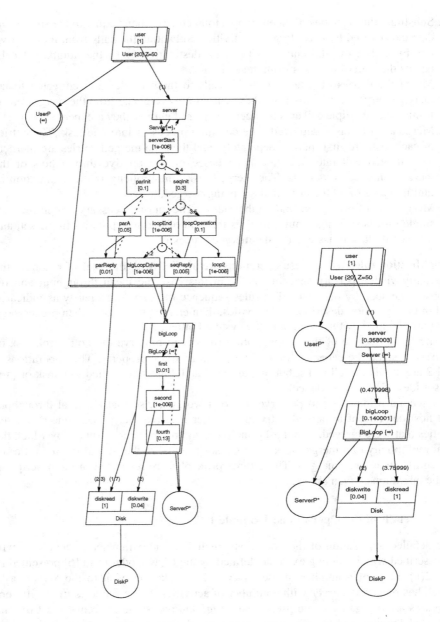

(a) Original model from [6]   (b) After aggregating the activities

Fig. 2. Aggregating activities in an example LQN model

The simplification operations are applied on this example and described as follows. Some calculations can take advantage of finding a single solution of the model being simplified, and this is assumed to be available.

Operation 1: Substituting Activities. In each task t, for each entry e that has activities in its definition, the activities are aggregated. For activity i, let:

s_e, s_i = execution demand of entry e (to be found), and activity i, (given)

λ_e, λ_i = throughput of entry e and activity i, in any solution of the model.

w_i = executions of activity i per request to entry e (this may be calculated by examining the activity graph, or from a model solution as $w_i = \lambda_i/\lambda_e$)

y_{ib} = mean calls from activity i to another entry b of another task

y_{eb} = aggregated mean calls from entry e to entry b (to be found).

Then the aggregated execution demand is

$$s_e = \Sigma_i\, w_i\, s_i \tag{1}$$

and the aggregated number of calls from entry e to another entry b is

$$y_{eb} = \Sigma_i\, w_i\, y_{ib} \tag{2}$$

where the sum in both cases is over the activities of entry e.

In the example, in entry *server* for each activity, the values of *(activity name, weight, execution demand)* are *(serverStart, 1, 1.e-6), (parinit, 0.6, 0.1), (parA, 0.6, 0.05), (parReply, 0.6, 0.01), (seqinit, 0.4, 0.3), (loopOperation, 1.4, 0.1), (loop2, 1.4, 1.e-6), (loopEnd, 0.4, 1.e-6), (bigLoopDriver, 0.48, 1.e-6), (seqReply, 0.4, 0.005)*. Applying Eq (1) we obtain $s_{server} = 0.358$. Applying Eq (2) for the call from *bigLoop-Driver* to *bigLoop*, the entry has the aggregated calls $y_{server,bigLoopDriver} = 0.48$. Fig. 2(b) represents the model after aggregating all the activities from Fig. 2(a).

Operation 2: Merging Entries. The second operation merges the entries of each task t having more than one entry. Let:

s_m, s_k = execution demand of the merged entry m (to be found), and of the original entry k of task t,

y_{kb}, y_{mb} = mean number of calls from entry k of task t to an entry b of another task, and from the merged entry m to entry b,

w_k = weight of original entry k = fraction of all calls to task t, that go to entry k. From any solution, w_k can be found as $\lambda_k/\Sigma_k\,\lambda_k$, where the sum is over the entries to be merged. Then the service demand of the merged entry is:

$$s_m = \Sigma_k\, w_k\, s_k \tag{3}$$

and the calls from entry m to another entry b are:

$$y_{mb} = \Sigma_k\, w_k\, y_{kb} \tag{4}$$

where the sums are over the entries to be merged in both equations.

In Fig. 2(b), only task *Disk* has more than one entry. So, the values of *(entry name, weight, execution demand)* are *(diskread, 0.797, 1), (diskwrite, 0.203, 0.04)*. Applying Eq (3), $s_m = 1 * 0.797 + 0.04 * 0.203 = 0.805$. There are no outgoing calls from *Disk*.

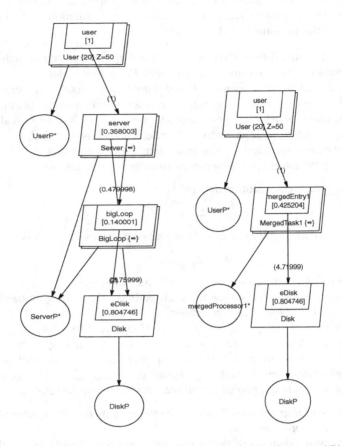

(a) after merging entries   (b) after merging the Server and BigLoop tasks

Fig. 3. More merging operations

The incoming calls are simply transferred to the merged entry (if this results in more than one call from a specific entry, the calls are merged and the numbers summed). Fig. 3(a) represents the model after merging entries.

Operation 3: Merging Tasks on the same Processor: We consider merging two tasks that share a host. Each task has a single entry (entries have been previously merged if necessary). If one task calls the other, we call it vertical merging, otherwise it is horizontal merging.

Vertical merging: Let

s_a, s_b, s_m = the service demands of the entries a and b of the two tasks, and the entry of the merged task, respectively.

y_{ab}, y_{ac}, y_{bc} = the number of calls from entry a to entry b, from a to entry c and from b to c, respectively.

y_{mc} = the number of calls from the merged entry m to entry c.

Then the service demand and number of calls for the entry of the merged task are:

$$s_m = s_a + y_{ab} s_b \qquad (5)$$

$$y_{mc} = y_{ac} + y_{ab} y_{bc} \qquad (6)$$

The incoming calls in vertically merged tasks are calculated as for merged entries. In Fig. 3(a), *Server* and *BigLoop* both are deployed on the same processor *ServerP*. They are merged in Fig. 3(b) as "*MergedTask1*" with an entry "*mergedEntry1*" with service demand of $0.3580003 + 0.479998 * 0.140001 = 0.425204$ (following Eq (5)). The number of outgoing calls from *mergedTask1* to *eDisk* is $= 3.7599926 + 0.479998 *2 = 4.71999$ (following Eq (6)).

Horizontal merging: We call it horizontal merging when there is no calling relationship between the tasks. Just as for merging two entries of the same task, the service demand and the calls of the merged task are computed by Eq (3) and (4), where the entry k designates the single entry of one of the tasks to be merged, and the sums are over this set of entries. As in merging entries, the calls into the separate entries are transferred to the merged entry m and if this results in multiple calls between a pair of entries, the calls are merged and the numbers summed. There are no additional sets of tasks sharing a processor in figure Fig. 3(b), so this calculation is not applied. For this example the last step would be to possibly merge some of the processors, each having a single task. This step will be discussed in the second example in Section 4.2.

Table 1. Performance results of three simplification operations of Example1

Model "SRVN"	Sys. Through put	Sys. Response time	U_{Server}	$U_{BigLoop}$	U_{Disk}	$U_{ServerP}$	U_{DiskP}	Relative error (%) in Sys. Throughput	Relative error (%) in Sys. Response time
Original model	0.261	26.511	6.669	1.296	0.993	0.111	0.993		
Activity simplification	0.261	26.622	6.604	1.282	0.991	0.111	0.991	0.144	0.415
Entry simplification	0.263	25.962	6.486	1.407	1.000	0.112	1	0.723	2.073
Task Simplification	0.257	27.801	6.809		0.976	0.109	0.976	1.657	4.864

In Example1, tasks *Server* and *BigLoop* are infinite servers (i.e., no thread limit), whereas *Disk* and the processors *ServerP* and *DiskP* are single servers.

The effect of the three levels of simplification on the model of Example1 can be seen in Table 1. On the first row of this table, the system throughput, system service time and resource utilizations of the original model are shown. In the subsequent rows, the same performance metrics are reported after activity, entry and task simplifications respectively. From the two rightmost columns of Table 1, it is observed that the amount of errors incurred by each simplification is relatively low comparing to the gain in the size of the models (discussed more in Section 5). Throughput error due to activity and entry simplifications are less than 1%, and due to task simplification less than 2%. The errors incurred by activity, entry and task simplifications on system response time is less than 1%, about 2% and almost 5% respectively. Moreover, along the simplifications steps, the utilizations of tasks and processors also remain almost same. The system bottleneck is *DiskP* (the disk hardware) for all cases. Although the *Disk* task is also saturated, its server *DiskP* is equally saturated, so *Disk* is not a software bottleneck (see [9] for techniques for identifying and mitigating software bottleneck).

4.2   Details of the Operations: Example 2

Fig. 4(a) represents another example of an LQN model called *eShop*, where a number of users' requests go through *StoreApp, CustomerDB, InventoryDB* and *FileServer* for read and write operations. This model has just one entry per task so it is ready for task-level simplification. Preliminary experiments show that the bottleneck is the task *StoreShopping*.

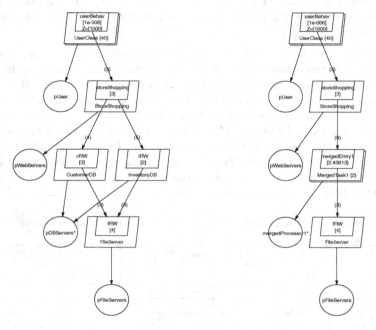

(a) Original model of eShop   (b) After merging CustomerDB and InventoryDB

Fig. 4. An LQN model of eShop

In this model, tasks *CustomerDB* and *InventoryDB* are merged since they are deployed on the same processor. The values of *(entry name, weight, execution demand)* are *(cRW, 0.439, 3)* and *(iRW, 0.561, 2)*. Thus, applying Eq (3) the service demand of the *mergedTask* we found $s_m = 3 * 0.439 + 2 * 0.561 = 2.439$ (where the throughputs of *cRW* and *iRW* are *0.03001* and *0.03833*, respectively). The number of incoming calls to the merged entry is 9 since the incoming calls from *storeShopping* should be directly summed up. For the number of outgoing calls, the values of *(entry name, weight, number of calls from merged entry of task to fRW)* are *(cRW, 0.439, 3)*, *(iRW, 0.561, 3)*. Thus, applying Eq (4), the number of calls from the merged entry to *fRW* is $3 * 0.439 + 3 * 0.561 = 3$. Fig. 4(b) represents the model after merging *CustomerDB* and *InventoryDB* tasks.

Operation 4: Merging Processors and Tasks

The next step of aggregation for this example will be merging different tasks that are deployed on different processors. In case of horizontal as well as vertical merging of such tasks, the service demands, incoming and outgoing calls and multiplicities of tasks are calculated as for horizontal and vertical merging of tasks on the same processor, as discussed in Operation 3 in Section 4.1. The merged processor's multiplicity is the aggregation of multiplicities of merging processors. In Fig. 5(a), *FileServer* and *MergedTask1* (originally deployed on different processors) are merged.

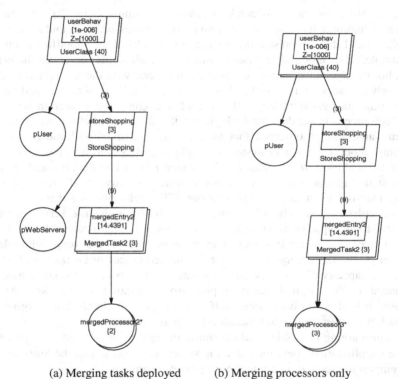

(a) Merging tasks deployed   (b) Merging processors only
on different processors

Fig. 5. More simplification operations on eShop

Table 2. Effects of the simplification operations on system Response Time and Throughput for Example2

Model "eShop"	Sys. Throughput (jobs/ms)	Sys. Response time (ms)	Relative error in Sys. Throughput (%)	Relative error in Sys. Response time(%)
Original model	0.002	12236.2	N/A	N/A
Merging CustomerDB and InventoryDB	0.002	12360.6	0.820	1.017
Merging MergedTask1 and FileServer	0.003	11816.4	5.738	3.431
Merging non-bottleneck processors	0.003	10139.9	17.213	17.132

Table 3. Effects of simplification operations on Utilizations of resources of Example2

Model "eShop"	$U_{StoreShopping}$	$U_{CustomerDB}$	$U_{InventoryDB}$	$U_{FileServer}$	$U_{pWebServers}$	$U_{pDBServers}$	$U_{pFileServers}$
Original model	0.999	0.457	0.520	0.812	0.022	0.167	0.812
Merging CustomerDB and InventoryDB	1.000	0.977{2}		0.811	0.023	0.164	0.811
Merging MergedTask1 and FileServer	0.999	0.977{3}			0.023	0.984{2}	
Merging non-bottleneck processors	0.999	0.977{3}			0.992{3}		

In Fig. 5(b), the non-bottleneck *pWebServers* (multiplicity 2) and *merged-Processor2* (multiplicity 1) are merged into *mergedProcessor3* with multiplicity 3.

Table 2 and Table 3 represent the performance results of the simplification process for Example2. The task and processor resources are all single servers in the original model, but the merged resources become multiservers, with the multiplicities shown in the curly brackets after the utilization. The saturation level of the merged resource is then (utilization)/(multiplicity). Because of averaging, the saturation level for a merged resource is lower than for the highest utilized resource before merging.

From Table 2, it is observed that merging tasks *CustomerDB* and *InventoryDB* incur only about 1% error in system throughput and system response time. Then, merging vertical tasks *MergedTask1* and *FileServer* incur less than 6% and 4% errors in system throughput and system response time respectively. However processor merging incurred much higher errors (about 17% each). We see that the database processor utilization is only 0.16, compared to 0.81 for the file server processor. When the total capacity is shared, the contention is significantly lower for the fileserver accesses, and this effect is even stronger after the very lightly loaded webserver processor is merged (note that the merged processor utilization of 0.992 is relative to a capacity of 3, so it is only 33% saturated). This effect would be much less pronounced if the original database processor utilization were lower. At 81% saturation, it is almost a bottleneck itself. So, it appears probable that merging near-bottleneck resources (tasks and processors) degrades accuracy.

It is worth noting in Table 3, which shows the utilizations of tasks and processors after the simplification operations, that the system bottleneck (i.e., the *StoreShopping* task) remains the same throughout the simplifcation process.

5   Case Study

The performance results reported in this section were obtained by simulation with the lqsim solver [5] with a confidence interval of ±1% of the mean at 95% confidence level. A Java application was created to generate a series of simplified models from the original LQN by merging activities, entries and tasks.

The Case study example is a Business Reporting System described in [7], which can retrieve reports and statistical data about various business processes from a data base. Fig. 6(a) shows the original LQN model (generated from a design specified as a Palladio Component Model) from [7]. The original model has 43 tasks with a large number of entries and activities, running on 43 processors.

Fig. 6(b) shows the final simplified model. The two highly utilized tasks are preserved in the simplified model: one is the software bottleneck of the system and the other is a direct caller of the bottleneck tasks, saturated due to pushback (waiting for services that are delayed by congestion). These two tasks remain the most highly utilized. All the other tasks are merged into a single task.

The performance results after different steps of the simplification process (i.e., activity, entry and task simplification) are comparable, as shown in Figures 7 and 8.

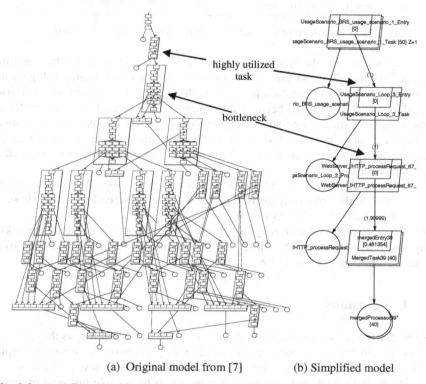

(a) Original model from [7]   (b) Simplified model

Fig. 6. Layered Queueing Network of the Business Reporting System generated from PCM

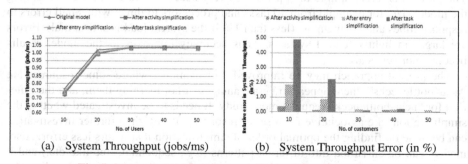

(a)   System Throughput (jobs/ms)   (b)   System Throughput Error (in %)

Fig. 7. System throughput after various simplification operations

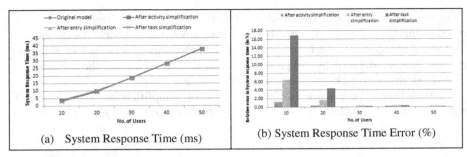

Fig. 8. System response time after various simplification operations

Fig. 7(a) shows that the system throughput is very little changed by the simplification. Fig. 7(b) further shows that the error caused by task simplification is higher than that of entry simplification, which in turn is higher than that of activity simplification. This relationship is expected because the simplifications done for larger model elements (e.g., task) make more approximations than simplifications for smaller model elements (e.g., activity and entry). Also, the simplification process incurs larger percentage errors for fewer customers, perhaps because the approximations involved are better in heavier traffic.

Fig. 8(a) shows similar effects on the response time accuracy. With more customers (e.g., N>20), the error in response time is less than 5% (see Fig. 8(b)). However, for small N (e.g., N=10), the error is larger.

What is important is that, throughout the simplification process, the bottlenecks of the system remain unchanged with similar utilizations. For example with N = 10 the two heaviest task utilizations (73% and 70%) changed by less than 1%.

6   Conclusions

Large performance models are problematic for human and computer, as they are difficult to maintain and take a long time to solve. This paper proposes a model simplification process that compacts a given LQN model to its smallest possible size by reducing non-bottleneck task and processor resources to a single task and processor. The process is rather simple to apply and is very successful in dramatically reducing the model size, for example from 40 tasks and processors to three, with small errors in response times and throughputs (less than 1% if the system load is substantial). Errors are larger in lightly loaded systems. Thus accuracy improves as the model results become more critical.

The present approach may be improved. For example, where some resources are "near-bottlenecks", there appear to be advantages in not merging them, and the criterion for "near-bottleneck" requires more study. It needs to be investigated whether the simplifications are associative for a set of resources. If not, then further investigation can be done on finding the optimal order of simplification that incurs less error. Also, models can be classified into different patterns (e.g., sequential, tree-like etc.) and it can be studied whether they need different rules for finding the optimal order. The position (e.g., at the top, middle or bottom) of the bottleneck resource as well as the bottleneck intensity in a model may also affect the optimal rule. Furthermore, traceability models can be developed to keep track of the simplification steps so that the modeler can go back to an intermediate simplification step and modify performance parameters if needed. The proposed simplification has been applied so far to systems with a single class of users. Further investigation is needed to find the effect of the simplification process on performance results for multiple classes of users.

Acknowledgement. This research was supported by grants from NSERC, the Natural Sciences and Engineering Research Council of Canada, through its Discovery and Strategic Projects programs.

References

1. Chandy, K.M., Herzog, U., Woo, L.: Parametric analysis of queuing networks. IBM Journal of Research and Development **19**(1), 36–42 (1975)
2. Kritzinger, P.S., Wyk, S.V., Krzesinski, A.E.: A generalization of Norton's theorem for multiclass queueing networks. Performance Evaluation **2**(2), 98–107 (1982)
3. Lazowska, E.D., Zahorjan, J., Graham, G.S., Sevcik, K.C.: Quantitative system performance - computer system analysis using queueing network models. Prentice Hall (1984). ISBN: 978-0-13-746975-8
4. Woodside, C.: The relationship of performance models to data. In: Kounev, S., Gorton, I., Sachs, K. (eds.) SIPEW 2008. LNCS, vol. 5119, pp. 9–28. Springer, Heidelberg (2008)
5. Layered Queuing Network homepage. http://www.sce.carleton.ca/rads/lqns/
6. Woodside, M.: Tutorial Introduction to Layered Modeling of Software Performance, Edition 4.0, RADS Lab. http://www.sce.carleton.ca/rads/lqns
7. Martens, A., Koziolek, H., Becker, S., Reussner, R.: Automatically improve software architecture models for performance, reliability, and cost using evolutionary algorithms. In: Proc. First Joint WOSP/SIPEW International Conference on Performance Engineering, pp. 105–116 (2010)
8. Franks, G., Al-Omari, T., Woodside, C.M., Das, O., Derisavi, S.: Enhanced Modeling and Solution of Layered Queueing Networks. IEEE Trans. on Software Eng. **35**(2) (2009)
9. Franks, G., Petriu, D., Woodside, M., Xu, J., Tregunno, P.: Layered bottlenecks and their mitigation. In: Proc of 3rd Int. Conference on Quantitative Evaluation of Systems QEST 2006, Riverside, CA, USA, pp. 103–114, September 2006

Moment-Generating Algorithm for Response Time in Processor Sharing Queueing Systems

Tiberiu Chis[✉] and Peter Harrison

Department of Computing, Imperial College London, SW7 2AZ, London, UK
{tiberiu.chis07,p.harrison}@imperial.ac.uk

Abstract. Response times are arguably the most representative and important metric for measuring the performance of modern computer systems. Further, service level agreements (SLAs), ranging from data centres to smartphone users, demand quick and, equally important, predictable response times. Hence, it is necessary to calculate moments, at least, and ideally response time distributions, which is not straightforward. A new moment-generating algorithm for calculating response times analytically is obtained, based on M/M/1 processor sharing (PS) queueing models. This algorithm is compared against existing work on response times in M/M/1-PS queues and extended to M/M/1 discriminatory PS queues. Two real-world case studies are evaluated.

1 Introduction

One could argue that performance is driving mobile [12,39,41] and cloud [1,2] technologies. For example, users wait, on average, just over nine seconds for a web page to load [27] before opting for more reliable performance from competitors. The same argument applies to delays in data centers [14,16] as part of quality of service (QoS) standards, which is incorporated, along with operational and energy costs [15,18], into service level agreements (SLAs). Whether it's using smartphones to download files using WiFi or streaming web content on the cloud, the delay principle still applies. With emerging technology companies selling increasingly more smartphones in 2015 – Xiaomi and Huawei are each aiming to sell 100 million handsets this year [47,48] – wireless communication via mobile devices will only intensify. Therefore, it is important to understand the effect of delay on asynchronous data transmission and how this impacts performance of millions of devices. From a queueing perspective, delay and response time (or latency, i.e. the time between a job arriving and leaving the system) are closely related. To meet QoS demands, application developers and content providers aim for quick response times to minimize performance bottlenecks. Modelling response times analytically requires a *fair* scheduling policy, such as processor-sharing (PS), which gives n incoming tasks an equal share of the processor (i.e. $1/n$ if service rate is 1). PS scheduling has relevant applications in web server designs and for bandwidth-sharing protocols in packet-switched networks [17,26]. PS queueing models provide an abstraction for such systems and allow analytical response time metrics to represent system delay. Minimising mean

© Springer International Publishing Switzerland 2015
M. Beltrán et al. (Eds.): EPEW 2015, LNCS 9272, pp. 80–95, 2015.
DOI: 10.1007/978-3-319-23267-6_6

response time alone is usually not acceptable nowadays because users tend to be equally frustrated with a highly variable service. They demand response time that is *predictable* [20, 23], which makes it important to calculate moments at least, and ideally response time distributions, which is not straightforward. In the past three decades, work has addressed response time in various ways using PS queues [19, 21, 22, 24, 36, 46]. In the present work, we introduce a novel moment-generating algorithm to calculate response times analytically. The algorithm is based on M/M/1-PS queues and offers the following contributions:

1. Iterative computation of moments, in terms of mean service rate (μ) and utilisation (ρ) of the system, using a partial differential equation for the Laplace transform of response time density.
2. Extension of the moment-generating algorithm to calculate response times for multiple job classes, which is automated for different job weights under discriminatory PS.
3. Applications include performance models dealing with smartphone data transfers, switching states for cloud servers given user demand, resource allocation for data centres, etc.

The rest of the paper is organised as follows: section 2 provides some background on queueing theory, PS scheduling and its applications and defines response times for different scheduling algorithms; in section 3, we describe related work on obtaining higher moments of response time in PS queueing systems; section 4 presents the moment-generating algorithm, which calculates response time in PS queueing systems analytically, with corresponding results under different scenarios; in section 5, we extend the moment-generating algorithm to support mutiple job classes and analyse two real-world case studies in section 6; we conclude and offer extensions in section 7.

2   Background

In this section, we introduce key queueing concepts and justify the importance of queueing models with respect to diverse applications such as servers in smartphones, data centres and networks. Queueing models allow us to abstract the dynamic processes governing modern, complex computer systems and obtain representative performance measures (i.e. response times) with minimal computational cost. Fundamentally, scheduling is an integral part of queueing models for obtaining such measures.

2.1   Scheduling

There exist many scheduling disciplines for servers in queueing models. The most well-known is the first-come first-served (FCFS) discipline, which serves jobs in order of arrival and the job that waits the longest is served first. The best example of the FCFS discipline is in the first-in first-out (FIFO) queue when organising a data buffer. Other scheduling disciplines include last-come first-served (LCFS), which selects the most recent job and serves it first. The most fundamental example of a data structure which implements LCFS is a stack. In terms of system utilisation (ρ), LCFS suffers from greater variability than FCFS as $\rho \to 1$ [7].

Organising servers under processor-sharing (PS) disciplines, such as egalitarian PS (EPS), allows for current jobs to be served at equal rates. Under EPS, if there are n jobs in a system with service rate 1, each job will be served at $1/n$ times the speed of the processor, which means there is no queueing and all jobs start immediately. One useful property of EPS is its *fairness*, where the expected response time of a job is directly proportional to its size. There are variants of PS such as discriminatory PS (DPS), where each job j in the system receives its own percentage of the server, therefore catering for multiple job classes. In DPS, a single processing system serving K job types is controlled by a vector of weights $(\alpha_j > 0, j = 1, \ldots, K)$. Further, assuming there are n_i class i jobs $(i = 1, \ldots, K)$ in the system, each class j job is served at rate:

$$r_j(n_1, \ldots, n_K) = \frac{\alpha_j}{\sum_{i=1}^{K} \alpha_i n_i}, \; j = 1, \ldots, K \qquad (1)$$

Note that when $\alpha_i = \alpha_j, i, j = 1, \ldots, K$, DPS scheduling becomes EPS as each job request has equal weight. Round robin (RR) scheduling offers equal time slices for each job, assigned in circular order and without priorities. The EPS algorithm is seen as an idealisation of RR scheduling in time-shared computer systems [43]. Hereinafter, we use the terms EPS and PS interchangeably. The following section summarises PS applications for a range of computer systems.

2.2   PS Applications

Within queueing systems, the PS server discipline has been of considerable interest for several decades. PS is applied to modelling performance of bandwidth-sharing protocols in packet-switched networks [17,26] and approximating the fair-queueing server disciplines used in communication network routers [45], where delays and congestion control are key measures. Further, PS has proved useful for modelling heavy-tailed service time distributions [31] and bulk arrivals [33]. Stochastic analysis of PS systems dealing with power management and energy consumption have also been of interest. More specifically, a queueing model with PS scheduling was employed when setting bounds on performance of dynamic speed scaling [29].

When predicting queueing delays, for example, the PS discipline is more complex to model than FCFS because the remaining response times in PS systems depend on future (i.e. uncertain) arrivals and dependent service requirements. Nonetheless, the simplicity of PS, coupled with fairness properties, has made it easily applicable to a variety of high-speed, computer systems that are abstracted by queueing systems.

Typically, modern servers are often difficult to replicate precisely in a numerically tractable way; to model such servers, PS scheduling is assumed for a number of reasons:

1. PS is popular for web server design [9] and evaluating flow-level performance of end-to-end flow control mechanisms like TCP [44].
2. Under PS, there is no queueing *per se* and arriving jobs start immediately to access server resources.
3. The implicit *fairness* means expected response time of a job is directly proportional to its size.
4. PS is effective for heavy-tailed service times, which may arise, for example, as short jobs are allowed to overtake long jobs. It also facilitates tractable asymptotic analysis of heavy-tailed service time distributions [31].

2.3   Queueing Models

The most fundamental queueing model is the M/M/1-FCFS queue, with Poisson mean arrival rate λ and exponential mean service time μ for one server with FCFS scheduling. Similarly, the M/M/1 queue under PS scheduling is written as M/M/1-PS using Kendall notation. Generalising such queues, the G/G/m queue offers generally distributed arrivals and service times for m parallel servers. Note that arrivals and service times may have specific distributions such as hyperexponential, phase-type, MAP-induced, etc.

One utilises underlying continuous time Markov chain (CTMC) properties of queueing models. Additionally, classes of product-form models exist, where state equilibrium probability is a scaled product of the marginal state probabilities of Markov processes that represent individual system components [30]. Therefore, queueing models approximate modern communication systems and their long-term behaviour, without the state explosion problem limiting modelling possibilities. Often, response time is a key measurement, which we define in subsequent sections, because it is useful for approximating performance and thus provides resource allocation on large-scale storage systems, mobile technology, wireless sensor networks (WSNs), etc. Typically, response times are obtained using aforementioned queues, given queueing theoretic assumptions.

2.4   Response Times

We refer to response time (or, sojourn time) as the time a customer spends in the system before completely departing from it. In queueing terms, response time T is the sum of the queueing time and the service time (i.e. duration of customer service). Let λ be the arrival rate, μ be the service rate, and $\rho = \lambda/\mu < 1$ be the equilibrium system utilisation. Of course, under FCFS queueing discipline, the response time probability density function is well known to be $f(t) = (\mu - \lambda)e^{-(\mu-\lambda)t}$, [8]. Under PS, the mean unconditional response time may be computed using Little's law. Let $L = \rho/(1 - \rho)$ be the mean number of jobs in the system and $\mathbb{E}[T]$ be the mean unconditional response time at equilibrium. Then, it follows from Little's law that $L = \lambda \mathbb{E}[T]$ and re-arranging for $\mathbb{E}[T]$ gives us:

$$\mathbb{E}[T] = \frac{L}{\lambda} = \frac{\rho}{\lambda(1 - \rho)} = \frac{\lambda/\mu}{\lambda(1 - \rho)} = \frac{1}{\mu(1 - \rho)} \tag{2}$$

When jobs require x units of service time, the mean conditional response time is given by $\mathbb{E}[T(x)] = x/(1-\rho)$. Therefore, $\mathbb{E}[T(x)]$ is linear in x, meaning that jobs with twice the size have double the response time, on average. Note that this fairness property only applies to means. As ρ approaches 1, the unconditional mean response time $\mathbb{E}[T]$ grows as $1/(1-\rho)$ and is independent of the variability of the service time distribution. Terms only affected by the mean of the service time distribution exhibit the *insensitivity* property [38].

Calculating higher moments of response time under PS scheduling requires, in general, an advanced understanding of layered branching of incoming jobs into the system [22]. Additionally, higher moments identify variability and skewness in time-series and approximate distributions, which may help to flatten heavy-tails, for instance. The next section describes existing methods in the literature for obtaining response time in PS queues.

3   Related Work on Response Times

There are a number of works on approximating response time under PS scheduling, but few which adopt analytical queueing theory, even for the Markovian M/M/1-PS queue. Some of the earliest significant work on PS queues is by Coffman *et al* in 1970 [4], which analysed waiting time means and variance of PS systems compared to FCFS. In 1980, Fayolle *et al* [37] summarised results of Kleinrock and Mitrani for DPS and also obtained average response time (both conditionally and unconditionally on job request sizes) in M/M/1-DPS queueing systems. Further, Laplace transforms provided average waiting time for multiple class types and asymptotic behaviour of service demand was also obtained, but no results on higher response time moments were given.

The abstraction of PS scheduling as a layered branching of incoming jobs into the system was first used by Yashkov in 1987 and led to a derivation of conditional response time moments a decade later [22,24]. The kth moment of response time of a job with service requirement x, $\mathbb{E}[T(x)^k]$, is given by:

$$\mathbb{E}[T(x)^k] = -\sum_{i=1}^{k} \binom{k}{i} (-1)^i \mathbb{E}[T(x)^{k-i}] \alpha_i(x)$$

$$\alpha_k(x) = \frac{k}{(1-\rho)^k} \int_{t=0}^{x} (x-t)^{k-1} F^{(k-1)*}(t) dt$$

$$F^{0*}(x) = 1$$

$$F^{n*}(x) = \int_{0}^{x} F^{(n-1)*}(x-u) dF(u) \quad \text{for} \ \ n \geq 2$$

$$F(x) = \frac{1}{\beta_1} \int_{0}^{x} \big(1 - B(u)\big) du$$

where $B(\cdot)$ is a general service time distribution with finite mean $\beta_1 < \infty$.

In 2003, Masuyama *et al* obtained a complementary response time distribution [32]. Specifically, for an M/M/1-PS queue with arrival rate λ, service rate μ and job size x, the complement of response time distribution $\bar{T}(x) = 1 - T(x)$ is defined recursively as:

$$\bar{T}(x) = \sum_{n=0}^{\infty} (1 - \rho)\rho^n \sum_{k=0}^{\infty} \frac{(\lambda + \mu)^k x^k}{k!} e^{-(\lambda+\mu)x} h_{n,k} \tag{3}$$

where $h_{n,k+1} = \frac{n}{n+1} \frac{\mu}{\lambda+\mu} h_{n-1,k} + \frac{\lambda}{\lambda+\mu} h_{n+1,k}$, $h_{n,0} = 1$ and $h_{-1,k} = 0$.

This computationally intensive recursion is more costly than Yashkov's iterative solution, although storing previous terms in a buffer would speed up calculations. However, truncating multiple infinite sums is a serious disadvantage of Masuyama's method.

In 2004, Kim *et al* offered a joint transform to obtain response time moments for K job classes with different shares of service. An M/M/1-DPS queueing system is considered, where $\rho_i = \lambda_i/\mu_i$, for all jobs $i = 1, \ldots, K$, subject to $\rho = \sum_{i=1}^{K} \rho_i < 1$. Let N_i be the number of jobs in the system at steady state and $Q(z_1, \ldots, z_K) = \mathbb{E}[z_1^{N_1} \cdots z_K^{N_K}]$ be the joint probability generating function for the numbers of jobs of each class in the queue at steady state. A job i with required service time greater than x is tagged such that when it attains service x, $S_i(x)$ and $N_{ij}(x)$ denote the elapsed response time and the number of class j jobs in the system, respectively $(j = 1, \ldots, K)$. The joint distribution of $S_i(x)$ and $N_{ij}(x)$ is then given by the transform $T_{ix}(s; z_1, \ldots, z_K) = \mathbb{E}[e^{-sS_i(x)} z_1^{N_{i1}(x)} \cdots z_K^{N_{iK}(x)}]$ for $|z_i| \leq 1$, $i = 1, \ldots, K$, and $s \geq 0$.

The joint transform $T_{ix}(s; z_1, \ldots, z_K)$ is governed by the following partial differential equation (PDE):

$$\frac{\partial}{\partial x} T_{ix}(s; z_1, \ldots, z_K)$$

$$= -\sum_{j=1}^{K} \frac{\alpha_j}{\alpha_i} \left\{ \left(s + \sum_{k=1}^{K} \lambda_k (1 - z_k) \right) z_j - \mu_j (1 - z_j) \right\} \frac{\partial}{\partial z_j} T_{ix}(s; z_1, \ldots, z_K)$$

$$- \left(s + \sum_{j=1}^{K} \lambda_j (1 - z_j) \right) T_{ix}(s; z_1, \ldots, z_K) \tag{4}$$

Deconditioning on x, which has exponential distribution with parameter μ_i requires only taking a Laplace transform.

We define $T_i(s; z_1, \ldots, z_K) = \int_0^{\infty} \mu_i e^{-\mu_i x} T_{ix}(s; z_1, \ldots, z_K) dx$ as the joint Laplace transform of the unconditional joint density of response time and probability generating function of class populations. It is easy to see this is given by the PDE, for $i = 1, \ldots, K$,

$$-\mu_i Q(z_1, \ldots, z_K) + \mu_i T_i(s; z_1, \ldots, z_K)$$

$$= -\sum_{j=1}^{K} \frac{\alpha_j}{\alpha_i} \left\{ \left(s + \sum_{k=1}^{K} \lambda_k (1 - z_k) \right) z_j - \mu_j (1 - z_j) \right\} \frac{\partial}{\partial z_j} T_i(s; z_1, \ldots, z_K)$$

$$- \left(s + \sum_{j=1}^{K} \lambda_j (1 - z_j) \right) T_i(s; z_1, \ldots, z_K) \tag{5}$$

Unconditional moments of response time are derived by differentiating equation (5). Kim *et al* solve $(K + 1)(K + 2)/2$ linearly independent equations to obtain unknown moments M_i^{jk}, $0 \leq j \leq k \leq K$, for each i, $i = 1, \ldots, K$, which are defined as:

$$M_i^{jk} = \left. \frac{\partial}{\partial z_j \partial z_k} T_i(s; z_1, \ldots, z_K) \right|_{s=0, z_1 = \cdots = z_K = 1} \tag{6}$$

where z_0 is taken to be the Laplace-parameter s. We illustrate the calculation of such moments in the case of one class ($K = 1$) in Appendix 7. The next section introduces a novel moment-generating algorithm that can *iteratively* calculate arbitrary moments of response time, thus improving an aspect of the Kims' method in this respect.

4   Moment-Generating Algorithm

In a PS queue with utilisation ρ, the response time T of an arriving customer that requires x units of service time is known to have a probability density function that has Laplace transform:

$$W^*(s \mid x) = \frac{(1 - \rho)(1 - \rho r^2) e^{-[\rho\mu(1-r)+s]x}}{(1 - \rho r)^2 - \rho(1 - r)^2 e^{-[1/r - \rho r]\mu x}} \tag{7}$$

where r is the smaller root of the equation $\rho r^2 - (\rho + 1 + s/\mu)r + 1 = 0$. The result is long known, see for example [4,7], and is derived by solving a partial differential equation for a certain generating function $G(z, s, x)$, viz.

$$(\mu z^2 - (\rho\mu + \mu + s)z + \rho\mu) \frac{\partial G}{\partial z} - \frac{\partial G}{\partial x} = (\rho\mu + s - \mu z)G \tag{8}$$

which yields $W^*(s \mid x) = (1 - \rho)G(\rho, s, x)$. We make the following observations:

1. The unconditional response time density for an arriving customer that has exponentially distributed service time requirement with mean $1/u$ is the product of u and the Laplace transform of $W^*(s|x)$ with respect to x, evaluated at Laplace-parameter u.
2. To calculate moments, the generating function's derivatives need only be computed at $s = 0$.

3. There is no need to solve the differential equation (8) for the generating function G since the moments are given by its derivatives evaluated at $s = 0$ and $z = \rho$, corresponding to the geometric equilibrium queue length probability distribution.
4. The Laplace transform of derivative $\partial G/\partial x$ yields the term $uG^{*x}(z, s, u) - G(z, s, 0)$, where G^{*x} denotes the Laplace transform of G with respect to x and the initial value $G(z, s, 0)$ is known to be $1/(1 - z)$.
5. At $s = 0$ and $z = \rho$, the coefficient of $\partial G/\partial x$ vanishes. Thus, by successive differentiation of the Laplace-transformed equation (8), we can determine the moments recursively.

In this way, we obtain the following unconditional moments for response time:

$$\mathbb{E}[T] = \frac{1}{\mu(1 - \rho)} \tag{9}$$

$$\mathbb{E}[T^2] = \frac{4}{\mu^2(2 - \rho)(1 - \rho)^2} \tag{10}$$

$$\mathbb{E}[T^3] = \frac{12(\rho + 2)}{\mu^3(2 - \rho)^2(1 - \rho)^3} \tag{11}$$

$$\mathbb{E}[T^4] = \frac{48(48 + 52\rho - 10\rho^2 - 6\rho^3 - 24\rho^4 + 9\rho^5)}{\mu^4(2 - \rho)^3(1 - \rho)^3(3 - 2\rho)(4 - 3\rho)} \tag{12}$$

In table 1, we summarise response time moments with fixed $\mu = 1$ whilst increasing ρ and also obtain moments with fixed $\rho = 0.5$ whilst increasing μ. After calculating response time moments analytically, approximating a full response time distribution is typically straightforward [11], for example via the general lambda distribution (GLD) [25,34]. As such approximations are not the main scope of this paper, we guide the reader to relevant material [35,40,42] for more information. We extend the moment-generating algorithm to multiple job types in the next section.

Table 1. Moments for varying ρ with fixed $\mu = 1$ (left) and varying μ with fixed $\rho = 0.5$ (right).

Moment	$\rho = 0.2$	$\rho = 0.5$	$\rho = 0.8$	Moment	$\mu = 0.5$	$\mu = 2.5$	$\mu = 8.5$
$\mathbb{E}[T]$	1.25	2.0	5.0	$\mathbb{E}[T]$	4.0	0.8	0.235
$\mathbb{E}[T^2]$	3.47	10.67	83.33	$\mathbb{E}[T^2]$	42.67	1.71	0.147
$\mathbb{E}[T^3]$	15.91	106.7	2.9e3	$\mathbb{E}[T^3]$	853.3	6.82	0.174
$\mathbb{E}[T^4]$	105.3	1.6e3	1.1e5	$\mathbb{E}[T^4]$	2.5e4	40.5	0.303

5   Multi-Class Algorithm

We build an automated moment-generating algorithm for multiple job classes, which supports DPS scheduling for K job classes and incorporates service weights α_i for each job class $i, i = 1, \ldots, K$. For simplicity of presentation, we use equal job weights (i.e. $\alpha_i = \alpha_j, i, j = 1, \ldots, K$), but this is not a requirement of our method. Adapting a multi-class version of the PDE given in equation (8), we apply similar methods of successive differentiation to determine moments recursively. Assuming two job classes (i.e. $K = 2$), with mean arrival rates λ_1 and λ_2, mean service rates μ_1 and μ_2, and utilisation $\rho_1 = \lambda_1/\mu_1$ and $\rho_2 = \lambda_2/\mu_2$ such that $\rho_1 + \rho_2 < 1$, we obtain respective mean response times $\mathbb{E}[T_1]$ and $\mathbb{E}[T_2]$ as:

$$\mathbb{E}[T_1] = \frac{1}{\mu_1(1 - \rho_1 - \rho_2)}; \; \mathbb{E}[T_2] = \frac{1}{\mu_2(1 - \rho_1 - \rho_2)} \tag{13}$$

Further, we derive second moments of response time $\mathbb{E}[T_1^2]$ and $\mathbb{E}[T_2^2]$ as:

$$\mathbb{E}[T_1^2] = \frac{4\big(\mu_1(1 + \rho_2) + \mu_2(1 - \rho_2)\big)}{\mu_1^2(1 - \rho_1 - \rho_2)^2(\mu_1(2 - \rho_1) + \mu_2(2 - \rho_1 - 2\rho_2))} \tag{14}$$

$$\mathbb{E}[T_2^2] = \frac{4\big(\mu_1(1 - \rho_1) + \mu_2(1 + \rho_1)\big)}{\mu_2^2(1 - \rho_1 - \rho_2)^2(\mu_1(2 - 2\rho_1 - \rho_2) + \mu_2(2 - \rho_2))} \tag{15}$$

```
K = 2;
L1sub = Solve[ (L# == λ# / μ# + Sum[αj (λj L# + λ# Lj) / (αj μj + α# μ#), {j, 1, K}]) & /@
      Range[K], L# & /@ Range[K]] // Simplify;
Do[M0,i = ((-Li / λi) /. L1sub) // Simplify, {i, 1, K}];
e1 =
      ( (μi + μ# α# / αi) M0,#,i - λ# Sum[M0,j,i αj / αi, {j, 1, K}] + Sum[M#,j,i αj / αi, {j, 1, K}] ==
      λ# M0,i - (1 + α# / αi) M#,i) & /@ Range[K];
e2 = ( (μi + 2 μ# α# / αi) M#,#,i - 2 λ# Sum[M#,j,i αj / αi, {j, 1, K}] ==
      μi L#,# + 2 (1 + α# / αi) λ# M#,i) & /@ Range[K];
e3 = Table[ (μi + μk αk / αi + μm αm / αi) Mm,k,i - λk Sum[Mm,j,i αj / αi, {j, 1, K}] -
      λm Sum[Mk,j,i αj / αi, {j, 1, K}] == μi Lm,k + λk (1 + αm / αi) Mm,i + λm (1 + αk / αi) Mk,i,
      {m, 1, K}, {k, 1, m-1}] // Flatten;
eM3 = Join[e1, e2, e3] /. M_x_,y_,i /; x > y → My,x,i /. L_x_,y_ /; x > y → Ly,x;
eM = ( (μi + μ# α# / αi) M#,i == λ# Sum[Mj,i αj / αi, {j, 1, K}] + μi L# + λ#) & /@ Range[K];
M2sub = Solve[(eM /. L1sub) // Flatten, M#,i & /@ Range[K]] // First;
eL2 = Table[Lj,k - Sum[αi (λj Lk,i + λk Li,j + λi Lj,k) / (αj μj + αk μk + αi μi), {i, 1, K}] ==
      (αj + αk) (λj Lk + λk Lj) / (αj μj + αk μk), {k, 1, K}, {j, 1, k}];
L2sub = Solve[(eL2 /. L1sub /. L_x_,y_ /; x > y → Ly,x) // Flatten,
      Table[La,b, {b, 1, K}, {a, 1, b}] // Flatten] // First // Simplify;
M0,0,i_ := (-2 M0,i - 2 Sum[M0,j,i αj / αi, {j, 1, K}]) / μi;
M3sub = Solve[eM3 /. L2sub /. M2sub, Join[Table[M0,b,i, {b, 1, K}],
      (Table[Ma,b,i, {b, 1, K}, {a, 1, b}] // Flatten)]] // First;
M0,1
```

Fig. 1. Mathematica code for K-class moments up to 2

These expressions were obtained by solving the moment equations obtained by repeatedly differentiating equation (5) up to two times. The algorithm to do this, written in Wolfram's Mathematica, is shown in figure 1. Obtaining the variance (i.e. $\sigma_i^2 = \mathbb{E}[T_i^2] - \mathbb{E}[T_i]^2$) of a class i job reveals the spread of the response time distribution from the mean. Further, calculating higher moments of response time is useful for predicting performance in a variety of multi-class applications where jobs have different priorities – or shares of a PS server. As with the second moment, higher moments are derived by differentiating equation (5) and defining the steady state generating function $Q(\cdot)$, which is straightforward to derive. This is the approach used in figure 1 for just two moments, but which is easy to extend to any higher moments. The difficulty that arises is the number of calculations needed, since every partial derivative up to p is required to calculate moment p – a rapidly increasing number, especially if there are many classes. A symbolic solution is surely intractable, but mathematical software could easily cope with a numerical solution when values are pre-set for the parameters of the model. Using such an automated multi-class algorithm, it is straightforward to estimate the probability distribution of response time for K job classes; good approximations can usually be found from the first four moments or so.

6   Case Studies

We obtained workload traces from two applications, which we abstract using M/M/1-PS queueing models, each with two job classes (i.e. $K = 2$). The first application is an HTC One (M7) smartphone transmitting data via 4G cell radio, where a time-stamped trace was recorded from a transmission period of 30 minutes. We summarise this HTC trace with the following mean service rates for each job class: $\mu_1 = 0.6$ and $\mu_2 = 2.4$. The second application is an Apache Cloud-Stack VM executing programs on an Intel Core i7-2600 CPU @ 3.40GHz host machine. The CloudStack trace was recorded with mean service rates $\mu_1 = 1.4$ and $\mu_2 = 6.1$. Using equation (13), we plot mean response times (i.e. $\mathbb{E}[T_i]$, for $i = 1, 2$) in figure 2 with increasing system load (i.e. $\rho_1 + \rho_2$) for the HTC and CloudStack traces. Further, using equations (14) and (15), we plot variance (i.e.

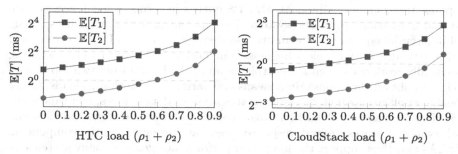

Fig. 2. $\mathbb{E}[T_1]$ and $\mathbb{E}[T_2]$ for HTC (left) and CloudStack (right) traces under increasing load.

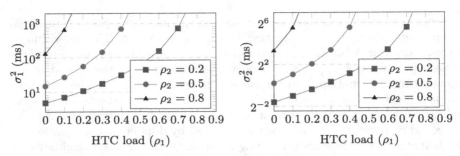

Fig. 3. σ_1^2 (left) and σ_2^2 (right) for the HTC trace under increasing load.

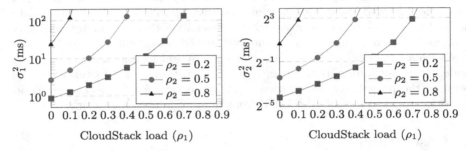

Fig. 4. σ_1^2 (left) and σ_2^2 (right) for the CloudStack trace under increasing load.

σ_i^2, for $i = 1, 2$) of response time for increasing values of ρ_1 and ρ_2 in figures 3 and 4. Note that for the variance, the total system load (i.e. $\rho_1 + \rho_2$) does not exceed 1. For systems with more than two job classes, it is important to measure performance via response time moments for resource provisioning whilst considering different system load. Indeed, the moment-generating algorithm allows such measurements for any K job classes.

7   Conclusion and Future Work

We proposed an automated moment-generating algorithm for calculating response times analytically in M/M/1-PS queues in terms of mean service rate (μ) and utilisation (ρ) of the system. This incremental algorithm uses a partial differential equation for recursively evaluating terms in a Laplace transform and is extended for multiple job classes. Further, we examined two case studies, specifically workloads from a smartphone and a VM exhibiting two job classes each, and obtained response time means and variance for both workloads. Other possible applications include resource allocation in data centres, run-time analysis of multi-class workload in storage systems, and online server provisioning with switching states. Indeed, response times have become essential components of SLAs and thus support the long-term performance goals of many systems.

Extensions include generalising response time analysis for G/G/1-DPS queues and catering for bursty arrivals through an online MMPP or HMM used for possible workload prediction. Further, incorporating energy cost into our performance models would match the SLA requirements more realistically. Indeed, battery models are popular in the literature [3,5,6,10,13,28], but there is scarce analysis on power consumption related to performance via higher response time moments for multiple job classes.

References

1. Curtis, J.: 10 top cloud computing providers for 2014. http://tinyurl.com/top-10-cloud-providers-2014
2. Velazco, C.: Google gives students unlimited cloud storage. http://www.engadget.com/2014/09/30/google-drive-for-education/
3. Huria, T., Ceraolo, M., Gazzarri, J., Jackey, R.: High fidelity electrical model with thermal dependence for characterization and simulation of high power lithium battery cells. In: Proc. IEEE IEVC, Greenville, pp. 1–8 (2012)
4. Coffman Jr., E.G., Muntz, R.R., Trotter, H.: Waiting Time Distributions for Processor-Sharing Systems. Journal ACM **17**, 123–130 (1970)
5. Prabhu, B.J., Chockalingam, A., Sharma, V.: Performance analysis of battery power management schemes in wireless mobile devices. In: Proc. IEEE WCNC, Orlando, vol. 2, pp. 825–831 (2002)
6. Open Battery. http://www.doc.ic.ac.uk/~gljones/openbattery/index.php
7. Harrison, P.G., Patel, N.M.: Performance Modelling of Communication Networks and Computer Architectures. Addison-Wesley (1993)
8. Stewart, W.J.: Probability, Markov Chains, Queues, and Simulation: The Mathematical Basis of Performance Modeling, p. 409. Princeton University Press (2009)
9. Kjaer, M.A., Kihl, M., Robertsson, A.: Response-time control of a processor-sharing system using virtualised server environments. In: Proc. IFAC, Korea, vol. 17, p. 3612–3618 (2008)
10. Rohner, C., Feeney, L.M., Gunningberg, P.: Evaluating battery models in wireless sensor networks. In: Tsaoussidis, V., Kassler, A.J., Koucheryavy, Y., Mellouk, A. (eds.) WWIC 2013. LNCS, vol. 7889, pp. 29–42. Springer, Heidelberg (2013)
11. Au-Yeung, S.W.M., Dingle, N.J., Knottenbelt, W.J.: Efficient approximation of response time densities and quantiles in stochastic models. In: Proc. ACM WOSP, Redwood Shores, vol. 4, pp. 151–155 (2004)
12. Shye, A., Scholbrock, B., Memik, G., Dinda, P.A.: Characterizing and Modeling User Activity on Smartphones, Technical Report, Northwest University (2010)
13. Rao, V., Singhal, G., Kumar, A., Navet, N.: Battery model for embedded systems. In: Proc. IEEE VLSID, Washington, DC, vol. 18, pp. 105–110 (2005)
14. Gao, P.X., Curtis, A.R., Wong, B., Keshav, S.: It's not easy being green. In: Proc. ACM SIGCOMM, Helsinki, vol. 44, pp. 211–222 (2012)
15. Wray, J.: Where's The Rub: Cloud Computing's Hidden Costs. http://tinyurl.com/cloud-computing-hidden-costs
16. Alawnah, R.Y., Ahmad, I., Alrashed, E.A.: Green and Fair Workload Distribution in Geographically Distributed Data. Journal Green Eng. **4**, 69–98 (2014)
17. Massoulie, L., Roberts, J.W.: Bandwidth sharing and admission control for elastic traffic. Telecomm. Systems **15**, 185–201 (2000)
18. AISO.net. http://www.aiso.net/index.html
19. Ott, T.J.: The Sojourn-Time Distribution in the M/G/1 Queue with Processor Sharing. Journal of Applied Probability **21**, 360–378 (1984)

20. Wierman, A.: Scheduling for Today's Computer Systems: Bridging Theory and Practice, PhD Thesis, School of Computer Science, Carnegie Mellon University (2007)
21. Kim, J., Kim, B.: Sojourn time distribution in the M/M/1 queue with discriminatory processor-sharing. Performance Evaluation **58**, 341–365 (2004)
22. Yashkov, S.F.: Processor-Sharing Queues: Some Progress In Analysis. Queueing Systems **2**, 1–17 (1987)
23. Wierman, A., Harchol-Balter, M.: Classifying scheduling policies with respect to higher moments of conditional response time. In: Proc. ACM SIGMETRICS (2005)
24. Zwart, A.P., Boxma, O.J.: Sojourn time asymptotics in the M/G/1 processor sharing queue. Queueing Systems **35**, 141–166 (2000)
25. Lakhany, A., Mausser, H.: Estimating the parameters of the General Lambda Distribution. Algo. Research Quarterly **3**, 47–58 (2000)
26. Roberts, J.W.: A survey on statistical bandwidth sharing. Computer Networks **45**, 319–332 (2004)
27. Lohr, S.: For Impatient Web Users, an Eye Blink Is Just Too Long to Wait. New York Times. http://tinyurl.com/eye-blink-too-long-to-wait
28. Jones, G.L., Harrison, P.G., Harder, U., Field, T.: Fluid queue models of battery life. In: Proc. IEEE MASCOTS, vol. 19, pp. 278–285 (2011)
29. Wierman, A., Andrew, L.L.H., Tang, A.: Power-aware speed scaling in processor sharing systems: Optimality and robustness. Performance Evaluation **69**(12), 601–622 (2012)
30. Casale, G., Harrison, P.G.: AutoCAT: automated product-form solution of stochastic models. In: Matrix-Analytic Methods in Stochastic Models, vol. 27, pp. 57–85 (2013)
31. Queija, R.N.: Sojourn times in non-homogeneous QBD processes with processor sharing. Stochastic Models **17**, 61–92 (2001)
32. Masuyama, H., Takine, T.: Sojourn time distribution in a MAP/M/1 processor-sharing queue. Op. Res. Letters **31**, 406–412 (2003)
33. Bansal, N.: Analysis of the M/G/1 processor-sharing queue with bulk arrivals. Op. Res. Letters **31**, 401–405 (2003)
34. Lebrecht, A.: Queueing network models of Zoned RAID system performance, PhD Thesis, Department of Computing, Imperial College London (2009)
35. Ramberg, J., Schmeiser, B.: An approximate method for generating asymmetrics random variables. Comm. ACM **17**, 78–82 (1974)
36. Ward, A.R., Whitt, W.: Predicting reponse times in processor-sharing queues. In: Proc. of Fields Institute Conference on Communication Networks (2000)
37. Fayolle, G., Iasnogorodski, R., Mitrani, I.: Sharing a Processor Among Many Job Classes. Journal ACM **27**(3), 519–532 (1980)
38. Kelly, F.: Stochastic Networks and Reversibility, vol. 1. Wiley (1979)
39. Embedded Microprocessor Benchmark Consortium (EEMBC). http://eembc.org/
40. Ramberg, J., Dudewicz, E., Tadikamalla, P., Mykytka, E.: A probability distribution and its uses in fitting data. Technometrics **21**, 201–214 (1979)
41. AndEBench-Pro. http://eembc.org/andebench/index_pro.php
42. Freimer, M., Mudholkar, G., Kollia, G., Lin, C.: A study of the generalized Tukey Lambda family. Comm. in Statistics **17**, 3547–3567 (1988)
43. Aalto, S., Ayesta, U., Borst, S., Misra, V., Nunez-Queija, R.: Beyond Processor Sharing. ACM SIGMETRICS Perform. Eval. Rev. **34**, 36–43 (2007)
44. Kherani, A.A., Kumar, A.: On processor sharing as a model for TCP controlled HTTP-like transfers. In: Proc. IEEE ICC, Paris, vol. 4, pp. 2256–2260 (2004)

45. Dukkipati, N., Kobayashi, M., Zhang-Shen, R., McKeown, N.: Processor sharing flows in the internet. In: de Meer, H., Bhatti, N. (eds.) IWQoS 2005. LNCS, vol. 3552, pp. 271–285. Springer, Heidelberg (2005)
46. Harrison, P.G.: Response time distributions in queueing network models. In: Donatiello, L., Nelson, R. (eds.) SIGMETRICS 1993 and Performance 1993. LNCS, vol. 729, pp. 147–164. Springer, Heidelberg (1993)
47. Huilgol, M.: Xiaomi aims to sell 100 million smartphones in 2015. http://tinyurl.com/xiaomi-100-million-smartphones
48. Moore, M.: Huawei Looks To Shift 100 Million Smartphones in 2015. http://tinyurl.com/huawei-100-million-smartphones

Appendix: The Kims' Method of Response Time Moments for $K = 1$

Conditional and unconditional joint transforms of response time are given in equations (4) and (5), respectively. This allows calculation of conditional and unconditional moments of response time [21], where there are K job classes. For the $K = 1$ case, let us assume the following conditions for the M/M/1-PS queueing model:

1. The mean arrival rate is λ and the mean service rate is μ.
2. Utilisation is $\rho = \lambda/\mu < 1$.
3. $z_1 = z/\rho$, where z is the parameter from equation (8).

Let $Q(z_1) = \mathbb{E}[z_1^{N_1}]$ be the probability generating function in the system at steady state for one job type, where N_1 is the number of jobs in the system at steady state. Note that $z_1 = z/\rho$ is the difference of deconditioning on ρ in Kim's method. Further, Kim $et\ al$ tag a job with required service time greater than x; when the tagged job attains service x, let $S_1(x)$ and $N_1(x)$ denote the elapsed response time and the number of jobs in the system, respectively. Then, Kim $et\ al$ use a joint transform to derive a relation on the joint distribution of $S_1(x)$ and $N_1(x)$:

$$T_x(s; z_1) = \mathbb{E}\big[e^{-sS_1(x)} z_1^{N_1(x)}\big]$$

which is defined for $|z_1| \leq 1$ and $s \geq 0$.

The proof of this relation is given in [21] and is omitted here. For the $K = 1$ case, we evaluate the expression for the PDE given in equation (4), such that we obtain equation (8) as follows:

$$\frac{\partial}{\partial x} T_x(s; z_1) =$$
$$-\frac{\alpha_1}{\alpha_1}\bigg(\big(s + \lambda(1 - z_1)\big)z_1 - \mu(1 - z_1)\bigg)\frac{\partial}{\partial z_1} T_x(s; z_1) - (s + \lambda(1 - z_1))T_x(s; z_1)$$

Simplifying terms gives us

$$\frac{\partial}{\partial x} T_x(s; z_1) = -\bigg(sz_1 + \lambda z_1 - \lambda z_1^2 - \mu + \mu z_1\bigg)\frac{\partial}{\partial z_1} T_x(s; z_1) - (s + \lambda - \lambda z_1)T_x(s; z_1)$$

Substituting z/ρ for z_1, we have

$$\tfrac{\partial}{\partial x}T_x(s;z_1) = -\rho\left(s\tfrac{z}{\rho} + \lambda\tfrac{z}{\rho} - \lambda\tfrac{z^2}{\rho^2} - \mu + \mu\tfrac{z}{\rho}\right)\tfrac{\partial}{\partial z}T_x(s;z_1) - (s + \lambda - \lambda\tfrac{z}{\rho})T_x(s;z_1)$$

Simplifying terms further and using relation $\rho = \lambda/\mu$ gives us

$$\tfrac{\partial}{\partial x}T_x(s;z_1) = \left(-sz - \lambda z + \lambda\tfrac{z^2}{\rho} + \rho\mu - \mu z\right)\tfrac{\partial}{\partial z}T_x(s;z_1) - (s + \lambda - \lambda\tfrac{z}{\rho})T_x(s;z_1)$$

$$\tfrac{\partial}{\partial x}T_x(s;z_1) = \left(-sz - \rho\mu z + \mu z^2 + \rho\mu - \mu z\right)\tfrac{\partial}{\partial z}T_x(s;z_1) - (s + \rho\mu - \mu z)T_x(s;z_1)$$

Replacing G for $T_x(s;z_1)$ and rearranging terms gives us equation (8) as follows:

$$\left(\mu z^2 - (\rho\mu + \mu + s)z + \rho\mu\right)\tfrac{\partial G}{\partial z} - \tfrac{\partial G}{\partial x} = (\rho\mu + s - \mu z)G$$

Obtaining unconditional moments of response time uses repeated differentiation of the PDE given in equation (5), where we use the joint transform $T(s;z_1)$ for the $K = 1$ case. To obtain the first moment of response time T (i.e. $\mathbb{E}[T]$), we use equation (2) and Little's law. The second moment requires derivation of $(K+1)(K+2)/2$ linearly independent equations with unknown moments $L^j, M^0, M^j, M^{00}, M^{0j}, j = 1, \ldots, K$, and $L^{jk}, M^{jk}, 1 \le j \le k \le K$. For $K = 1$, the moments are defined as follows:

$$L^1 = \left.\frac{\partial}{\partial z_1}Q(z_1)\right|_{z_1=1}, \quad M^0 = \left.\frac{\partial}{\partial s}T(s;z_1)\right|_{s=0,z_1=1}, \quad M^1 = \left.\frac{\partial}{\partial z_1}T(s;z_1)\right|_{s=0,z_1=1},$$

$$M^{00} = \left.\frac{\partial^2}{\partial s^2}T(s;z_1)\right|_{s=0,z_1=1}, \quad M^{01} = \left.\frac{\partial^2}{\partial s\partial z_1}T(s;z_1)\right|_{s=0,z_1=1},$$

$$L^{11} = \left.\frac{\partial^2}{\partial z_1^2}Q(z_1)\right|_{z_1=1}, \quad M^{11} = \left.\frac{\partial^2}{\partial z_1^2}T(s;z_1)\right|_{s=0,z_1=1}$$

$$\tag{16}$$

Evaluating derivatives for these moments gives us

$$L^1 = \left.\mathbb{E}\left[N_1 z_1^{(N_1-1)}\right]\right|_{z_1=1},$$

$$M^0 = \left.\mathbb{E}\left[-S_1 e^{-sS_1} z_1^{N_1}\right]\right|_{s=0,z_1=1}, \quad M^1 = \left.\mathbb{E}\left[e^{-sS_1} N_1 z_1^{(N_1-1)}\right]\right|_{s=0,z_1=1},$$

$$M^{00} = \left.\mathbb{E}\left[S_1^2 e^{-sS_1} z_1^{N_1}\right]\right|_{s=0,z_1=1}, \quad M^{01} = \left.\mathbb{E}\left[-S_1 e^{-sS_1} N_1 z_1^{(N_1-1)}\right]\right|_{s=0,z_1=1},$$

$$L^{11} = \left.\mathbb{E}\left[N_1(N_1-1)z_1^{(N_1-2)}\right]\right|_{z_1=1}, \quad M^{11} = \left.\mathbb{E}\left[e^{-sS_1} N_1(N_1-1)z_1^{(N_1-2)}\right]\right|_{s=0,z_1=1}$$

$$\tag{17}$$

Substituting values for s and z_1, we have

$$L^1 = \mathbb{E}[N_1], \; M^0 = \mathbb{E}[-S_1], \; M^1 = \mathbb{E}[N_1], \; M^{00} = \mathbb{E}[S_1^2], \; M^{01} = \mathbb{E}[-S_1 N_1],$$

$$L^{11} = \mathbb{E}[N_1(N_1 - 1)], \; M^{11} = \mathbb{E}[N_1(N_1 - 1)]$$

$$\tag{18}$$

Note that $L^1 = M^1$ and $L^{11} = M^{11}$ such that these terms are used interchangeably hereinafter. Further, it is known that $\mathbb{E}[N_1] = \rho/(1 - \rho)$ and $\mathbb{E}[-S_1] = -1/\mu(1 - \rho)$. In the $K = 1$ case, taking partial derivatives of equation (5) gives us three linearly independent equations from which we obtain the moments. The first equation is obtained by taking partial derivatives twice in equation (5) with respect to s and evaluating at $s = 0$, $z_1 = 1$:

$$\mu M^{00} + 2M^{01} = -2M^0 \tag{19}$$

Then, we take partial derivatives of equation (5) with respect to s and z_1 and evaluate at $s = 0$, $z_1 = 1$:

$$(2\mu - \lambda)M^{01} + M^{11} = \lambda M^0 - 2M^1 \tag{20}$$

Again, we take partial derivatives twice in equation (5), but this time with respect to z_1 and evaluate at $s = 0$, $z_1 = 1$:

$$(\mu - \lambda)M^{11} = 2\lambda M^1 \tag{21}$$

Solving equations (19), (20) and (21), we obtain the following values for the moments:

$$M^{00} = \frac{4}{\mu^2(2 - \rho)(1 - \rho)^2}; \; M^{01} = \frac{-\lambda(3 - \rho)}{\mu^2(2 - \rho)(1 - \rho)^2}; \; M^{11} = \frac{2\rho^2}{(1 - \rho)^2} \tag{22}$$

Therefore, we verify that values for M^{00} from equation (22) and $\mathbb{E}[T^2]$ from equation (10) are indeed the same for the $K = 1$ case. Extending analysis to the third moment is computationally more complex and Kim *et al* do not provide explicit values for M^{000} as we do for $\mathbb{E}[T^3]$ in equation 11. Hence, this is an advantage of the moment-generating algorithm proposed in our work over existing work.

On Generalized Processor Sharing and Objective Functions: Analytical Framework

Jasper Vanlerberghe[1]([✉]), Joris Walraevens[1], Tom Maertens[1],
Stijn De Vuyst[2], and Herwig Bruneel[1]

[1] Stochastic Modelling and Analysis of Communication Systems Research Group,
Department of Telecommunications and Information Processing (TELIN),
Ghent University (UGent), Sint-Pietersnieuwstraat 41, B-9000 Gent, Belgium
{jpvlerbe,jw,tmaerten,hb}@telin.UGent.be
[2] Supply Networks and Logistics Research Center (SNLRC),
Department of Industrial Management, Ghent University (UGent),
Technologiepark 903, B-9052 Zwijnaarde, Belgium
Stijn.DeVuyst@UGent.be

Abstract. Today, telecommunication networks host a wide range of heterogeneous services. Some demand strict delay minima, while others only need a best-effort kind of service. To achieve service differentiation, network traffic is partitioned in several classes which is then transmitted according to a flexible and fair scheduling mechanism. Telecommunication networks can, for instance, use an implementation of Generalized Processor Sharing (GPS) in its internal nodes to supply an adequate Quality of Service to each class. GPS is flexible and fair, but also notoriously hard to study analytically. As a result, one has to resort to simulation or approximation techniques to optimize GPS for some given objective function. In this paper, we set up an analytical framework for two-class discrete-time probabilistic GPS which allows to optimize the scheduling for a generic objective function in terms of the mean unfinished work of both classes without the need for exact results or estimations/approximations for these performance characteristics. This framework is based on results of strict priority scheduling, which can be regarded as a special case of GPS, and some specific unfinished-work properties in two-class GPS. We also apply our framework on a popular type of objective functions, i.e., convex combinations of functions of the mean unfinished work. Lastly, we incorporate the framework in an algorithm to yield a faster and less computation-intensive result for the optimum of an objective function.

Keywords: Generalized processor sharing (GPS) · Optimization · Queueing · Scheduling · Objective function

1 Introduction

Times when telecommunication networks were used for one single service like telephony or television are long gone. Nowadays, telecommunication systems host

© Springer International Publishing Switzerland 2015
M. Beltrán et al. (Eds.): EPEW 2015, LNCS 9272, pp. 96–111, 2015.
DOI: 10.1007/978-3-319-23267-6_7

a wide collection of services. Amongst those services are the traditional services like internet, telephony, and television, but modern telecommunication networks also support more demanding interactive multimedia services such as online gaming and video conferencing. Every service desires other network requirements in order to deliver a certain Quality of Service (QoS) or Quality of Experience (QoE) to the end user [4,8]. Hence, the network needs a way to differentiate services. This can be achieved by dividing the network traffic into several classes and implementing some kind of priority scheduling amongst those classes.

Giving strict priority to the different classes in a hierarchical way may not be flexible enough. Additionally, strict priority is not fair since a high load of a higher-priority class can lead to starvation of lower-priority classes [3,5,6]. One scheduling mechanism able to deliver fairness and flexibility is Generalized Processor Sharing (GPS) [10,11]. With GPS, each class is given a certain weight and the available link capacity is shared according to the weights of the backlogged classes. In this way, no capacity goes to waste and each class gets a minimum capacity. Starvation is thus not an issue for GPS. The biggest drawback of all GPS-like scheduling mechanisms is the complexity of obtaining analytical results for their performance characteristics, such as (mean) delays, queue contents or unfinished work. As a consequence, it is hard to analytically determine the optimal weights minimizing an objective function that depends on these characteristics.

In this paper, we consider a discrete-time, probabilistic emulation of single-server, two-class GPS. The weights of the two classes are normalized such that they sum up to one. Then setting the weight of one of the classes to 1 implies that this class has strict priority over the other class; so strict priority scheduling can be seen as a special case of GPS. Strict priority scheduling is well-studied and allows, in many important cases, for an explicit analytical solution. In the remainder, we first show how to use (i) results of strict priority scheduling and (ii) some specific properties of the unfinished work in this GPS system, to transform a generic objective function as to determine its behaviour. This transformation leads to a format which does not require exact results or estimations/approximations for performance characteristics. It allows to find the number of (local) extrema and inflection points and determine the values of the objective function in these points. This analytical framework can thus be invaluable to network operators as it provides an opportunity for them to quickly estimate the need for the possible time-consuming quest for the optimal weight. Furthermore, in case such a quest is recommended, the framework helps to make it more efficient.

Secondly, we extend the results of [13], where we have considered the same GPS-like model. Our theoretical study there concentrated on the behaviour of a specific type of objective functions, i.e., convex combinations of increasing functions (both convex, concave or linear) of the mean unfinished work of both classes. Here, we consider a more general type of objective functions, i.e., convex combinations of (more) general functions of the mean unfinished work of both classes. This removes the requirement for the functions of the unfinished work to be increasing and both either convex, concave or linear in the relevant

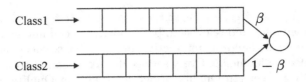

Fig. 1. GPS system at hand

domain. We show how to study the behaviour of this popular type of objective functions, based on some new theorems and by using the analytical framework described in the first part of our paper. Finally, we show how to exploit our theoretical results in a sensitivity analysis on the objective function parameters of the optimum. This is very useful for network operators, as it allows them to avoid simulating the system or use complex approximation techniques.

In the next section, we unfold our analytical framework, for a generic objective function. We show some illustrative examples of our findings in Sect. 3. In Sect. 4, we apply this framework to study a more specific type of objective functions. We prove a theorem to easily carry out the higher order derivative test on this type of objective functions and demonstrate how to study the behaviour of these objective functions with respect to (w.r.t.) the GPS weights and the coefficients in these functions. Before summarizing the most important conclusions, we illustrate the gains of the framework with a practical example in Sect. 5.

2   Analytical Framework

As already mentioned, we consider a discrete-time GPS model with one server and two traffic classes. We denote these classes by class 1 and class 2. The weights of both classes are normalized: class 1 is assigned weight β and class 2 is assigned weight $1 - \beta$, with $0 \le \beta \le 1$. So when both classes are backlogged, the server will choose a class-1 packet with probability β and a class-2 packet with probability $1 - \beta$. In case one of the classes is not backlogged, the server picks a packet of the other class. For the sake of convenience, we assume that both classes have their own queue. This GPS system is depicted in Fig. 1. The cases $\beta = 0$ and $\beta = 1$ reduce to strict priority scheduling.

Next, we define the unfinished work in a queue at the beginning of a slot as the sum of the residual service/transmission times of the packets present in the queue at that moment. It is obvious that the unfinished work in both queues depends on the parameter β: the lower (higher) the value of β, the less capacity for class 1 (2). Therefore, the mean unfinished work in queue j at the beginning of a random slot in steady state is denoted as $\bar{w}_j(\beta)$ $(j = 1, 2)$. We assume that the arrival and service processes of the system at hand are such that $\bar{w}_1(\beta)$ and $\bar{w}_2(\beta)$ satisfy two important properties, i.e.,

Property 1. Function $\bar{w}_1(\beta) + \bar{w}_2(\beta)$ is independent of β.

Property 2. $\bar{w}_2(\beta)$ and $\bar{w}_1(\beta)$ are analytic and strictly monotonic (increasing and decreasing, respectively) w.r.t. β, i.e., on the interval $[0, 1]$.

These properties basically follow (i) from the observation that GPS is a work-conserving scheduling mechanism and (ii) that class 2 is given less capacity with increasing β, respectively (see, e.g., [13,14,18] for more formal proofs). (In fact, only strict monotonicity of one of the $\bar{w}_j(\beta)$ functions is required as Property 1 implies the other is stricly monotone as well.) The first property states that the mean total unfinished work is a constant w.r.t. β. This constant, say $\bar{w}_T$, can be calculated explicitly for a whole range of arrival and service processes. Indeed, since the scheduling mechanism is work-conserving, in every time slot one unit of work (if any) is executed. So to study the total unfinished work, we can consider the system to be a single queue consisting of units of work which are, for instance, executed according to a First-In-First-Out (FIFO) scheduling. Discrete-time, single-queue systems with a FIFO scheduling are much easier to study analytically than multi-queue systems (see, e.g., [2]).

The second property, furthermore, has important consequences as well. Since $\bar{w}_2(\beta)$ is continuous on the interval $[0, 1]$ and takes the values $\bar{w}_2(0)$ and $\bar{w}_2(1)$ at each end of that interval, we can apply the intermediate value theorem to conclude that $\bar{w}_2(\beta)$ takes any value between $\bar{w}_2(0)$ and $\bar{w}_2(1)$ at minimum one point within $[0, 1]$. From the *strict* monotonic increasing property, finally, we have that $\bar{w}_2(\beta)$ is bijective on $[0, 1]$, i.e., that there is a one-to-one correspondence between all values in $[0, 1]$ and all values in $[\bar{w}_2(0), \bar{w}_2(1)]$. For $\bar{w}_1(\beta)$, we can set up a similar reasoning. However, since $\bar{w}_1(\beta)$ is decreasing w.r.t. β, the image of $\bar{w}_1(\beta)$ is the interval $[\bar{w}_1(1), \bar{w}_1(0)]$. For ease of notation, we define the intervals $[\bar{w}_1(1), \bar{w}_1(0)]$ and $[\bar{w}_2(0), \bar{w}_2(1)]$ as Ω_1 and Ω_2, respectively. For a whole range of arrival and service processes, Ω_1 and Ω_2 can be determined explicitly, as they arise from results of strict priority systems (see, e.g., [16,17]).

Now we turn to optimization. The optimal β is defined as the β-value that minimizes some objective function. In the context of scheduling mechanisms, objective functions are often constructed in terms of (mean) delays or holding times. Here, we assume the objective function to be a generic function of the mean unfinished work in both queues, i.e., $f(\bar{w}_1(\beta), \bar{w}_2(\beta))$.[1] It is clear that the objective function $f(\bar{w}_1(\beta), \bar{w}_2(\beta))$ can be seen as a function in terms of β, say $F(\beta) \triangleq f(\bar{w}_1(\beta), \bar{w}_2(\beta))$, with domain $[0, 1]$. The objective function $f(\bar{w}_1(\beta), \bar{w}_2(\beta))$, however, can also be expressed in terms of another single variable. In particular, the first unfinished-work property states that $\bar{w}_1(\beta) = \bar{w}_T - \bar{w}_2(\beta)$, implying that $f(\bar{w}_1(\beta), \bar{w}_2(\beta))$ can be expressed in terms of $\bar{w}_2(\beta)$ only. With a slight abuse of notation, we define this format as $f^*(\bar{w}_2)$. It is obvious that $F = f^* \circ \bar{w}_2$. As mentioned earlier, analytical results for $\bar{w}_2(\beta)$ are notoriously complex to obtain, so we need estimations/approximations for this performance measure to study the objective function $f(\bar{w}_1(\beta), \bar{w}_2(\beta))$ in

---

[1] In some specific cases, it is possible to find easy relations between the mean unfinished work in both queues and the corresponding mean queue contents and/or delays (see [13] for such a case). Then one can initially consider an objective function in terms of, e.g., mean packet delays, which may be practically most relevant in the context of heterogeneous services, translate it to an objective function in terms of the mean unfinished work in both queues, and apply our framework to study the latter.

terms of β. However, since we know the image ($\Omega_2 = [\bar{w}_2(0), \bar{w}_2(1)]$) and behaviour (continuous and strictly increasing) of $\bar{w}_2(\beta)$, we can already study $f(\bar{w}_1(\beta), \bar{w}_2(\beta))$ in terms of $\bar{w}_2(\beta)$ instead of β, with domain Ω_2 instead of $[0, 1]$ (i.e., studying $f^*(\bar{w}_2)$).

Consequently, we can observe the number of extrema and inflection points and determine the values of $f(\bar{w}_1(\beta), \bar{w}_2(\beta))$ in these points without running simulations or relying on possibly inaccurate approximate expressions. Obviously, we do not know the β-values corresponding to these points (except when they coincide with the endpoints). To determine these β-values, we still need estimations/approximations. Now some preliminary conclusions can be drawn from the behaviour of $f^*(\bar{w}_2)$. For instance, the minimum can be in the endpoints $\beta = 0$ or $\beta = 1$. In that case, strict priority is optimal and we do not have to simulate. Another possible conclusion is that the difference in the objective function between the minimum and one of the endpoints is too small to justify a time-consuming quest for the β-value corresponding to the minimum. Summarized, from the analysis of $f^*(\bar{w}_2)$, an interval in Ω_2 with an acceptable value for the objective function can be selected. The optimization problem then reduces to finding a value of β for which the continuous and monotonic function $\bar{w}_2(\beta)$ reaches a value in the selected interval (stopping criterium). In the next section, we demonstrate these findings by means of some illustrative examples.

3   Some Illustrative Examples

For the examples, we consider one-slot service times and a two-dimensional binomial arrival process characterized by the joint probability generating function

$$A(z_1, z_2) \triangleq \left(1 + \frac{\lambda_1}{N}(z_1 - 1) + \frac{\lambda_2}{N}(z_2 - 1)\right)^N, \tag{1}$$

of the independently and identically distributed number of class-1 and class-2 arrivals in a slot. Here, λ_j ($j = 1, 2$) is the arrival rate of class-j packets. This is the arrival process in a queue of an NxN output-queueing switch with Bernoulli arrivals at its inlets and with independent and uniform routing towards the outlets. Parameter N expresses the maximum total number of arrivals in a queue during a slot. For the sake of convenience, we also introduce the parameters λ_T and α, indicating the total arrival rate (i.e., $\lambda_T = \lambda_1 + \lambda_2$) and the fraction of class-1 packets in the overall arrival stream (i.e., $\alpha = \frac{\lambda_1}{\lambda_T}$), respectively. A queueing model with this type of arrival process, one-slot service times and strict priority scheduling is, for instance, studied in [15]. Adopting some concrete values for the parameters of the arrival process, the results of [15] can be used to calculate $\bar{w}_j(0)$ and $\bar{w}_j(1)$ ($j = 1, 2$). For $N = 16$, $\lambda_T = 0.9$, and $\alpha = 0.8$, for example, we find that

$$\bar{w}_1(0) = 4.50, \qquad \bar{w}_2(0) = 0.20,$$
$$\bar{w}_1(1) = 1.59, \qquad \bar{w}_2(1) = 3.11. \tag{2}$$

Then $\bar{w}_T = \bar{w}_1(\cdot) + \bar{w}_2(\cdot) = 4.70$ and the intervals Ω_1 and Ω_2, defined in the previous section, equal $[1.59, 4.5]$ and $[0.2, 3.11]$, respectively.

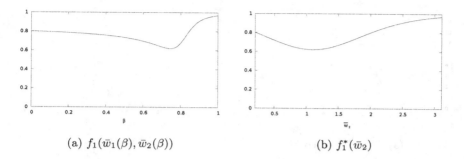

(a) $f_1(\bar{w}_1(\beta), \bar{w}_2(\beta))$ (b) $f_1^*(\bar{w}_2)$

Fig. 2. Comparison between the objective function $f_1(\bar{w}_1(\beta), \bar{w}_2(\beta))$ and $f_1^*(\bar{w}_2)$

As objective function for our first example, we add two logistic functions:

$$f_1(\bar{w}_1(\beta), \bar{w}_2(\beta)) \triangleq \frac{1}{1 + e^{-2\bar{w}_1(\beta)+8}} + \frac{1}{1 + e^{-2\bar{w}_2(\beta)+3}}. \tag{3}$$

Applying the framework of the previous section and using the values of (2) then yields $f_1(\bar{w}_1(\beta), \bar{w}_2(\beta))$ in terms of $\bar{w}_2(\beta)$ only, i.e.,

$$f_1^*(\bar{w}_2) = \frac{1}{1 + e^{2\bar{w}_2-1.4}} + \frac{1}{1 + e^{-2\bar{w}_2+3}}. \tag{4}$$

This function is plotted in Fig. 2b. In Fig. 2a, $f_1(\bar{w}_1(\beta), \bar{w}_2(\beta))$ is plotted as a function of β; this figure results from a Monte-Carlo simulation over one million slots (see further for more details), as it is notoriously complex to find analytical results for $\bar{w}_2(\beta)$. We see from the figures that both graphs have equal characteristics. Both graphs, for instance, have the same number of minima (i.e., one). Also, the ranges of both graphs are the same. So, $f_1^*(\bar{w}_2)$ helps identifying the minimum value of the objective function and determining how much this value differs from the values in the endpoints. As opposed to $f_1(\bar{w}_1(\beta), \bar{w}_2(\beta))$, however, we can plot $f_1^*(\bar{w}_2)$ right away.

Nevertheless, we cannot conclude from the behaviour of $f_1^*(\bar{w}_2)$ at what β-value this minimum occurs (say $\beta_{\min}$). We only know that $\bar{w}_2(\beta_{\min}) = 1.1$ at $F = 0.62$. So for instance if we are satisfied with the objective function within 2% of its minimum (i.e. F smaller than 0.6324), we then argue that we need $\bar{w}_2$ to be in the interval $[0.9, 1.3]$. This subsequently is the stopping criterium for a simulation or approximation procedure on the function $\bar{w}_2(\beta)$. As can be seen from Fig. 2a, the procedure should result in a β_{opt} value in the interval $[0.71, 0.77]$, as to have a value for the objective function within 2% of the minimum.

As a second example, we examine the objective function

$$f_2(\bar{w}_1(\beta), \bar{w}_2(\beta)) \triangleq (0.5\bar{w}_1(\beta))^2 + (1.3\bar{w}_2(\beta))^2. \tag{5}$$

Applying the framework of Sect. 2 and using the values in (2) then leads to

$$f_2^*(\bar{w}_2) = (2.35 - 0.5\bar{w}_2)^2 + (1.3\bar{w}_2)^2. \tag{6}$$

We know from the theory of [13] that $f_2(\bar{w}_1(\beta), \bar{w}_2(\beta))$ will have a minimum different from the endpoints $\beta = 0$ or $\beta = 1$, because of the convex character of

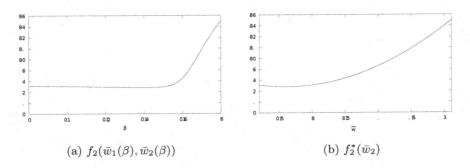

(a) $f_2(\bar{w}_1(\beta), \bar{w}_2(\beta))$ (b) $f_2^*(\bar{w}_2)$

Fig. 3. Comparison between the objective function $f_2(\bar{w}_1(\beta), \bar{w}_2(\beta))$ and $f_2^*(\bar{w}_2)$

the squaring in the objective function. Plots for this second example are found in Fig. 3. Function $f_2^*(\bar{w}_2)$ provides us with the value of the objective function in its minimum (see Fig. 3b). The difference between the value of the objective function at $\beta = 0$ (5.13) and the minimum value (4.81) is perhaps not worth the effort to search for $\beta_{\min}$ (the difference is 6.6%). Fig. 3b can thus be used in advance to decide whether significant gain can be won by searching $\beta_{\min}$ compared to using the priority cases $\beta = 0$ or $\beta = 1$.

If, on the other hand, we make the same reasoning as in the previous example, thus allowing at most 2% deviation of the minimum, we need the objective function F to be smaller than 4.90 and consequently $\bar{w}_2$ in $[0.38, 0.83]$. An accompanying simulation/approximation method should then result in a β_{opt}-value in the interval $[0.44, 0.69]$.

The reasoning done here can be done for an arbitrarily complex objective function in $\bar{w}_j(\beta)$. Indeed, every objective function minimization problem can be translated to a problem of finding the corresponding β for a certain performance vector (with or without some error margin). This effectively simplifies procedures; we will come back to this in Sect. 5.

Note: Critical for this method is the availability of the values for $\bar{w}_j(0)$ and $\bar{w}_j(1)$ $(j = 1, 2)$. However, it does not matter how these values are obtained. For strict priority systems, a lot of analytical results are available; this is, for instance, the case for the arrival and service processes we have used in the examples above (see [15]). For more complex arrival and/or service processes, this is not necessarily the case. To still obtain accurate values for $\bar{w}_j(0)$ and $\bar{w}_j(1)$ $(j = 1, 2)$, we can, for example, simulate the system for $\beta = 0$ and $\beta = 1$ only.

4   Framework Application

In this section, we address another important issue, namely the selection of an objective function and in particular the influence of this selection on the optimum. In a first step, a network operator chooses the type of relation of each performance characteristic in the objective function. When equal increments for high or low values should have an equal influence on the objective function, a linear relation can be used. The behaviour of other types of relations can easily be derived from a plot of the corresponding function. Other examples are a squared relation (as in f_2, see (5)) or a logistic one (as in f_1, see (3)). This choice of relation is closely related to the relation between QoS and QoE [4] and the choice of utility functions [7,9]. A second question for the operator is how to weigh the performance of both classes. An answer to this question is less clear and in most cases more arbitrarily chosen. In the next paragraphs, we derive a method the operator can use to do a sensitivity analysis on these weights. This way, he can assess the impact of his choice on the behaviour and the resulting minimum of the objective function.

Following the reasoning above, we propose the following template for the objective function:

$$F(\gamma, \beta) = \gamma g_1(\bar{w}_1(\beta)) + (1 - \gamma)g_2(\bar{w}_2(\beta)), \tag{7}$$

whereby other characteristics can be achieved by incorporating them in g_j as noted before. The parameter γ, assumed to be in the interval $[0, 1]$, serves as weight parameter, on which we want to do our sensitivity analysis. When $\gamma = 0$, the objective function only takes into account $\bar{w}_2(\beta)$; when $\gamma = 1$, only $\bar{w}_1(\beta)$ plays a role.

The exposition in the previous sections allows us to study this objective function for a specific value of γ. In [13], we already studied a subclass of this kind of objective functions. The analysis there was limited to increasing g_j that were either both linear, convex or concave. We proved that for linear or concave g_j the optimal β-value is always one of the endpoints (i.e., $\beta = 0$ or $\beta = 1$). For convex g_j, on the other hand, we have found that for certain values of γ the objective function reaches a minimum for some β different from 0 and 1.

In [13], we were able to study the objective function for all γ in one effort, but we were limited to increasing functions g_j. Now we generalize the results of [13], allowing all functions g_j. The function

$$\phi(\beta) \triangleq \frac{g_2'(\bar{w}_2(\beta))}{g_2'(\bar{w}_2(\beta)) + g_1'(\bar{w}_1(\beta))}, \tag{8}$$

defined in [13], plays a key role.

Theorem 1. *Assume g_j is continuously differentiable in Ω_j. Then $\gamma = \phi(\beta)$ if and only if (iff) $\frac{\partial F}{\partial \beta}(\gamma, \beta) = 0$.*[2]

Proof. From (7), we find that $\frac{\partial F}{\partial \beta}(\gamma, \beta) = 0$ is equivalent with

$$[g_2'(\bar{w}_2(\beta)) - \gamma(g_1'(\bar{w}_1(\beta)) + g_2'(\bar{w}_2(\beta)))]\bar{w}_2'(\beta) = 0, \tag{9}$$

where we have used that $\bar{w}_1'(\beta) = -\bar{w}_2'(\beta)$, see Property 1. According to Property 2 ($\bar{w}_2'(\beta) > 0$) and (9), this leads to

$$\gamma = \phi(\beta). \tag{10}$$

---

[2] In fact, Theorem 1 is a generalization of Lemma 1 in [13].

β-values for which $\phi(\beta) = \gamma$ are the critical points of the objective function $F(\gamma, \beta)$. Critical points can be either extrema or inflection points with a horizontal asymptote.

For simplicity, we assume $\phi(\beta)$ to be continuous. Discontinuities only occur for β-values for which $g_2'(\bar{w}_2(\beta)) = -g_1'(\bar{w}_1(\beta))$. For these β-values,

$$\frac{\partial F}{\partial \beta}(\gamma, \beta) = g_2'(\bar{w}_2(\beta))\bar{w}_2'(\beta). \tag{11}$$

As $\bar{w}_2'(\beta)$ is positive, the objective function will increase or decrease like g_2. We will disregard these cases in the remainder, as the discontinuities in $\phi(\beta)$ do not lead to special cases for $F(\gamma, \beta)$. The extensions are straightforward but only result in more involved expressions.

To be able to distinguish extrema from inflection points, we need higher-order derivatives of the objective function. These will allow us to perform the higher-order derivative test on the objective function. Therefore, we extend Theorem 1.

Theorem 2. *Assume g_j is n times continuously differentiable in Ω_j. Then $\phi^{(i-1)}(\beta) = 0, \forall i = 2, ..., n$, and $\phi(\beta) = \gamma$ iff $\frac{\partial^i F}{\partial \beta^i}(\phi(\beta), \beta) = 0, \forall i = 1, ..., n$. Here, $\phi^{(j)}(\beta)$ denotes the j-th derivative of $\phi(\beta)$.*

Proof. (by induction) The base case $n = 1$ follows from Theorem 1. For the induction hypothesis, assume that $\phi(\beta) = \gamma$ and $\phi^{(i-1)}(\beta) = 0$ for $i = 2, ..., n-1$ iff $\frac{\partial^i F}{\partial \beta^i}(\phi(\beta), \beta) = 0$ for $i = 1, ..., n-1$. To complete the theorem, we prove that $\phi^{(n-1)}(\beta) = 0$ iff $\frac{\partial^n F}{\partial \beta^n}(\phi(\beta), \beta) = 0$. We find that

$$\frac{\partial^n F}{\partial \beta^n} = \frac{\partial^{n-1}}{\partial \beta^{n-1}}\left(\frac{\partial F}{\partial \beta}\right) = \frac{\partial^{n-1}}{\partial \beta^{n-1}}\left(\bar{w}_2'(\beta)\big(g_1'(\bar{w}_1(\beta)) + g_2'(\bar{w}_2(\beta))\big)\big(\phi(\beta) - \gamma\big)\right),$$

where we have used (9). Define, furthermore, $\Delta(\gamma, \beta)$ as $\phi(\beta) - \gamma$ and $\chi(\beta)$ as $\bar{w}_2'(\beta)\big(g_1'(\bar{w}_1(\beta)) + g_2'(\bar{w}_2(\beta))\big)$. Then we can write

$$\frac{\partial^n F}{\partial \beta^n} = \frac{\partial^{n-1}}{\partial \beta^{n-1}}\big(\chi(\beta)\Delta(\gamma, \beta)\big) = \sum_{r=0}^{n-1}\binom{n-1}{r}\chi^{(n-1-r)}(\beta)\frac{\partial^r \Delta}{\partial \beta^r}(\gamma, \beta). \tag{12}$$

We know from the induction hypothesis that $\frac{\partial^r \Delta}{\partial \beta^r}(\gamma, \beta) = \frac{\partial^r(\phi(\beta)-\gamma)}{\partial \beta^r} = \frac{\partial^r \phi(\beta)}{\partial \beta^r} = \phi^{(r)}(\beta) = 0, r = 1, ..., n-2$, and that $\Delta(\gamma, \beta) = 0$. This yields

$$\frac{\partial^n F}{\partial \beta^n} = \binom{n-1}{n-1}\chi^{(0)}(\beta)\frac{\partial^{n-1}\Delta}{\partial^{n-1}\beta}(\gamma, \beta) = \bar{w}_2'(\beta)\big(g_1'(\bar{w}_1(\beta)) + g_2'(\bar{w}_2(\beta))\big)\phi^{(n-1)}(\beta).$$

Strict monotonicity of $\bar{w}_2(\beta)$ and the continuity assumption of $\phi(\beta)$ that we made earlier prove that $\phi^{(n-1)}(\beta) = 0$ iff $\frac{\partial^n F}{\partial \beta^n}(\phi(\beta), \beta) = 0$.

The next corollary follows directly from Theorem 2:

Corollary 1. *If $\gamma = \phi(\beta)$, $\phi^{(1)}(\beta) = \cdots = \phi^{(n)}(\beta) = 0$ and $\phi^{(n+1)}(\beta) \neq 0$, then $F(\gamma, \beta)$ has a local extremum at β if n is even and an inflection point at β if n is odd.*

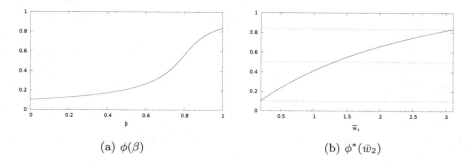

(a) $\phi(\beta)$ (b) $\phi^*(\bar{w}_2)$

Fig. 4. For F_3: Comparison between $\phi(\beta)$ and $\phi^*(\bar{w}_2)$

With this corollary, we can determine the behaviour of $F(\gamma, \beta)$ by studying the behaviour of $\phi(\beta)$. Suppose, for instance, that $\phi(\beta)$ has one inflection point $\hat{\beta}$ with horizontal tangent. Then $\phi^{(2)}(\hat{\beta}) = \phi^{(1)}(\hat{\beta}) = 0$ and $\phi^{(3)}(\hat{\beta}) \neq 0$, and, as a consequence, $F(\gamma, \beta)$ has an extremum at $\hat{\beta}$ when $\gamma = \phi(\hat{\beta})$.

Unfortunately, we do not have a formula for $\phi(\beta)$, as we do not have explicit analytical results for the functions $\bar{w}_j(\beta)$. This is where the framework of Sect. 2 comes into play. In particular, the function $\phi(\beta)$ can be translated into a function in terms of $\bar{w}_2(\beta)$ instead of β (i.e. $\phi^*(\bar{w}_2)$ in the remainder). As there is a one-to-one mapping between the values in Ω_2 and the values in $[0, 1]$, we find that

$$\phi^*(\bar{w}_2) = \frac{g_2'(\bar{w}_2)}{g_2'(\bar{w}_2) + g_1'(\bar{w}_T - \bar{w}_2)}. \tag{13}$$

Using the framework, the previous corollary is reformulated as follows:

Corollary 2. *If* $\gamma = \phi^*(\bar{w}_2)$, $\phi^{*(1)}(\bar{w}_2) = \cdots = \phi^{*(n)}(\bar{w}_2) = 0$ *and* $\phi^{*(n+1)}$ $(\bar{w}_2) \neq 0$, *then* $F^*(\gamma, \bar{w}_2)$ *has a local extremum at* $\bar{w}_2$ *if* n *is even and an inflection point at* $\bar{w}_2$ *if* n *is odd. As* $\bar{w}_2(\beta)$ *is bijective on* $[0, 1]$, $F(\gamma, \beta)$ *also has a local extremum at* β *if* n *is even and an inflection point at* β *if* n *is odd.*

Hereby, we defined $F^*(\gamma, \bar{w}_2)$ analogously to the other functions marked with a star, using the framework presented in Sect. 2.

As in Sect. 3, we have composed figures to compare $\phi(\beta)$ (see Fig. 4a) with $\phi^*(\bar{w}_2)$ (see Fig. 4b). For these figures, we have used the objective function

$$F_3(\gamma, \beta) = \gamma(0.5\sqrt{2}\bar{w}_1(\beta))^2 + (1 - \gamma)(1.3\sqrt{2}\bar{w}_2(\beta))^2. \tag{14}$$

and the same arrival and service processes as in Sect. 3. It should be noticed that $F_3(0.5, \beta) = f_2(\bar{w}_1(\beta), \bar{w}_2(\beta))$ (see (5)); we refer to this equivalence later. We can draw similar conclusions as in the previous section. For instance, we can see that both graphs have the same range.

Using the aforementioned corollary, we see that for $\gamma \in [0.1, 0.83]$ (obtained using strict priority scheduling results only, see [13]), $F_3(\gamma, \beta)$ reaches an extremum at a β-value different from 0 or 1. In particular for $\gamma \in [0.1, 0.83]$, there is a β-value and corresponding $\bar{w}_2(\beta)$-value, say $\hat{\beta}$ and $\bar{w}_2(\hat{\beta})$, respectively, for

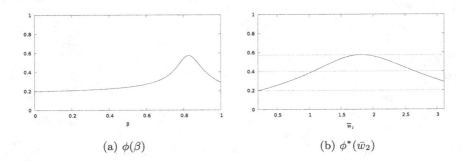

(a) $\phi(\beta)$ (b) $\phi^*(\bar{w}_2)$

Fig. 5. For F_4: Comparison between $\phi(\beta)$ and $\phi^*(\bar{w}_2)$

which $\gamma = \phi(\beta) = \phi^*(\bar{w}_2(\beta))$. Visually, this $\hat{\beta}$ and $\bar{w}_2(\hat{\beta})$ can be presented in the Cartesian coordinate systems $(\beta, \phi(\beta))$ and $(\bar{w}_2, \phi^*(\bar{w}_2))$, by drawing a horizontal line at the chosen value of γ (see Fig. 4b); the intersection points of the horizontal lines and curves of $\phi(\beta)$ and $\phi^*(\bar{w}_2)$ then yield $\hat{\beta}$ and $\bar{w}_2(\hat{\beta})$, respectively.

Now according to Theorem 2, $\frac{\partial F_3}{\partial \beta}(\gamma, \hat{\beta}) = 0$ and we have an extremum at $\hat{\beta}$ if $\frac{\partial^2 F_3}{\partial \beta^2}(\gamma, \hat{\beta}) \neq 0$ or, equivalently, if $\phi^{(1)}(\hat{\beta}) \neq 0$. From Fig. 3, where we have depicted $F_3(\gamma, \beta)$ for $\gamma = 0.5$, we can see that $F_3(\gamma, \beta)$ is decreasing at $\beta = 0$ for $\gamma \in [0.1, 0.83]$. As we have a $\hat{\beta}$ for which $\gamma = \phi(\beta)$ in that interval, $F_3(\gamma, \beta)$ has an extremum, which is necessarily a minimum. Summarized, the couples $(\phi(\beta), \beta)$, indicated by the curve in the figure, are parameter combinations for (γ, β) that minimize the objective function. For $\gamma < 0.1$, there is no β for which $\gamma = \phi(\beta)$ and thus, according to Theorem 1, $F_3(\gamma, \beta)$ has no extremum between 0 and 1. Since $F_3(\gamma, \beta)$ is increasing at $\beta = 0$, $F_3(\gamma, \beta)$ is increasing w.r.t. all β. Analogously, for $\gamma > 0.83$, the objective function is decreasing w.r.t β.

As a last example, we look at the objective function

$$F_4(\gamma, \beta) = \frac{\gamma}{1 + e^{-3\bar{w}_1(\beta)+9}} + \frac{1 - \gamma}{1 + e^{-4\bar{w}_2(\beta)+7}}. \tag{15}$$

Using the same arrival and service processes and the same arrival process parameters as before, we depict the corresponding $\phi(\beta)$ in Fig. 5a and $\phi^*(\bar{w}_2)$ in Fig. 5b. Remember that the former is obtained via simulations, while the latter can be drawn directly. We can make the same reasoning as before. $\phi(\beta)$ and $\phi^*(\bar{w}_2)$ will have intersection points with horizontal lines at $\gamma \in [0.2, 0.58]$ only. For a γ-value in this interval, Theorem 1 dictates that $F_4(\gamma, \beta)$ will have extrema or inflection points. Using the corollary of Theorem 2, we know that inflection points only occur when also $\phi'(\beta) = 0$, so when $\phi(\beta)$ and $\phi^*(\bar{w}_2)$ have an extremum. For the example at hand, this occurs for $\gamma = 0.58$ and $\bar{w}_2 = 1.8$ (and from simulation, $\beta = 0.82$).

At $\bar{w}_2 = 0.2$ ($\beta = 0$), $F_4^*(\gamma, \bar{w}_2)$ is decreasing if $\gamma > 0.2$ and increasing if $\gamma < 0.2$ (this can easily be seen from (15)). So if we take $\gamma = 0.4$, we find that $F_4^*(\gamma, \bar{w}_2)$ is decreasing at $\bar{w}_2 = 0.2$. Furthermore, $F_4^*(0.4, \bar{w}_2)$ reaches an extremum at $\bar{w}_2 = 1.1$ because at that $\bar{w}_2$-value $\phi^*(\bar{w}_2)$ intersects with a horizon-

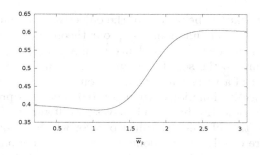

Fig. 6. Objective function $F_4^*(\gamma, \bar{w}_2)$ for $\gamma = 0.4$

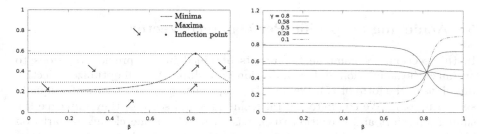

Fig. 7. Annotated version of Fig. 5a   **Fig. 8.** Simulation of F_4 for several γ

tal line at 0.4 (see Fig. 5b). This extremum is necessarily a minimum. For higher values of $\bar{w}_2$, $F_4^*(\gamma, \bar{w}_2)$ is increasing again. At $\bar{w}_2 = 2.7$, we once more have a point of intersection between $\phi^*(\bar{w}_2)$ and the horizontal line at 0.4, and, hence, $F_4^*(\gamma, \bar{w}_2)$ has a second extremum, in this case a maximum. These conclusions can be verified in Fig. 6, where we have plotted $F_4^*(0.4, \bar{w}_2)$.

Using similar arguments for other values of γ, we constructed an annotated version of Fig. 5a in Fig. 7. In this figure, the arrows indicate the behaviour of $F_4(\gamma, \beta)$ for the (γ, β)-values in that area. We see that $F_4(\gamma, \beta)$ is decreasing above the curve of $\phi(\beta)$ and increasing under the curve. In fact, if we draw a path in the unit square (a collection of couples for (γ, β)), we know that the behaviour of the objective function is indicated by the arrows in the figure. Furthermore, the sign (and thus the behaviour) of $\frac{\partial F_4}{\partial \beta}(\gamma, \beta)$ will only change if the path intersects with $\phi(\beta)$. At the intersection point, the sign will change and the objective function will have an extremum or an inflection point. We can thus conclude that from a plot of $\phi^*(\bar{w}_2)$, we can determine the complete behaviour of the objective function without running any simulation or relying on possibly inaccurate approximate expressions for the performance characteristics.

In Fig. 8, finally, we have graphs of $F_4(\gamma, \beta)$, for several γ; they are obtained through simulation (see further for more details). We have chosen a γ-value from each area in Fig. 7 and we see that the graphs in Fig. 8 confirm our analysis. For $\gamma = 0.5$, for instance, we see that $F_4(\gamma, \beta)$ first reaches a minimum and later a maximum; for $\gamma = 0.58$, $F_4(\gamma, \beta)$ has an inflection point. We see from this study that the behaviour of the objective function largely depends on the coefficient γ in that function. Using our theorems presented here, these different behaviours can be seen at a glance from a graph of $\phi^*(\bar{w}_2)$.

Let us now return to the sensitivity analysis. From Fig. 5b, an analist can see what the impact of a variation in γ will be on $\bar{w}_2$ (and subsequently also on the value of the objective function). An annotated version as presented in Fig. 7 easily shows how the objective function behaves for different γ. This figure can be used to make a selection of different values of γ for which the objective function can be studied more closely, as was done in Fig. 8. The mapping from $\bar{w}_2$ to β, however, is still unknown. To get this information, one needs to resort to, e.g., simulation. In the next section, we show how this can be done efficiently.

5   Achieving a Specific Performance Vector

In this section, we combine the results obtained in the previous sections to optimize the simulation and optimization procedure. We will optimize objective functions of the form of (7). We do this by optimizing for 101 values of γ equally spaced in $[0, 1]$. We compare several techniques and see how they influence the simulation effort and execution time. Lastly, we give some closing remarks on how the procedure could be sped up even further. If you only need to optimize a simple objective function (without changing the γ parameter) the optimization is even faster, though completely analog to the one presented here.

For all simulations, we used one slot service times and the arrival process presented in Sect. 3. We used Monte-Carlo simulations over 10^8 slots. This length of trajectory is long enough to minimize bias and variance from the transient behaviour (before reaching steady state) and the selection of the specific trajectory. To guarantee, however, that Property 2 is fulfilled we use identical arrival and decision variable trajectories, i.e., in each M-th slot of every simulation the same number of packets arrives and the same decision variable to choose a queue to serve (which then needs to be compared to β) is used. We achieve this by initializing the random generator with the same seed for each simulation. This is the well known method of common random numbers (CRN) [1,12].

For the numerical results in this section, we optimized the objective function F_3. During the simulation runs for different γ, we only do one simulation for a given value of β. A table in memory holds the already simulated values for β and their results. This is a first way to speed up the process. As we use the CRN-method, each simulation for the same weight β will give us the same result.

A first, albeit naive, method is to just simulate equally spaced β's in $[0, 1]$. We call this the *brute force method*. In a second method, we use a golden section search on the objective function, like we used in [13]. This method, however, is only usable for objective functions known to have a single extremum and this extremum being a minimum. This is, for instance, the case for F_3 with convex g_j.

A third method follows from Sect. 2. We know that it is not needed to work directly with the objective function. From the objective function, the optimal value of $\bar{w}_2$ can be calculated. Subsequently, the corresponding value of β

Table 1. Simulation times

Method	Stopping criterium	Needed simulations
Brute force	β-precision: 0.0001	10001
Golden section on F	β-precision: 0.0001	897
Golden section on $\bar{w}_2$	β-precision: 0.0001	617
Golden section on $\bar{w}_2$	β-precision: 0.0001 or F-precision: 1%	12

to achieve this value of $\bar{w}_2$ needs to be obtained by simulation. Knowing that $\bar{w}_2(\beta)$ is monotonically increasing, we can use a simplified version of the golden section search method (mentioned in the previous paragraph). In this method, the algorithm maintains an interval $[A, B]$ for β (starting from $[0, 1]$) wherein the optimal solution can be found. Each iteration, the algorithm chooses a point C in the interval (according to the golden ratio). This point is subsequently simulated whereafter the algorithm updates the interval to either $[A, C]$ or $[C, B]$.

We can use two different stopping criteria. The first one is the β-precision. This is the value $B - A$; if this value is small enough, we stop the algorithm and use the average $(A + B)/2$ as value for β_{opt}. Another option, called the F-precision, is to stop the simulation once we found a β that leads to a *reasonable* value of F. This F-precision is the percentage deviation from the minimum of the objective function F (which we can calculate in advance, as shown in Sect. 2).

The computational effort of the different simulation methods for this specific case can be found in Table 1. Cases can be engineered where the efficiency order of these methods is different; however, these cases are exceptional and need to be tailor-made. Furthermore, golden section search on F has a limited usability as it can only be used for objective functions with a single extremum, specifically a minimum. It is clear that in general the more information and knowledge you have about the queueing system and objective function, the less simulations are needed. This often leads to complex algorithms that are only usable in a limited number of cases. The framework presented here, being as simple and general as it is, leads to large simulation gains without significant increase in complexity.

The methods presented here can be improved even further. Instead of using the golden section search in its purest form, the table with the already simulated β's could be used to select a starting interval $[A, B]$ after which golden section search could be used to further refine the result. This would lead to faster convergence. Another method (variation on golden section search), is to use multiple cores and select multiple C's. This way the interval $[A, B]$ will shrink much faster. Lastly, one could also vary the number of simulated slots as we get further into the algorithm, simulating less slots (and having a rougher estimation) when the interval $[A, B]$ is still large. A word of caution however is in order here, as this also induces extra variance. This variance could cause the algorithm to exclude the minimum. Using the presented insights the algorithm can easily detect when this happens and act accordingly.

6   Conclusions

In this paper, we have shown that no simulations or complex approximation techniques are needed for two-class GPS to study a generic objective function in terms of mean performance characteristics. With our framework, we are able to calculate the number of local extrema and the values of the objective function in these extrema. In this way, we are able to characterize the entire behaviour of the objective function w.r.t. the GPS weights. Our framework is based on results of strict priority scheduling and some specific properties of two-class GPS. To find the weights that optimize GPS, we still need to resort to a simulation approach. However, knowing the behaviour of the objective function beforehand can aid immensely in this optimization process.

Acknowledgments. This research has been co-funded by the Interuniversity Attraction Poles (IAP) Programme initiated by the Belgian Science Policy Office.

References

1. Asmussen, S., Glynn, P.W.: Stochastic Simulation: Algorithms and Analysis, vol. 57. Springer (2007)
2. Bruneel, H., Kim, B.G.: Discrete-time models for communication systems including ATM. Kluwer Academic Publishers (1992)
3. Demoor, T., Walraevens, J., Fiems, D., Bruneel, H.: Performance analysis of a priority queue: Expedited forwarding PHB in diffserv. AEU-International Journal of Electronics and Communications **65**(3), 190–197 (2011)
4. Fiedler, M., Hossfeld, T., Tran-Gia, P.: A generic quantitative relationship between quality of experience and quality of service. IEEE Network **24**(2), 36–41 (2010)
5. Homg, M.F., Lee, W.T., Lee, K.R., Kuo, Y.H.: An adaptive approach to weighted fair queue with QoS enhanced on IP network. In: Proceedings of IEEE Region 10 International Conference on Electrical and Electronic Technology, TENCON 2001, vol. 1, pp. 181–186. IEEE (2001)
6. Huang, T.Y.: Analysis and modeling of a threshold based priority queueing system. Computer Communications **24**(3), 284–291 (2001)
7. Lee, H.W., Kim, C., Chong, S.: Scheduling and source control with average queue-length control in cellular networks. In: IEEE International Conference on Communications, ICC 2007, pp. 109–114. IEEE (2007)
8. van Moorsel, A.: Metrics for the internet age: quality of experience and quality of business. In: Fifth International Workshop on Performability Modeling of Computer and Communication Systems, vol. 34, pp. 26–31. Arbeitsberichte des Instituts für Informatik, Universität Erlangen-Nürnberg, Germany (2001)
9. Neely, M.J.: Delay-based network utility maximization. IEEE/ACM Transactions on Networking (TON) **21**(1), 41–54 (2013)
10. Parekh, A.K., Gallager, R.G.: A generalized processor sharing approach to flow control in integrated services networks: the single-node case. IEEE/ACM Transactions on Networking (TON) **1**(3), 344–357 (1993)
11. Parekh, A.K., Gallagher, R.G.: A generalized processor sharing approach to flow control in integrated services networks: the multiple node case. IEEE/ACM Transactions on Networking (TON) **2**(2), 137–150 (1994)

12. Spall, J.C.: Introduction to stochastic search and optimization: Estimation, simulation and control, vol. 65. John Wiley & Sons (2005)
13. Vanlerberghe, J., Maertens, T., Walraevens, J., De Vuyst, S., Bruneel, H.: A hybrid analytical/simulation optimization of generalized processor sharing. In: Proceedings of The 25th International Teletraffic Congress (ITC 25), Shanghai, September 2013
14. Verloop, I.M., Ayesta, U., Borst, S.: Monotonicity properties for multi-class queueing systems. Discrete Event Dynamic Systems **20**(4), 473–509 (2010)
15. Walraevens, J., Steyaert, B., Bruneel, H.: Performance analysis of a single-server ATM queue with a priority scheduling. Computers & Operations Research **30**(12), 1807–1829 (2003)
16. Walraevens, J., Steyaert, B., Bruneel, H.: Performance analysis of a GI-Geo-1 buffer with a preemptive resume priority scheduling discipline. European Journal of Operational Research **157**(1), 130–151 (2004)
17. Walraevens, J., Steyaert, B., Bruneel, H.: Analysis of a discrete-time preemptive resume priority buffer. European Journal of Operational Research **186**(1), 182–201 (2008)
18. Walraevens, J., Vanlerberghe, J., Maertens, T., De Vuyst, S., Bruneel, H.: Strict monotonicity and continuity of mean unfinished work in two queues sharing a processor (forthcoming)

Software Performance

Comparing the Accuracy of Resource Demand Measurement and Estimation Techniques

Felix Willnecker[1]([⊠]), Markus Dlugi[1], Andreas Brunnert[1], Simon Spinner[2], Samuel Kounev[2], Wolfgang Gottesheim[3], and Helmut Krcmar[4]

[1] Fortiss GmbH, Guerickestr. 25, 80805 München, Germany
{willnecker,dlugi,brunnert}@fortiss.org
[2] Universität Würzburg, Am Hubland, 97074 Würzburg, Germany
{simon.spinner,samuel.kounev}@uni-wuerzburg.de
[3] Dynatrace Austria GmbH, Freistädter Str. 13, 4040 Linz, Austria
wolfgang.gottesheim@dynatrace.co
[4] Technische Universität München, Boltzmannstr. 3,
85748 Garching, Germany
krcmar@in.tum.de

Abstract. Resource demands are a core aspect of performance models. They describe how an operation utilizes a resource and therefore influence the systems performance metrics: response time, resource utilization and throughput. Such demands can be determined by two extraction classes: direct measurement or demand estimation. Selecting the best suited technique depends on available tools, acceptable measurement overhead and the level of granularity necessary for the performance model. This work compares two direct measurement techniques and an adaptive estimation technique based on multiple statistical approaches to evaluate strengths and weaknesses of each technique. We conduct a series of experiments using the SPECjEnterprise2010 industry benchmark and an automatic performance model generator for architecture-level performance models based on the Palladio Component Model. To compare the techniques we conduct two experiments with different levels of granularity on a standalone system, followed by one experiment using a distributed SPECjEnterprise2010 deployment combining both extraction classes for generating a full-stack performance model.

Keywords: Performance model generation · Resource demand measurements · Resource demand estimations

1 Introduction

Performance models can be used to predict the performance of application systems. Resource demands are an important parameter of such performance models. They describe how an operation utilizes the available resources. A busy resource increases the time an operation needs to execute, therefore increasing the response time of the operation and ultimately the time for the user

M. Beltrán et al. (Eds.): EPEW 2015, LNCS 9272, pp. 115–129, 2015.
DOI: 10.1007/978-3-319-23267-6_8

accessing the system. When performance models are applied for capacity management, such information is essential as the available hardware must be sized according to the demand of the operations for a certain workload. Demands can be extracted from different sources. Expert guesses are used, especially when no running application artifact is available, to forecast the application's performance behavior. If running artifacts are available (e.g., in a test environment), measurement and estimation techniques can be applied. This work compares two direct measurement techniques and an adaptive estimation technique based on multiple statistical approaches and compares strengths and weaknesses of each technique.

Manually creating performance models often outweighs their benefits [6]. Therefore, automatic performance model generator (PMG) frameworks for running applications have been introduced in the scientific community [3,6]. Such PMGs create performance models, which include the software architecture, control flow and the resource demand of the application. These PMGs use either direct measurements by instrumenting the operations that are executed or resource demand estimations calculated from coarse-grained measurement data like total resource utilization and response time per transaction invocation.

Applying direct measurements requires to alter the installation of the system that is instrumented by applying an agent that intercepts invocations. This allows for extracting the software architecture and control flow, but causes overhead on the system running for every instrumented operation that is invoked [5]. Furthermore, such measurements require that for each instrumented technology and resource type, a dedicated measurement approach must be available. A number of industry solutions for direct measurements are already available and have been integrated into such a PMG previously [17].

As an alternative to direct measurements, resource demand estimation techniques can approximate the demand of a resource from coarse-grained monitoring data like Central Processing Unit (CPU) utilization of a system and response time of a transaction. Such data can be collected for a wide range of systems and technologies and requires no in-depth measurement of the application's technology stack. This coarse-grained monitoring data causes less overhead, produces less data to collect, and to process. However extracting the control flow of an application is not possible with such an approach.

The Library for Resource Demand Estimation (LibReDE)[1] provides different resource demand estimation approaches [15]. In order to do the estimations, LibReDE requires information about the resource utilization as well as about the response times of an operation or transaction during the same time frame. This work integrates LibReDE with the PMG introduced by Brunnert et al. [6] in order to be able to generate models based on direct resource demand measurements or estimations. This integration allows to compare the direct measurement and estimation approaches and to determine strengths and weaknesses for

---

[1] http://se.informatik.uni-wuerzburg.de/tools/librede/

extracting resource demands using the SPECjEnterprise2010[2] industry benchmark as representative enterprise application for the evaluation.

We compare these two extraction classes for resource demands in a series of experiments evaluating the accuracy of automatically generated performance models in terms of CPU utilization and response times. Therefore, the main contributions of this work are as follows:

(i) An integration of resource demand estimation in a PMG.
(ii) A comparison of the accuracy of two direct measurement techniques with the most common resource demand estimation approaches used in practice.
(iii) An evaluation of an integrated PMG, utilizing the benefits of direct measurement and estimation techniques.

This work begins with an introduction to the performance model generation workflow followed by introducing measurement technologies. We continue with an introduction to LibReDE and the approaches used to estimate resource demands including the selection of the most accurate estimation approach for meaningful resource demands. The experiment for comparing all three approaches is described and evaluated, followed by a hybrid setup where a combination of direct measurements and resource demand estimations is used. The work closes with related work, followed by the conclusion and future work section.

2   Extracting Resource Demands

In order to support resource demand measurement and estimation approaches, we extend the previously introduced Performance Management Work (PMW)-Tools' automatic PMG with LibReDE [6,15]. Generating a performance model is divided into three separate steps depicted in Figure 1. First monitoring data is gathered. This monitoring data is, in a second step, aggregated per operation and stored in a monitoring database (DB). The last step is the actual model generation, which uses the aggregated data and generates an architecture-level performance model based on the Palladio component Model (PCM) [1].

The PMG supports data from different data sources:

(i) PMW-Tools monitoring, a monitoring solution for Java Enterprise Edition (EE) applications to measure CPU, memory, and network demands and response times of Java EE components and its operations [4,6].
(ii) Dynatrace[3] Application Monitoring (AM), an industry monitoring solution for Java, .NET, PHP and other technologies [17].
(iii) System Activity Reporter (SAR), an Unix/Linux based tool to display various system loads like CPU utilization.

---

[2] SPECjEnterprise is a trademark of the SStandard Performance Evaluation Corp. (SPEC). The SPECjEnterprise2010 results or findings in this publication have not been reviewed or accepted by SPEC, therefore no comparison nor performance inference can be made against any published SPEC result. The official web site for SPEC-jEnterprise2010 is located at http://www.spec.org/osg/Enterprise2010.
[3] http://www.dynatrace.com

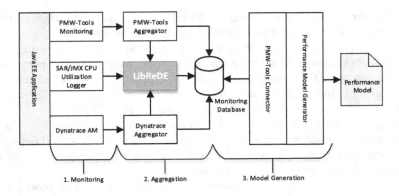

Fig. 1. Performance model generator framework (adapted from [5, 17])

(iv) Java Management Extensions (JMX) Logger, a command line tool that
     reads CPU utilization values from Java Virtual Machines (JVMs) using
     the JMX interface.

The first two data sources are able to collect direct measurement data, but
also response times for estimation techniques. The demand estimation is realized using LibReDE [15]. This library uses response times of an operation or
transaction and utilization of a resource, collected by one of the last two data
sources, to estimate the resource demands of an operation [15].

2.1 Performance Management Work - Tools Monitoring

PMW-Tools monitoring provides a Servlet Filter, an Enterprise JavaBean (EJB)
Interceptor, a SOAP-Handler and a Java Database Connectivity (JDBC)-Wrap-
per for Java EE applications [4,6]. The aforementioned technologies allow to
collect CPU time, heap allocation and network demand on the level of single
operation invocations [4–6]. Furthermore, the PMW-Tools monitoring allows to
collect information about the transaction control flow and about an application architecture on the level of components and their operations. All public
operations within the instrumented system are extracted and combined to one
transaction. The PMW-Tools monitoring agent is able to measure the response
time of an operation. The start and end time of each operation invocation is
measured. Subinvocations are removed from this time interval, so the actual
response time of one operation invocation is calculated.

2.2 Dynatrace Application Monitoring

The Dynatrace AM solution allows for measurements on different levels of granularity. This ranges from measuring the response time on the system entry point
level, through fine-grained measurements per operation invocation. Dynatrace
AM uses, depending on the host system, various timers that measure the CPU

utilization in different time intervals [7]. It furthermore traces a transaction throughout the instrumented system and can therefore determine the control flow as the PMW-Tools monitoring does [17]. The Representational State Transfer (REST) interface of this solution provides, among other metrics, the ability to access CPU time and response times of the instrumented operations. Thus, this approach, as well as the PMW-Tools monitoring approach can be used for direct measurements and estimation techniques.

2.3   Library for Resource Demand Estimation

Demand Estimation Approaches. While the monitoring tools described in subsection 2.1 and subsection 2.2 are able to directly measure the CPU time per operation invocation, their usage is infeasible in certain situations, e.g., when using third-party or legacy applications that cannot provide the required instrumentation. For other scenarios, the costs for fine-grained instrumentation can be considered too high. Therefore, different statistical approaches have been proposed in the literature to estimate resource demands for individual operations based on aggregated measurements such as average response time or CPU utilization. These aggregated measurements are often collected by default in applications (e.g., in access log files) and in the operating system (OS). Therefore, resource demand estimation techniques can be applied in many situations where the usage of direct measurements is prohibitive.

LibReDE is a Java library providing different ready-to-use implementations of statistical approaches for resource demand estimation [15]. The library currently comes with implementations of six commonly used approaches: response time approximation [3], service demand law [3], linear regression [13], two variants of a Kalman filter [16,18] and an optimization-based approach [12]. Previous work [14] showed that the accuracy of the individual techniques strongly depends on the characteristics of the observations and the modeled system resulting in significant differences in the estimates. In order to evaluate the accuracy of the estimated resource demands, LibReDE supports the evaluation of the results using k-fold cross-validation: the input data is randomly partitioned into k equally large subsets and the estimation is repeated k times, each time using a different one of the k subsets as validation set and the others as training set. As the actual values of the resource demands are unknown, the estimation error is evaluated using the observed utilization U_{act} and the observed response times $R_{act,r}$ of operation r. The observed values are compared to the calculated ones, U_{calc} and $R_{calc,r}$, which are obtained using equations from operational analysis of queuing networks. Using the estimated resource demands, U_{calc} is determined based on the Utilization Law [8, Chap.6]:

$$U_{calc}(\lambda) = \frac{1}{p} \sum_{r=1}^{n} \lambda^r D^r \qquad (1) \qquad R_{calc}^r(\lambda) = D^r(1 + \frac{P_Q}{1 - U_{calc}(\lambda)}). \quad (2)$$

Assuming a M/M/k/PS queue for Equation 2 [8, Chap.14]: n is the number of operations, D_r is the estimated resource demand of operation r, $\lambda = (\lambda_1, \dots, \lambda_n)$

is a vector of arrival rates, p is the number of processor cores and P_Q is the probability that an arrival finds all servers busy (calculated using the Erlang-C formula [8, Chap.14]).

The mean relative errors E_{util} for the utilization and $E_{rt,r}$ are then determined on the validation set $V=\{(\lambda_1^{(i)}, \ldots, \lambda_n^{(i)}, R_{act,1}^{(i)}, \ldots, R_{act,n}^{(i)}, U_{act}^{(i)}) : i = 1 \ldots m\}$:

$$e_{util} = \frac{1}{m} \sum_{i=1}^{m} \frac{|U_{act}^{(i)} - U_{calc}(\lambda^{(i)})|}{U_{act}^{(i)}} \quad (3) \qquad e_{rt}^{r} = \frac{1}{m} \sum_{i=1}^{m} \frac{|R_{act}^{(r,i)} - R_{calc}^{r}(\lambda^{(i)})|}{R_{act}^{(r,i)}} \quad (4)$$

The relative errors are calculated for each of the k validation sets and the result of the cross-validation is the mean relative error over all validation sets. Based on the relative errors, the PMG dynamically chooses an approach as described in the next section.

Estimation Approach Selection. Selecting the right estimation approach for LibReDE makes a huge difference (in our experiments we observed differences in the range of 6% to 6000% relative response time error). Each approach has strengths and weaknesses depending on the application in place [14, 15].

$$\begin{bmatrix} e_{util}^{(1)} \\ \vdots \\ e_{util}^{(i)} \\ \vdots \\ e_{util}^{(m)} \end{bmatrix} + \left(\begin{bmatrix} e_{rt}^{(1,1)} & \cdots & e_{rt}^{(1,j)} & \cdots & e_{rt}^{(1,n)} \\ \vdots & & \vdots & & \vdots \\ e_{rt}^{(i,1)} & \cdots & e_{rt}^{(i,j)} & \cdots & e_{rt}^{(i,n)} \\ \vdots & & \vdots & & \vdots \\ e_{rt}^{(m,1)} & \cdots & e_{rt}^{(m,j)} & \cdots & e_{rt}^{(m,n)} \end{bmatrix} \times \begin{bmatrix} \lambda^{(1)} & \cdots & 0 & \cdots & 0 \\ \vdots & & \vdots & & \vdots \\ 0 & \cdots & \lambda^{(i)} & \cdots & 0 \\ \vdots & & \vdots & & \vdots \\ 0 & \cdots & 0 & \cdots & \lambda^{(n)} \end{bmatrix} \right) \times \begin{bmatrix} 1 \\ \vdots \\ 1 \\ \vdots \\ 1 \end{bmatrix} \quad (5)$$

We are looking for the approach that calculates the most accurate resource demands, therefore we use both validators and select the one with the lowest relative error when combining both validation results provided by LibReDE. The utilization law validator provides a vector E_{util}, as we only use one resource, with the length of m, where m is the number of estimation approaches used. Each row in this vector contains the relative utilization error of one approach. The response time validator provides a $m \times n$ matrix E_{rt}, where m is the number of estimation approaches used and n the number of operations to estimate resource demands for. Each row i contains all relative response time errors of one approach and each column j contains the relative response time error of one operation. Therefore, the value at index i,j is the relative response time error of operation j using approach i.

Some operations might get a small amount of calls, misleading the approach selection when just selecting the approach with the smallest relative error. We weight the relative error of each operation according to the arrival rates of the input data as the number of values used for the estimation varies due to different workload on each operation. We therefore multiply the arrival rates matrix λ with the relative response time error matrix E_{rt}. The result is a weighted matrix that considers the operation call probability. To select the best suited approach we need to reduce this matrix to a vector, where each value contains a meaningful relative error for one approach considering all operations. We calculate the sum over each row of the matrix resulting in a relative response time error vector. Both vectors, containing either the response times or the CPU utilization error, are added up as shown in Equation 5.

We finally select the approach with the minimum total error in the resulting vector. The resource demands D_r of this approach are stored in the monitoring DB of the PMG. The model generation then uses these resource demands for building an architecture-level performance model.

3 Evaluation

In order to evaluate the accuracy of resource demand measurement and estimation approaches, we used two environments. The first evaluation compares the three presented approaches (PMW-Tools monitoring, Dynatrace AM and LibReDE) with each other on two levels of granularity in a virtualized environment. In the second evaluation, we use a distributed bare-metal installation and combine direct measurement and estimation approaches.

For both evaluations, we use the orders domain application of the SPECjEnterprise2010 (Version 1.03) industry standard benchmark as exemplary enterprise application. Since the benchmark defines a workload and a dataset for the test execution, the results are reproducible for others. The orders domain application is a Java EE web application comprised of servlet, JavaServer Pages (JSPs) and EJB components. The application represents a platform for automobile dealers to sell and order cars; the dealers (henceforth called users) interact with the platform using the Hypertext Transfer Protocol (HTTP). There are three basic business transactions which describe how users interact with the system: Browse, Manage and Purchase.

3.1 Standalone Evaluation

For the standalone evaluation, we installed the SPECjEnterprise2010 benchmark and its corresponding load test driver on two Virtual Machines (VMs), each deployed on separate hosts (IBM System X3755M3) to avoid interferences between the two systems. The system under test (SUT) VM contains the application server, hosting the orders domain application. The other VM executes load tests on the SUT using the Faban[4] harness driver of the benchmark. Both virtual machines run openSUSE 12.3 64-bit as OS and have access to 40 gigabytes of Random Access Memory (RAM). The application server VM uses six CPU cores while the driver VM has access to four CPU cores.

The benchmark is deployed on a JBoss Application Server (AS) 7.1 in the Java EE 7.0 full profile. The DB on the test system VM is an Apache Derby DB

---

[4] https://java.net/projects/faban/

in version 10.9.1.0. The JBoss AS and the Apache Derby DB are both executed in the same 64-bit Java OpenJDK VM (JVM version 1.7.0_17).

The first step of the evaluation is to obtain the relevant performance metrics (response time, utilization and throughput) of the SUT under different workloads by performing measurement runs. As the network overhead between the Faban harness and the SUT is not considered in the first step, the response time measurements are conducted by measuring the system entry point response times with the PMW-Tools monitoring. For this purpose, a workload of 600, 800, 1000 and 1200 concurrent users is put on the SUT, resulting in a mean CPU utilization of 39%, 56%, 69% and 79% on the server. Each measurement run lasts for sixteen minutes while data is only collected between a five minute ramp-up and a one minute ramp-down phase.

The standalone evaluation is conducted on two levels of granularity. We compare system entry point level, where only the boundaries of the system are monitored, with a component operation level monitoring, where each public operation of each used component is instrumented. This results in different performance models as resource demands are only measured or estimated for either servlet invocations (system entry point) or servlet calls and EJB operation invocations. For both cases we execute a load test with 600 concurrent users and collect monitoring data. Depending on the approach selected, this monitoring data contains either fine-grained measurements of CPU demanded time per operation invocation or only response times and total CPU utilization of the VM.

The performance models generated with this monitoring data are used for simulating the same and higher amounts of concurrent users (800 - 1200). We compare the simulated CPU utilization and the response times with actual measurements on the system. For the utilization we compare the measured mean CPU utilization (MMCPU) with the simulated mean CPU utilization (SMCPU) and calculate the relative CPU utilization prediction error (CPUPE).

When examining the CPU utilization prediction results shown in Table 1, it is visible that LibReDEs prediction is very accurate, especially in the replay case with 600 concurrent users and the upscaled case with 1200 concurrent users. The two monitoring solutions only measure the CPU time of the actual request thread while LibReDE also takes the overhead of the application server and CPU time for other processing like garbage collection (GC) into account. Dynatrace AM can use different CPU timers optimized for specific environments (i.e., VM, Windows OS, etc.) and the here used POSIX Hi-Res timer produces more accurate results than the PMW-Tools monitoring [7].

Table 1. Measured and simulated CPU utilization for system entry point level

System		PMW-Tools monitoring		Dynatrace AM		LibReDE - estimation	
Users	MMCPU	SMCPU	CPUPE	SMCPU	CPUPE	SMCPU	CPUPE
600	39,33%	36.66%	6.80%	38.73%	1.53%	39.73%	1.01%
800	55,69%	48.68%	12.58%	51.41%	7.68%	52.69%	5.37%
1000	69,28%	60.92%	12.06%	64.02%	7.58%	65.56%	5.36%
1200	79,31%	73.21%	7.69%	77.33%	2.50%	78.66%	0.82%

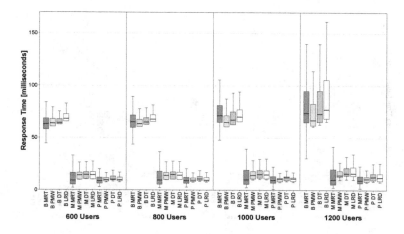

Fig. 2. Measured and simulated response times on system entry point level

Figure 2 shows the response times for system entry point level granularity using box plots. Each box depicts one measurement/simulation series. The figure is divided into four sections, distinguishing between different user amounts. In each section, three measured response time (MRT) box plots are shown, one for each business transaction: Browse (B), Manage (M), Purchase (P). The sections are completed by nine simulation box plots, one for each of the three business transactions times the three techniques: PMW-Tools monitoring (PMW), Dynatrace AM (DT) and LibReDE (LRD).

We see that LibReDE tends to overestimate the resource demands, leading to a higher median and broader Interquartile range (IQR) for the Browse and Manage transaction, but delivers good results in general. The differences between PMW-Tools monitoring and Dynatrace AM are minimal in most cases. All approaches have in common that they cannot predict the lower quartiles. However, this is most likely caused by the fact, that only mean values for CPU demands are represented in the resource demands of the generated performance models.

The CPU utilization results and errors are similar for component operation level compared to system entry point level. Table 2 shows that LibReDE again produces the most accurate resource demands when simulating and comparing

Table 2. Measured and simulated CPU utilization for component operation level

System		PMW-Tools monitoring		Dynatrace AM		LibReDE - estimation	
Users	MMCPU	SMCPU	CPUPE	SMCPU	CPUPE	SMCPU	CPUPE
600	39,33%	36.39%	7.49%	37.21%	5.39%	39.61%	0.69%
800	55,69%	48.42%	13.04%	49.83%	10.51%	52.77%	5.24%
1000	69,28%	60.26%	13.01%	61.89%	10.67%	65.71%	5.15%
1200	79,31%	71.78%	9.49%	74.07%	6.60%	79.32%	0.01%

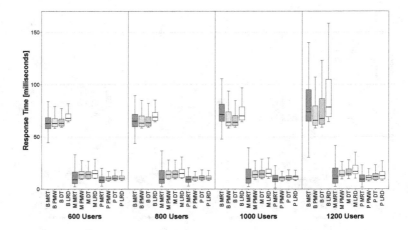

Fig. 3. Measured and simulated response times on component operation level

the CPU utilization with actual measurements. Dynatrace again is more accurate than PMW-Tools monitoring but the differences are smaller compared to the system entry point level.

The response time errors presented in Figure 3 are best predicted with direct measurements. The differences between the two monitoring approaches are rather small. LibReDE overestimates in most of the cases. The upper quartiles are better predicted using estimation than direct measurements, but the median and IQR are worse with estimation approaches. Again all approaches have in common that they cannot predict the lower quartile.

3.2   Distributed Setup

The previous evaluation showed that resource estimation techniques provide sufficiently accurate results for most of the evaluated scenarios. However, in order to use these estimations, it is important to be able to measure control flows and response time on the level of granularity that needs to be represented in a model. Furthermore, estimations work only as long as response time and throughput values for all requests are available for a measurement interval. Therefore, there are a lot of cases in which it is desirable to mix direct measurements with resource estimation techniques.

This evaluation validates a distributed deployment scenario for SPECjEnterprise2010 in which direct measurements and estimations are used in combination. This is necessary to be able to properly account for the resource demands and times spent on different layers of the architecture (e.g., what portion is spent in the DB tier). It is important to note that the following models also account for network resource demands which was not done for the previous evaluations as the standalone setup was deployed on a single server. The models for this evaluation are automatically generated using the PMG by providing input from multiple sources (PMW-Tools monitoring, Dynatrace AM, SAR and LibReDE).

The SPECjEnterprise2010 benchmark is deployed in a multi-tier architecture consisting of a presentation, application and a data tier. As we do not have an in-depth monitoring for the data tier, we use estimation here while the presentation and application tier are instrumented using the PMW-Tools monitoring as well as the Dynatrace AM. The resulting resource demands are used to build a performance model based on PCM. In order to model the data tier, the data collection solution (i.e., PMW-Tools monitoring, Dynatrace AM) gathers the tier's response times, CPU utilization on the DB is gathered using SAR. These values are used as input for a resource demand estimation using LibReDE [15]. The generated performance model is then enriched with the data tier's estimated resource demands. Finally, the model is used to perform simulations with increasing workloads; the results are then compared to measurements of the real system to gauge the prediction performance of the approach.

To obtain a multi-tier architecture, the standard orders domain application is modified by converting the EJB components to web services. This allows for the application's deployment on two different machines. In addition, the application tier is connected to a PostgreSQL DB located on a third machine.

The different tiers of the application are deployed on three different machines which in the following will be called User Interface (UI) server, Web Service (WS) server and DB server. Additionally, a benchmark driver is deployed on one VM to generate load on the whole system by accessing the UI server using the three business transactions. To achieve a moderate load on each system, the CPU core count of each system has been modified by disabling some cores. All of the systems' technical specifications are listed in Table 3.

The distributed evaluation also begins with performing similar measurement runs using minimal instrumentation. Executing the same workload (600 - 1200 users), as in the previous evaluation results in a maximum CPU utilization of 77%, 59% and 68% on the UI, WS and DB server, respectively. The benchmark driver has been modified to collect the response time of the three business transactions for each invocation, instead of measuring them directly on the SUT as in the previous evaluation.

Table 3. Software and hardware configuration of the SUT

Server	UI Server	WS Server	DB Server
Application	SPECjEnterprise2010 (version 1.03) orders domain		
AS/DB	GlassFish 4.0 (build 89)	JBoss AS 7.1.1	PostgreSQL 9.2.7
JVM	64-bit Java HotSpot JVM version 1.7.0_71	64-bit Java OpenJDK JVM version 1.7.0_40	-
OS	openSUSE 12.2		openSUSE 12.3
CPU Cores	2 x 2.1 GHz	6 x 2.1 GHz	4 x 2.4 GHz
CPU Sockets	4 x AMD Opteron 6172		2 x Intel Xeon E5645
RAM	256 GB		96 GB
Hardware System	IBM System X3755M3		IBM System X3550M3
Network	1 gigabit-per-second (GBit/s)		

Table 4. Measured and simulated CPU utilization using PMW-Tools monitoring

	UI server			WS server			DB server		
Users	MMCPU	SMCPU	CPUPE	MMCPU	SMCPU	CPUPE	MMCPU	SMCPU	CPUPE
600	39.97%	40.36%	0.96%	30.96%	26.93%	14.96%	34.51%	40.77%	15.35%
800	53.11%	54.05%	1.74%	41.86%	36.11%	15.94%	45.89%	54.54%	15.86%
1000	65.27%	67.37%	3.11%	48.39%	44.99%	7.57%	56.51%	68.02%	16.93%
1200	77.01%	80.52%	4.36%	59.71%	53.81%	10.96%	68.38%	81.42%	16.01%

Table 5. Measured and simulated CPU utilization using Dynatrace AM

	UI server			WS server			DB server		
Users	MMCPU	SMCPU	CPUPE	MMCPU	SMCPU	CPUPE	MMCPU	SMCPU	CPUPE
600	39.97%	33.29%	20.06%	30.96%	30.54%	1.36%	34.51%	34.25%	0.77%
800	53.11%	44.47%	19.43%	41.86%	40.82%	2.55%	45.89%	45.80%	0.20%
1000	65.27%	55.55%	17.49%	48.39%	51.03%	5.17%	56.51%	57.20%	1.21%
1200	77.01%	66.82%	15.25%	59.71%	61.34%	2.66%	68.38%	68.92%	0.79%

Afterwards, the UI and WS server are instrumented and another benchmark run with a workload of 600 concurrent users is performed. The collected data is used to generate a performance model using the PMG. Simultaneously, the response times per invocation and aggregated utilization of the DB server are collected. These are automatically used by the PMG as input for the LibReDE resource demand estimation. The model is further enhanced by adding latency and throughput values of the network connecting the individual servers as shown in [4]. These values are gathered using the lmbench[5] benchmark suite. Finally, the finished model is used to simulate the SUT with a workload of 600, 800, 1000 and 1200 concurrent users; the duration and steady state times correspond to the ones used for the measurements.

When examining the CPU utilization values in Table 4 and Table 5, we see that the SMCPU of the DB server is predicted with very high accuracy using Dynatrace AM, with the highest error being 1.21% at 1000 concurrent users. The PMW monitoring does not intercept all JDBC calls, leading to an overestimation of CPU demands on the calls that are intercepted. Furthermore, the accounting of this calls is also missing in the WS server, leading to an underestimation of the CPU demands in the business tier. The CPU utilization of the WS server is predicted very well using Dynatrace AM, while the UI server's utilization is predicted too low. Dynatrace distributes the processing time to all active operations. We have more running operations on the WS server, leading to better results for this tier compared to the UI server. The PMW monitoring instruments the CPU demands of the UI server better, because its servlet interceptor measures each operation individually. Overall, the results show that the approach is well suited for predicting the performance of a multi-tier application.

The response time values are illustrated in the box plots in Figure 4. The figure is divided into four sections, one section for each user amount. Each section again contains three MRT series (Browse, Manage, Purchase) and six simulation box plots. Three plots for the combination PMW-Tools monitoring and LibReDE

---

[5] http://lmbench.sourceforge.net/

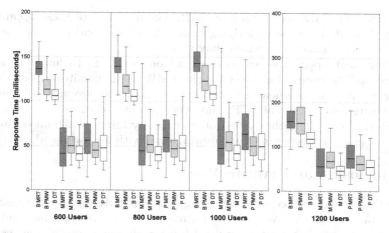

Fig. 4. Measured and simulated response times

(PMW) and three plots for the combination Dynatrace AM and LibReDE (DT). Note that the last section uses another scale as the first three sections, as the response times are significantly higher with 1200 concurrent users.

The comparison shows that the combination of resource demand measurement and estimation techniques leads to a good representation of the real system. The median of the simulated response time is close to the actual measurements. The prediction error for the median response time values is at most 25.02% for the browse transaction at 1200 concurrent users. The IQR prediction using PMW is usually a bit closer to the real system measurements than DT.

4   Related Work

This section presents related work that is concerned with measurement accuracy in different environments or the overhead caused by such measurements.

CPU accounting on VMs can be error prone due to sharing the same physical resource over multiple machines. Hofer et al. [9] discovered that malicious accounting, so called steal time, can be detected and calculated in a VM. If not corrected, CPU utilization measurements produce wrong resource demands. Wrong CPU utilization accounting decreases the quality of performance models created either using direct measurement or estimation methods. We avoid this by isolating the SUT VM on a single host. However, virtualized environments need to correct this steal time in order to calculate accurate resource demands.

Estimating the overhead of virtualized environments has been described by Brosig et al. [2] and Huber et al. [10]. These approaches estimate, among others, virtualization overhead based on monitoring data using a queuing network. Such calculations can increase the accuracy of resource demands of such environments.

Kuperberg compared different timers and measurement approaches for a number of systems [11]. While the Dynatrace AM already offers different timers to select the most suitable one, the other two approaches rely on either the

ThreadMXBean, JMX monitoring or SAR. The accuracy of these approaches can vary depending on the underlying system monitored and therefore the calculated resource demands accuracy may vary.

Measurement approaches cause overhead on the SUT. Brunnert et al. [5] measured and discussed this effect for the PMW-Tools monitoring solution in previous work. This overhead effect turns out to be at around 0.003 ms for each measurement when only CPU no other resource demands are collected. This overhead can effect the system at its capacity limits, while an estimation approach can use coarse-grained monitoring data with less overhead.

5   Conclusion and Future Work

This work compared three different techniques for deriving resource demands for performance models. We compared a monitoring approach from academia, an industry monitoring solution and a library combining six different estimation approaches. These techniques have been integrated into a single automatic PMG. The evaluation compared all techniques in a standalone and a distributed setup, as well as in a virtualized and a bare-metal environment for two levels of granularity: system entry point level and component operation level.

All techniques deliver good results for both granularity levels and in all environments. Estimation techniques deliver better results for the system entry point level, but fall short behind direct measurements for the component operation level. Furthermore, direct measurements can extract resource demands on any level of detail, while estimation techniques must calculate demands for the complete system to distribute the measured utilization among the components. Estimation techniques can be applied to a broad variety of technologies as the requirements for data collection are lower. We demonstrated accurate results using a hybrid setup, where measurement approaches are used to extract resource demands for the UI and WS combined with estimations for the DB.

The evaluation uses a Java EE application. Industry monitoring like Dynatrace AM are capable of observing other technologies. Demonstrating the applicability of the framework for other technology stacks as well as extending the monitored resources are interesting challenges for further research.

Acknowledgement. This work has been supported by the Research Group of the Standard Performance Evaluation Corporation (SPEC).

References

1. Becker, S., Koziolek, H., Reussner, R.: The Palladio Component Model for Model-Driven Performance Prediction. Journal of Systems and Software **82**(1), 3–22 (2009). special Issue: Software Performance - Modeling and Analysis
2. Brosig, F., Gorsler, F., Huber, N., Kounev, S.: Evaluating approaches for performance prediction in virtualized environments. In: 2013 IEEE 21st International Symposium on Modeling, Analysis Simulation of Computer and Telecommunication Systems (MASCOTS), pp. 404–408, August 2013

3. Brosig, F., Kounev, S., Krogmann, K.: Automated extraction of palladio component models from running enterprise java applications. In: Proceedings of the 1st International Workshop on Run-time Models for Self-managing Systems and Applications (ROSSA 2009). ACM, New York (2009)
4. Brunnert, A., Krcmar, H.: Continuous Performance Evaluation and Capacity Planning Using Resource Profiles (under review). Journal of Systems and Software (2015)
5. Brunnert, A., Neubig, S., Krcmar, H.: Evaluating the Prediction Accuracy of Generated Performance Models in Up- and Downscaling Scenarios, pp. 113–130 (2014)
6. Brunnert, A., Vögele, C., Krcmar, H.: Automatic performance model generation for java enterprise edition (EE) applications. In: Balsamo, M.S., Knottenbelt, W.J., Marin, A. (eds.) EPEW 2013. LNCS, vol. 8168, pp. 74–88. Springer, Heidelberg (2013)
7. Dynatrace: Dynatrace Agent Timers. https://community.compuwareapm.com/community/display/DOCDT60/Agent+Timers (accessed: May 14, 2015)
8. Harchol-Balter, M.: Performance Modeling and Design of Computer Systems. Cambridge University Press, New York (2013)
9. Hofer, P., Hörschläger, F., Mössenböck, H.: Sampling-based steal time accounting under hardware virtualization. In: Proceedings of the 6th ACM/SPEC International Conference on Performance Engineering (ICPE 2015), pp. 87–90. ACM, New York (2015)
10. Huber, N., von Quast, M., Hauck, M., Kounev, S.: Evaluating and modeling virtualization performance overhead for cloud environments. In: Proceedings of the 1st International Conference on Cloud Computing and Services Science (CLOSER 2011), pp. 563–573. SciTePress, May 2011
11. Kuperberg, M.: Quantifying and Predicting the Influence of Execution Platform on Software Component Performance, vol. 5. KIT Scientific Publishing (2010)
12. Menascé, D.A.: Computing missing service demand parameters for performance models. In: Proceedings of the 2008 Computer Measurement Group Conference (CMG 2008), Las Vegas, NV, USA, pp. 241–248 (2008)
13. Rolia, J., Vetland, V.: Parameter estimation for performance models of distributed application systems. In: Proceedings of the 1995 Conference of the Centre for Advanced Studies on Collaborative Research (CASCON 1995), p. 54. IBM (1995)
14. Spinner, S.: Evaluating Approaches to Resource Demand estimation. Master's thesis, Karlsruhe Institute of Technology (KIT) (2011)
15. Spinner, S., Casale, G., Zhu, X., Kounev, S.: LibReDE: A library for resource demand estimation. In: Proceedings of the 5th ACM/SPEC International Conference on Performance Engineering (ICPE 2014), pp. 227–228. ACM, New York (2014)
16. Wang, W., Huang, X., Qin, X., Zhang, W., Wei, J., Zhong, H.: Application-level CPU consumption estimation: towards performance isolation of multi-tenancy web applications. In: Proceedings of the 5th International Conference on Cloud Computing (CLOUD), pp. 439–446. IEEE (2012)
17. Willnecker, F., Brunnert, A., Gottesheim, W., Krcmar, H.: Using dynatrace monitoring data for generating performance models of java EE applications. In: Proceedings of the 6th ACM/SPEC International Conference on Performance Engineering (ICPE 2015), pp. 103–104. ACM, New York (2015)
18. Zheng, T., Woodside, M., Litoiu, M.: Performance Model Estimation and Tracking Using Optimal Filters. IEEE Transactions on Software Engineering 34(3), 391–406 (2008)

Estimating the Impact of Code Additions
on Garbage Collection Overhead

Peter Libič[✉], Lubomír Bulej, Vojtěch Horký, and Petr Tůma

Department of Distributed and Dependable Systems,
Faculty of Mathematics and Physics, Charles University in Prague,
Malostranské náměstí 25, 118 00 Prague 1, Czech Republic
{libic,bulej,horky,tuma}@d3s.mff.cuni.cz

Abstract. In managed memory environments, code changes influence performance both through time spent executing the code and time spent collecting garbage generated by the code. This complicates decision making when considering performance impact of code changes—while the impact on execution time is usually easy to assess in isolation, the impact on garbage collection time depends on the memory allocation behavior of the code surrounding the changes. In our paper, we describe a method to estimate the impact of code changes with additional allocations on garbage collection time, which can be applied, e.g., when assessing the overall performance impact of alternative changes. The method is demonstrated on experiments with the HotSpot virtual machine.

Keywords: Garbage collection · Performance · Modeling

1 Introduction

Performance is intuitively related to execution time, which is why execution time is often chosen as a performance metric. Unfortunately, the relationship between execution time and performance of a particular software artifact (application, component, function) is not always straightforward. Besides having its own code executed, a software artifact can also cause the platform it executes on to perform additional work—and therefore consume additional time—when the artifact code is not running.

The technical reasons for the additional work vary. Among typical examples, leaving dirty data in caches may cause later writes, loading code may cause later compilation and optimization, allocating memory on heap may cause additional garbage collections. In all these situations, tasks such as performance debugging or performance optimization become difficult when the additional execution time is not properly attributed.

Here, we focus on additional execution time due to garbage collection (GC). So far, methods of predicting GC time are only coarse grained [12]. It has also been shown that generational GC time can vary significantly even with minute workload changes [8]. Furthermore, GC time is related to program allocation

© Springer International Publishing Switzerland 2015
M. Beltrán et al. (Eds.): EPEW 2015, LNCS 9272, pp. 130–145, 2015.
DOI: 10.1007/978-3-319-23267-6_9

behavior, which is technically difficult to observe in detail due to associated overhead [5,10], and may even change due to observation [8].

In light of these obstacles, we devise a method that does not attempt to predict GC time of a complete application, but instead estimates the change in GC time after code modifications that insert additional allocations. The estimate is made under some limiting assumptions which make it possible to use only easily obtained input, in particular the GC log and the information on heap occupation in the modified locations, as provided by the standard HotSpot virtual machine.

We believe the method can find application in situations where modifications to an application with stringent GC overhead budget are considered, for example to choose between alternative modifications or alternative locations where the modifications are applied. The method also provides some insight on how objects survive individual collections and migrate between generations, which can be useful in GC tuning tasks. Finally, we illustrate the expressive limits of easily obtained input as far as the allocation behavior reconstruction is concerned.

The presentation structure is as follows. First, we review essential GC features related to our estimation method in Section 2. We follow with the description of the method itself in Section 3. Evaluation and discussion are given in Section 4. Related work and conclusion close the paper.

2 Garbage Collection Essentials

Our overhead estimation method has been developed in the context of the default (parallel) garbage collector of the HotSpot virtual machine. We present the essential GC features we rely on, the reader can find more details in [11].

Architecture. The collector architecture is generational. It separates objects into *young generation* and *tenured generation* and uses two configurable collectors, one to collect the young generation, one to collect both generations.

The default young generation collector (sometimes called *parallel scavenging*) is a copying collector. The memory allocated for the young generation is separated into one *eden* area and two *survivor* areas. New objects are allocated (mostly) in eden, each GC copies reachable objects from eden and one survivor into the other survivor. Objects that survive more than a particular young collection count, called *tenuring threshold*, are promoted to the tenured generation. (§1)

The default full collector (sometimes called *parallel mark and sweep*) is a mark and sweep collector with support for optional compaction. Each GC traverses reachable objects and releases objects that were not traversed. The young generation is evacuated into the tenured generation on each collection. (§2)

Dimensioning. The generations have default dimensions derived from the memory capacity of the execution platform. A suite of ad hoc rules, sometimes called *ergonomics*, is used to dynamically scale the generations. This can sometimes lead to performance anomalies [9], which is why manual sizing is recommended for production deployment.

A young collection is triggered whenever the eden in the young generation is full. During collection, the survivor that the objects are copied to may overflow,

(§3) leading to *premature promotion* of the remaining objects to the tenured gen-
(§4) eration. These facts can be used to configure the young generation—it should
be big enough to avoid excessive overhead due to frequent collections but small
enough to prevent individual collections taking too much time, and the tenuring
threshold should be small enough to prevent frequent premature promotions.

A full collection is triggered whenever the tenured generation is close to full.
(§5) Some reserve is maintained to prevent promotion failures in young collections,
the size of this reserve is derived dynamically as a weighted average of the amount
of promoted objects.

Monitoring. The HotSpot virtual machine provides support for displaying
information about heap occupation at GC events. The information is recorded
(§6) in the GC log, whose abbreviated example follows:

```
[GC (Alloc Failure) [PSYoungGen: 131072K->64672K(131072K)] 556149K->511381K
 ↪  (589824K), 0.5975770 secs] [Times: user=0.58 sys=0.01, real=0.60 secs]
[Full GC (Ergonomics) [PSYoungGen: 64672K->0K(131072K)] [ParOldGen:
 ↪  446709K->333843K(458752K)] 511381K->333843K(589824K), ...
```

For each young collection (first line above), we have the collection reason, the
size of the young generation before and after the collection, the size of the entire
heap before and after collection, and the collection time. For each full collection
(second line above), we additionally have the size of the tenured generation
before and after collection.

Outside the garbage collection events, an application can also use a standard
interface to query information on free memory in the virtual machine. Although
the exact meaning of the provided values is not documented, subtracting consecutive values provides an estimate on the amount of object allocations.

3   Modeling Garbage Collection Overhead

Besides the obvious time spent traversing and collecting heap content, GC may
impose overhead for example by trashing memory cache content, adding barriers
to memory access operations, enforcing particular object layout or reference
structure, and so on. Although the overhead can be measured by comprehensive
experiments [4], the interactions involved are too many to be captured in a white
box model of reasonable complexity.

Rather than modeling GC overhead for an entire application—a task that
requires detailed input on application allocation behavior even for partial tasks
such as modeling GC frequency [8]—we focus on modeling the impact of certain
application modifications on the total GC time. We consider modifications that
(§7) add allocations of short lived objects into particular application locations—in
practice, these are for example minor code patches, insertion or activation of
features such as logging, modifications that change optimization decisions and
turn stack allocations into short lived heap allocations, and so on.

In contrast, we do not consider modifications that allocate significant
amounts of long lived objects. We also assume applications that have a rel-
(§8) atively stable allocation behavior, rather than passing through phases whose
allocation behavior varies significantly. We discuss these assumptions in more

detail together with the model description—we believe they represent reasonably minimal constraints for a model that does not require expensive inputs.

In the limited context of this paper, the problem of estimating the total GC time can be decomposed into estimating the time of each collection and estimating the collection frequency. To estimate the time, we use gray box modeling with assumptions about algorithmic complexity of the GC algorithms involved. For estimating frequency, we look at the partial application allocation behavior that can be reconstructed from the information provided in the GC log (§6).

In the following derivations, we use verbose variable and function names to make the formulas more readable. For example, the heap dimensions are denoted as $max.size.eden$, $max.size.survivor$, $max.size.tenured$. The information from the GC log is $log.young.before$ and $log.young.after$ for the young generation occupation, $log.heap.before$ and $log.heap.after$ for the whole heap occupation. The current occupation of the heap is denoted $in.eden$, $in.survivor$, $in.tenured$, obviously $in.young = in.eden + in.survivor$ and so on.

Some symbols refer to information concerning a particular garbage collection. When presented without additional specification, the symbols refer to the current collection in the discussion context. Symbols with subscript refer to a particular collection index or rank. We index young collections following a full collection starting from 1, and also define collection rank r as the collection index i capped one collection above the tenuring threshold, $r = \min(i, tenuring.threshold + 1)$.

3.1   Reconstructing Allocation Behavior

Central to our model is the construction of a function that approximates how objects survive young collections. The function is directly related to object lifetime—only objects whose lifetimes exceed that of the particular young collection survive, objects whose lifetimes are shorter are collected.

We use the information from the GC log, specifically the sizes of the young generation and the entire heap before and after each young collection (§6). Obviously, we also have $log.tenured.before = log.heap.before - log.young.before$ and $log.tenured.after = log.heap.after - log.young.after$.[1]

After a full collection, the young generation is empty (§2). The first young collection following a full collection (rank 1) is triggered when the eden is full and both survivors are empty, we therefore have $log.young.before_1 = max.size.eden$ as the amount of objects allocated during one young collection period. Denoted as $surviving_1$, $log.young.after_1$ is the amount of surviving objects allocated during one young collection period.

The second young collection following a full collection (rank 2) is again triggered when the eden is full, we therefore have $log.young.before_2 = max.size.eden + log.young.after_1$. The amount of surviving objects allocated during two young collection periods, $surviving_2 = log.young.after_2$, consists of objects that have

---

[1] Some configurations of HotSpot can display object lifetime distribution at GC events, however, that feature is not available in the default collector and not sufficiently complete in other collectors.

survived two young collections (allocated before the first young collection) and objects that have survived one young collection (allocated after the first young collection).

We can proceed inductively for as long as no objects are promoted. When young collection with index i promotes some objects, the savings in the young generation will not match the savings in the entire heap:

$$log.young.before_i - log.young.after_i \neq log.heap.before_i - log.heap.after_i$$

When this happens, it no longer holds that $surviving_i = log.young.after_i$, because $log.young.after_i$ does not include the promoted objects. We can, however, still salvage the computation of $surviving_i$:

$$surviving_i = log.heap.after_i - log.heap.before_i + log.young.before_i$$

After a promotion in young collection with index i, we no longer have enough information to compute the amount of surviving objects $surviving_j$ for $j > i$, simply because the liveness of promoted objects is only examined during a full collection. However, if the promotion is due to objects reaching the tenuring threshold (§1), we can still use the young collections with index $j > i$ to compute additional estimates for $surviving_i$:

$$surviving_i = log.heap.after_j - log.heap.before_j + log.young.before_j$$

Because every full collection empties the young generation, we can repeat the same computations as many times as there are full collections, obtaining multiple estimates $surviving_i$ for each $i \in 1 \ldots tenuring.threshold + 1$. We note that a premature promotion due to survivor overflow (§3) may introduce inflation in the estimate of the amount of surviving objects, we therefore omit such estimates and average over the remaining ones:

$$valid.surviving_i = \{surviving_i : log.young.after_i < max.size.survivor\}$$
$$surviving.average_i = \text{average}(valid.surviving_i)$$

3.2   Considering Additional Allocations

We now consider modifications that add allocations of short lived objects (§7). In the text, we refer to the application without modifications as the *original* application (and original allocations, original collections and so on for artifacts present in the original application), and the application with envisioned modifications as the *modified* application (and modified allocations, modified collections and so on for artifacts not present in the original application).

(§9) To describe where the modified allocations happen, we execute the original application with minimalistic instrumentation that records information on free memory (§6) in all locations where the modified allocations are to be added. Merged with the record of the original GC behavior (§6), this forms the record of the original allocation behavior as the input of our model, which then estimates the modified GC behavior.

Triggering Young Collections. We pass sequentially through the record of the original allocation behavior, keeping track of the aggregate size of original and modified objects in eden, $in.eden.original$ and $in.eden.modified$, and the aggregate size of objects in survivors, $in.survivor.original$. We only need to consider the original objects in survivors, because the modified objects are short lived and therefore unlikely to survive (§7). We denote $in.young.original = in.eden.original + in.survivor.original$.

The modified allocations will cause the eden to fill up faster until a modified young collection is triggered, this is simply the moment when $in.eden.original + in.eden.modified$ reaches $max.size.eden$.

Estimating Surviving Amount. When the modified young collection is triggered, the young generation will contain both original and modified objects. To estimate the modified amount of surviving objects, we rely on the knowledge of the average amount of surviving objects in original collections of rank r, or $surviving.average_r$. We define $surviving.interpolated(x)$ as an interpolation of $surviving.average$ for allocated amount $x \in 0 \dots max.eden.size \cdot (tenuring.threshold + 1)$:

$$surviving.average_0 = 0$$
$$r.complete = x \operatorname{div} max.size.eden$$
$$r.partial = x \operatorname{mod} max.size.eden$$

$$surviving.interpolated(x) = surviving.average_{r.complete} +$$
$$+ \frac{r.partial}{max.size.eden} \cdot \left(surviving.average_{r.complete+1} - surviving.average_{r.complete}\right)$$

Given an entirely stable allocation behavior (§8), we could set $surviving.modified = surviving.interpolated(in.young.original)$, using the interpolation directly. To support some fluctuations in survival behavior, however, we further adjust the estimate by looking at the original survival behavior in the nearest young collection. Specifically, for a modified young collection of rank r, we find the nearest original young collection of the same rank r. We then look at how the amount of surviving objects in this original collection differs from the average amount of surviving objects and adjust the estimate accordingly:

$$surviving.original = log.heap.after_r - log.heap.before_r + log.young.before_r$$
$$scale = \frac{surviving.original}{surviving.average_r}$$
$$surviving.modified = surviving.interpolated(in.young.original) \cdot scale$$

Estimating Promoted Amount. After estimating the modified amount of surviving objects, we estimate the modified amount of promoted objects. Premature promotions aside, a modified young collection can only promote objects if its index exceeds the tenuring threshold. For such collections, we compute the promotion rate of the nearest original young collection:

$$promotion.rate = \frac{log.tenured.after_r - log.tenured.before_r}{surviving.original}$$

When the nearest original young collection involved premature promotions, we instead use the average promotion rate computed from $surviving.interpolated$:

$$surviving.all.lifetimes = surviving.interpolated(in.young.original_r)$$
$$surviving.except.oldest = surviving.interpolated(in.young.original_{r-1})$$
$$promotion.rate = \frac{surviving.all.lifetimes - surviving.except.oldest}{surviving.all.lifetimes}$$

Finally, we adjust the aggregate size of objects in survivors and in the tenured generation. The survivors will hold $min(surviving.modified \cdot (1 - promotion.rate), max.size.survivor)$ bytes, the rest is promoted.

Triggering Full Collections. After estimating the promoted amount for the modified young collection, we update the weighted average of the promoted amount, which serves as the reserve for triggering full collection (§5). For this, we simply reproduce the formulas used in the virtual machine sources.

Finally, we test whether a modified full collection is triggered. If it is, we find the nearest original full collection and use the size of the tenured generation after this collection as the size of the tenured generation after the modified full collection. This estimate is possible because the modified objects are short lived and the tenured generation therefore contains mostly original objects (§7).

3.3 Estimating Collection Time

Given the estimate of the modified GC behavior, we complete the model with estimates of the modified GC time. This is trivial for the full collections— although the modified collections may differ from the original collections in frequency, they traverse and collect mostly original objects in similar amounts. As a consequence, we estimate that each modified full collection takes about as much time as the nearest original full collection.

For the young collections, we rely on the empirical observation that the young collection time is strongly correlated with the number of live objects. We use total amount in place of total number of objects and estimate the modified time based on the time of the nearest original young collection with the same rank:

$$surviving.original_r = log.heap.after_r - log.heap.before_r + log.young.before_r$$
$$time.modified_r = time.original_r \cdot \frac{surviving.modified_r}{surviving.original_r}$$

4   Evaluation and Discussion

The primary goal of our evaluation is to understand and explain what makes or breaks the model. Towards this, we present and discuss the results of using the model on two workloads. With one workload, the model works reasonably well, especially given the quality of the inputs. With the other workload, the results are notably worse. This is interesting because the workloads represent similar applications, but differ in how they satisfy the model assumptions on application behavior. In particular, the first workload has a stable allocation behavior, while the second workload has two alternating phases, each allocating objects with significantly different demographics.

4.1   Methodology and Metrics

Our evaluation is based on comparing the measured and predicted values of metrics that capture the amount of GC work. The key metrics are the total number of young and full collections and the total time spent doing young and full collections (in seconds). We also collect internal model metrics which serve as the basis for the high-level metrics—the average amount of data surviving young collections, the average amount of data promoted in young collections, and the average tenured generation occupancy before and after full collections (all in bytes). These enable better understanding of the results, especially in the cases where the model loses accuracy.

In each experiment, we first execute the original workload with a workload-specific GC configuration[2] that conforms to (§4). The JVM is instructed to produce a GC log (§6). The planned modification locations are instrumented per (§9), the instrumentation is carefully designed to avoid allocating any heap memory. The results of this run provide inputs for the model as well as baseline data for evaluating the real effect of the added allocations.

In the second step, the workload is modified to allocate more data at the designated locations and run using the same JVM configuration. The allocated data is a single integer array of configurable length, and is only used in the scope of the modified code, thus increasing the allocation rate of the workload without increasing the steady state live size. The results from this run provide data for establishing the ground truth regarding the effect of the added allocations.

Third, we solve the model using data from the original workload execution.

4.2   Workloads

Ideally, our evaluation workloads would be standard benchmarks. Unfortunately, this runs into difficulties—SPECjvm2008 does not exhibit an interesting allocation behavior in our context, SPECjEnterprise2010 is extremely unwieldy and not well supported with open source technologies (proprietary platforms restrict

---

[2] We manually fix the min and max heap sizes, the size of the young generation space, the ratio of the eden and survivor spaces, and the tenuring threshold.

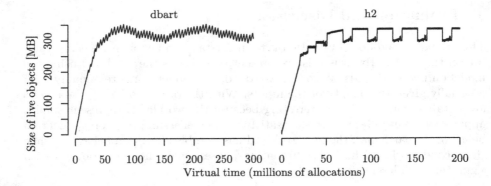

Fig. 1. Live size, dbart and DaCapo h2 workload, partial

result publication), and the DaCapo [3] suite breaks our requirements with often rather low heap occupancy and distinct sawtooth patterns in the live size profile (owing to naive use of iterations) [8].

The h2 workload from DaCapo comes closest to the live size stability assumption (§8), but is still rather uninteresting in a single iteration—with a reasonable[3] heap size, the workload does not trigger any full collection. We therefore use the h2 workload with 400 iterations, default input size, 1 thread, and no forced GC. We also create a modified benchmark that performs additional allocations in each transaction, with a total of 13 479 600 added allocations. The JVM configuration uses 256 MB tenured generation, 96 MB young generation (32 MB eden, two 32 MB survivor spaces), and a tenuring threshold of 7.

Even in the above configuration, the h2 workload still violates some of the model assumptions—the most notable being the presence of two distinct phases in every iteration. In the work phase, h2 allocates objects representing database records that contribute to the global state. In the cleanup phase, which restores the initial state of the database, h2 allocates only very short-lived objects that do not survive even one collection. Both phases are clearly visible in the GC log, and in the live size trace shown in Figure 1. Moreover, the workload design (together with multiple iterations) causes the full collections to synchronize with the end of some iterations (every second one in our case). The full collections always happen approximately in the middle of the cleanup phase, and changing the size of the tenured generation only influences their frequency, but not the point at which they occur during the iteration.

To evaluate the model on a workload that better conforms to the underlying assumptions, we created a benchmark simulating a simple university information system accessing an in-memory database. The benchmark, called dbart, operates on entities such as students, courses, schedules, and grades. Its live size trace is more stable, as shown in Figure 1. The modified benchmark performs additional allocations in the operation that records courses for a student, with a total of

---

[3] Heap size 10 % above the minimal heap size required by the workload.

5 935 084 added allocations. The JVM configuration uses 448 MB tenured generation, 192 MB young generation (64 MB eden, two 64 MB survivor spaces), and a tenuring threshold of 4.

4.3   Measurement Platform and Results

We conducted the experiments on two different hardware platforms. The h2 workload was executed on a Dell PowerEdge M620 system[4] and OpenJDK 1.8.0_25 JVM[5]. The dbart workload was executed on a Dell PowerEdge 1955 system[6] and Oracle HotSpot 1.8.0_11 JVM[7]. In both cases, the JVM was configured to use a single GC thread to make the GC times stable.

To evaluate the model for different amounts of additional allocations, the modified workloads were executed in two configurations, using arrays of 2^{11} (2K) and 2^{12} (4K) elements as a base allocation unit. This corresponds to allocating 8 208 and 16 400 extra bytes (including object header and alignment), respectively, at each instrumented workload location.

The measurement and model evaluation results are summarized in Table 1. For each workload, the table shows the key metrics corresponding to the execution of the original workload, followed by results for the two modified workloads. For each modified workload, the table shows metrics obtained by measurement and by evaluating the model. Table 2 summarizes the accuracy of the model in form of prediction errors for the metrics that serve as a basis for calculating collection counts and durations. For each modified workload configuration, the table shows two prediction errors. The *error.wrt.base* is calculated as $\left|1 - \frac{modified.model}{modified.measurement}\right|$, and expresses the relative difference between the measured and predicted metrics for the modified workload. The *error.wrt.change* is calculated as $\left|1 - \frac{original.measurement - modified.model}{original.measurement - modified.measurement}\right|$, and expresses the error made in predicting the change in GC work. This error is not calculated for the tenured amounts before and after full collection, because the high-level metrics such as collection counts and times are influenced by the difference between the amounts before and after full GC, but not by the difference between the original and modified workloads.

4.4   Results Discussion

The results for the dbart workload are encouraging. The predicted young collection counts are very close to the measured values, which is expected [8]. The prediction accuracy for the survived and promoted amounts, summarized as averages in Tables 1 and 2, can be considered reasonable, as illustrated in Figures 2 and 3, which plot the individual predicted values.

---

[4] 48 GB RAM, two Intel Xeon E5-2660 (Sandy Bridge) chips, 16 processors, NUMA.

[5] OpenJDK Runtime (1.8.0_25-b18) and OpenJDK 64-Bit VM (25.25-b02).

[6] 24 GB RAM, two Intel Xeon E5345 (Clovertown) chips, 8 processors.

[7] Java(TM) SE Runtime (1.8.0_11-b12), and Java HotSpot[TM]64-Bit VM (25.11-b03).

Table 1. Measured and modeled results

Configuration	Young collection Time [s]	Count	Full collection Time [s]	Count	Survived young[1] [MB]	Promoted[2] [MB]	Tenured at full GC start[3] [MB]	Tenured after full GC end[4] [MB]
dbart								
Original	1418.8	2128	1556.2	203	55.48	11.18	428.73	313.13
2k – measurement	1629.7	2856	1563.1	206	48.64	8.92	431.88	309.81
2k – model	1604.9	2854	1557.8	203	47.96	8.99	438.20	313.38
4k – measurement	1744.7	3587	1617.0	214	40.15	7.74	436.52	308.67
4k – model	1641.6	3579	1770.9	231	38.65	8.11	437.38	313.19
h2								
Original	189.6	9762	196.4	200	8.54	1.05	249.08	198.54
2k – measurement	234.0	13395	197.4	200	6.59	0.81	252.30	198.53
2k – model	70.2	13057	196.4	200	2.42	0.87	254.73	198.54
4k – measurement	255.2	16405	197.2	200	5.47	0.70	254.94	198.53
4k – model	68.2	16348	196.4	200	1.84	0.75	258.98	198.54

[1] Total size of objects that survived young collection and stayed in survivor space (averaged over all young GCs).
[2] Total size of objects that were promoted to tenured space in the young collection (averaged over all young GCs).
[3] Total size of objects in tenured space at the moment when full collection started (averaged over all full GCs).
[4] Total size of objects in tenured space right after the full collection (averaged over all full GCs).

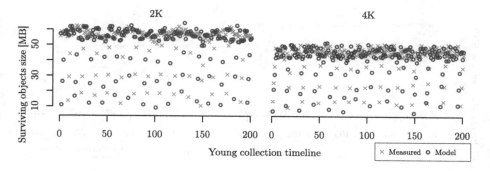

Fig. 2. Size of objects in survivor space after young collections, dbart workload, partial

The lower prediction accuracy for the young collection times and full collection counts can be primarily attributed to errors in predicting the survived and promoted amounts—these errors are additive and accumulate over all young collections, of which there are thousands. The errors also influence each other, e.g., over-estimating the promoted amount causes under-estimating the survived amount, which in turn results in under-estimating the total young collection time (which has a linear dependency on live size).

The accuracy of full collection count estimates is also influenced by the estimates of free space in the tenured generation, which is based on the estimated amounts of tenured objects before and after a full collection. We again consider these estimates reasonably accurate, as illustrated in Figure 4.

To explain why we consider the high-level results generally encouraging, consider the 4.82 % error in the prediction of the promoted amount for the 4K variant of the dbart workload. To predict the full collection count exactly, the promoted amount would need to be predicted with error no more than 0.47 %, which we consider impossible given the model inputs.

The results for the h2 workload are considerably less accurate. Similarly to the dbart workload, the individual predicted values of selected metrics are shown in Figures 5–7. The predicted full collection counts and times basically match the measured values, which we consider a coincidence of two errors canceling each other (over-estimating the promoted amount by 7 % and compensating by over-estimating the free space in the tenured generation).

The match in the tenured amount after full collection is not really surprising—the model estimates the value using data from the nearest full collection occurring in the original workload, and due to the full collections always occurring at the same point in the cleanup phase, there is basically no room for observing different values. The tenured amount before full collection is predicted with reasonably low error, given the complexity of the tenured space reserve calculation.

The estimate of the young collection time is rather inaccurate. This is due to significant under-estimation of the survived amount, which is in turn

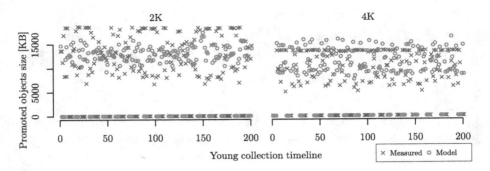

Fig. 3. Size of promoted objects in young collections, dbart workload, partial

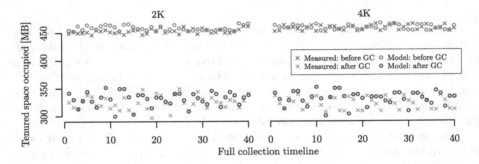

Fig. 4. Tenured space sizes before and after full collections, dbart workload, partial

caused by the presence of the alternating workload phases and the fact that a full collection always occurs at the same point in the cleanup phase. Because the cleanup phase allocates extremely short-lived data, the model observes $surviving_i \approx 0, i \in \{1 \ldots 5\}$, which severely disrupts the $surviving.average$ calculation—a major contributing factor to the survived amount estimation.

5   Related Work

Our work complements research on predictive performance models in software development, where frameworks such as [1] can be extended with platform specific models of GC overhead to increase prediction accuracy. In this context, we are aware of no models that would estimate GC overhead of a production VM accurately—as explained in our earlier work [8,9], GC overhead is sensitive to many minute details and accurate estimates may be infeasible in practice.

GC overhead modeling is also useful in the context of VM configuration, where appropriate GC settings are often tuned heuristically. Vengerov [12] introduced an analytical GC model to optimize the generation sizes and the tenuring threshold. The model is to be incorporated in the VM runtime, obtaining

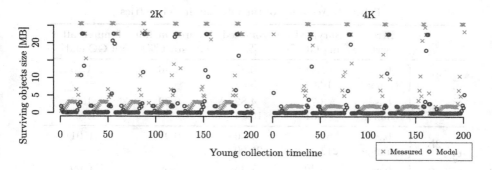

Fig. 5. Size of objects in survivor space after young collections, h2 workload, partial

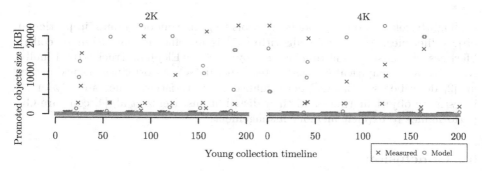

Fig. 6. Size of promoted objects in young collections, h2 workload, partial

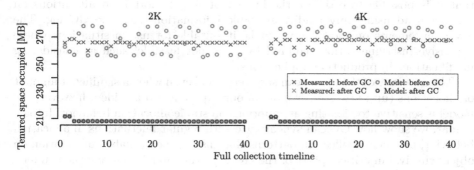

Fig. 7. Tenured space sizes before and after full collections, h2 workload, partial

its inputs by observing application execution. In a related modeling work, Li et al. [7] define a windowed liveness metric, which can be used to derive optimal heap sizes and minimal survival rates in applications. This rate estimate is, however, significantly lower than the actual rate.

Table 2. Accuracy of the internal model metrics

		Error wrt	Survived young GC [%]	Promoted [%]	Tenured at full GC start [%]	Tenured after full GC end [%]
dbart	2K	base	1.41	0.85	1.46	1.15
		change	10.04	3.34	—	—
	4K	base	3.75	4.82	0.20	1.47
		change	9.81	10.84	—	—
h2	2K	base	63.22	7.36	0.97	0.01
		change	213.86	25.00	—	—
	4K	base	66.40	7.45	1.59	0.00
		change	118.52	14.60	—	—

Tightly related to GC models are tools to gather model inputs, in particular object lifetimes. The Merlin algorithm [5] is an efficient algorithm to collect lifetimes, the most current tool inspired by Merlin is ElephantTracks [10]. Finally, research often empirically investigates general GC performance. Blackburn et al. [2] describe essential GC performance characteristics, Jones and Ryder [6] investigate object lifetimes and their distributions, Hertz et al. [4] compare GC performance to explicit memory management.

6   Conclusion

Although GC overhead is generally difficult to model accurately, we demonstrate that it is possible to estimate the impact of added short lived allocations on GC overhead using only readily available information from the GC log. The low requirements on model input make it suitable for use in situations where expensive instrumentation required by [8] is not feasible, such as considering modifications in production applications.

The accuracy of the model can stay very good even when a significant amount of allocations (up to two thirds more in our experiments) is added, however, the model is sensitive to violating assumptions on stable allocation behavior. In particular, we show how GC can synchronize with regular fluctuations in allocation behavior, producing collection patterns that distort the available information on object survival in critical parts of the GC log—the model accuracy then suffers.

Our work is provided together with complete data and tools, available at http://d3s.mff.cuni.cz/resources/epew2015.

Acknowledgments. This work has been supported by Charles University institutional funding SVV-2015-260222, by EU project 257414 ASCENS and by the Research Group of the Standard Performance Evaluation Corporation (SPEC).

References

1. Becker, S., Koziolek, H., Reussner, R.: The Palladio component model for model-driven performance prediction. J. Syst. Softw **82**(1) (2009)
2. Blackburn, S.M., Cheng, P., McKinley, K.S.: Myths and realities: The performance impact of garbage collection. Perform. Eval. Rev. **32**(1), 25–36 (2004). http://doi.acm.org/10.1145/1012888.1005693
3. Blackburn, S.M., et al.: The DaCapo benchmarks: Java benchmarking development and analysis. In: OOPSLA, pp. 169–190 (2006). http://doi.acm.org/10.1145/1167473.1167488
4. Hertz, M., Berger, E.D.: Quantifying the performance of garbage collection vs. explicit memory management. In: OOPSLA, pp. 313–326 (2005). http://doi.acm.org/10.1145/1094811.1094836
5. Hertz, M., Blackburn, S.M., Moss, J.E.B., et al.: Generating object lifetime traces with Merlin. ACM Trans. Program. Lang. Syst. **28**(3), 476–516 (2006). http://doi.acm.org/10.1145/1133651.1133654
6. Jones, R.E., Ryder, C.: A study of Java object demographics. In: ISMM, pp. 121–130 (2008). http://doi.acm.org/10.1145/1375634.1375652
7. Li, P., Ding, C., Luo, H.: Modeling heap data growth using average liveness. In: ISMM, pp. 71–82 (2014). http://doi.acm.org/10.1145/2602988.2602997
8. Libič, P., Bulej, L., Horký,V., Tůma, P.: On the limits of modeling generational garbage collector performance. In: ICPE, pp. 15–26 (2014). http://doi.acm.org/10.1145/2568088.2568097
9. Libič, P., Tůma, P., Bulej, L.: Issues in performance modeling of applications with garbage collection. In: QUASOSS, pp. 3–10 (2009). http://doi.acm.org/10.1145/1596473.1596477
10. Ricci, N.P., Guyer, S.Z., Moss, J.E.B.: Elephant tracks: portable production of complete and precise GC traces. In: ISMM, pp. 109–118 (2013)
11. Sun Microsystems Inc: Memory management in the Java HotSpot virtual machine (2006). http://www.oracle.com/technetwork/java/javase/memory management-whitepaper-150215.pdf
12. Vengerov, D.: Modeling, analysis and throughput optimization of a generational garbage collector. In: ISMM, pp. 1–9 (2009). http://doi.acm.org/10.1145/1542431.1542433

Performance Awareness in Java EE Development Environments

Alexandru Danciu[1]($\boxtimes$), Alexander Chrusciel[2], Andreas Brunnert[1],
and Helmut Krcmar[2]

[1] fortiss GmbH, Guerickestrasse 25, 80805 München, Germany
{danciu,brunnert}@fortiss.org
[2] Technische Universität München, Boltzmannstr. 3, 85748 Garching, Germany
{alexander.chrusciel,krcmar}@in.tum.de

Abstract. The earlier performance problems are detected, the easier
they can be solved. Performance evaluations during the implementation phase of software projects cause overhead for developers. Unless
performance evaluations are highly automated, they are not adopted
in practice. This paper presents an approach to introduce performance
awareness in Java Enterprise Edition (EE) integrated development environments (IDE) by providing automated model-based performance evaluations. The approach predicts response times of Java EE component
operations and provides feedback to the developer within the IDE.
Response time predictions are performed based on the component implementation and the response time of required services. The source code of
the component to be evaluated is parsed and represented as an abstract
syntax structure. This structure is then converted into a performance
model representing the control flow of component operations and calls
to required services. The response time of external calls is parameterized using monitoring data acquired by application performance monitoring (APM) tools from production systems. Developers are provided
with immediate feedback, if the estimated response time of a component
operation exceeds a predefined threshold.

Keywords: Performance awareness · Performance evaluation · Performance modeling · Palladio Component Model · Java Enterprise Edition

1 Introduction

Evaluating the performance of enterprise application systems in terms of
response time, resource utilization and throughput is a complex task which
requires deployable software artifacts and a realistic testing environment. Thus,
performance tests are often conducted late during the development phase. Evaluating the performance of their software artifacts continuously during implementation highly impacts the productivity of developers. Unless these activities are
highly automated, they are not adopted in practice. The concept of performance
awareness [16] addresses these challenges. It describes the availability of insights

© Springer International Publishing Switzerland 2015
M. Beltrán et al. (Eds.): EPEW 2015, LNCS 9272, pp. 146–160, 2015.
DOI: 10.1007/978-3-319-23267-6_10

on the performance of software systems and the ability to act upon them [16]. From a developer's perspective, the concept aims at providing methods and tools supporting them in improving the performance of the code they are currently developing.

Obtaining insights on the performance of enterprise application systems becomes increasingly difficult due to continuous developments in their architecture, governance, and life cycle [3]. Application system architectures evolve from monolithic structures to complex system of systems architectures implying a considerable amount of components, dependencies, environments, and deployments. The transparency on system structures cannot be easily obtained by individuals. The technical, organizational, and cultural diversity of architectures also leads to a more complex IT governance. The rights and obligations associated with individual components and functions are distributed across many organizational units. Accessing data and coordinating activities related to application system performance is, thus, more difficult. Finally, application systems are nowadays subject to a continuous shift between life cycle phases. Individual teams and components undergo life cycle phases in an independent rhythm. Ensuring timeliness of data and the existence of ever-new versions complicate the assessment of application systems.

Developers of component-based software systems specify loosely coupled components fostering separation of concerns and supporting reuse across the system. Existing components are, thus, reused by developers for implementing new functionality. Reuse thereby highly impacts the performance of components. The factors influencing the performance of software components are [11]: component implementation, required services, deployment platform, usage profile, and resource contention.

For gathering insights on the performance of components before these can be deployed, developers would have to investigate the implementation, required services, and the usage profile. The steps and tools required for developers to investigate the performance manually is shown in figure 1. In a first step, the developer needs to employ the integrated development environment (IDE) to extract the control flow of the component and to identify external calls from the

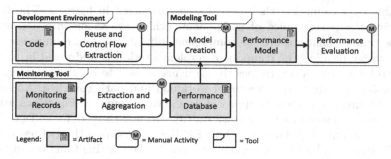

Fig. 1. Manual steps for evaluating the response time of a component which is currently being implemented

source code. For each external call, developers have to gather monitoring records and aggregate them to expectancy values. By employing a performance modeling tool, the developer needs to combine the knowledge about the control flow and the performance metrics for external calls to a performance model. Using this model, performance evaluations can then be performed.

This work presents an approach for integrating performance awareness in Java Enterprise Edition (EE) IDEs. The approach provides developers with response time estimations for the component operations they are currently implementing. Component developers are not required to have any expertise in the performance engineering domain or to apply corresponding tools. By automatically deriving component dependencies and providing access to the latest monitoring data for reused components, the approach attempts to overcome the challenges mentioned above. The contribution of this work is to provide and evaluate an actual implementation of the approach outlined in [6] only conceptually.

The remainder of this paper is organized as follows. Section 2 provides an overview of the approach and describes the phases of data collection and aggregation, performance model generation, and feedback provision in detail. The accuracy of the response time estimations provided by the approach is presented in section 3. Section 4 describes similar research performed in this area. Section 5 concludes this article and presents future research directions.

2   Performance Awareness Approach

The aim of the performance awareness approach presented in this paper is to support developers of Java EE components with response time estimations for individual component operations they are currently implementing without requiring additional effort. The Java EE specification distinguishes several types of components such as applets, Enterprise JavaBeans (EJB), servlets and JavaServer Pages (JSP) [7]. Components either run on the client or on the server. The approach is intended to support all types of server components which are developed using the Java programming language and reuse existing services.

The approach is intended to support developers with performance estimations for components they are currently implementing and which are not yet deployable by automating the manual steps displayed in figure 1. The factors influencing the performance of components which are taken into account for response time estimations are the component implementation, required services, and the usage profile. The component implementation is addressed in terms of the control flow, which determines the sequence of calls to required services. The usage profile is addressed in terms of workload and path coverage. An outline of the performance awareness approach is shown in figure 2. Functions are distributed over two separate tools, the Java EE IDE and the tool for collecting and aggregating monitoring data. The Java EE IDE is based on the Eclipse IDE for Java EE developers[1] adding several extensions to the existing functionality using the Eclipse plugin architecture. The user interface provided by the approach is

---

[1] http://eclipse.org/

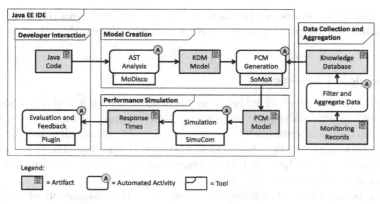

Fig. 2. Overview of the performance awareness approach

integrated in the existing source code editor. Developers are able to request response time estimations for any component within the project workspace. The corresponding source code is passed to the model creation module. MoDisco [2] is an open source Eclipse plugin used to parse the source code and to extract the control flow of the component operations. As a result, MoDisco creates a Knowledge Discovery Model (KDM) containing the abstract syntax tree (AST) of operations and an inventory of all software artifacts discovered in the code. This KDM model serves as input for the SOftware MOdel eXtractor (SoMoX) [12]. SoMoX is an open source Eclipse plugin used to reverse engineer the component architecture of software systems based on static analysis. The Palladio Component Model (PCM) [14] serves as meta-model for the output of SoMoX. PCM enables analyzing the performance of a software system without actually having to run the system. Several model layers are used to represent the user workload, the system behavior and the component deployment. Response times of reused components are also modeled and parameterized with values provided by the monitoring tool.

The monitoring tool represents a central server application responsible for the provision of performance measurements. Running Java EE applications are monitored using the Kieker framework [9]. Kieker supports the instrumentation of Java classes and stores performance measurements in local monitoring records. These records are transferred to a central database application. Within the database application, these measurements are filtered and aggregated. Aggregated performance data is provided to clients via a web-service interface.

The resulting PCM model serves as input for the performance simulation module. SimuCom [1] implements an engine for simulating the execution of a system specified using PCM. During a simulation run, the response time of component operations is simulated. Both the creation of performance models and the simulation are performed in the background of the Java EE IDE so that the workflow of the developer is not interrupted considerably. The collected measurements are then evaluated and used to provide feedback on the expected response time of component operations to the developer within the code editor. The modules of the performance awareness approach are described in detail below.

2.1 Developer Interaction

There are different types of interaction between developers and our approach. Developers are able to request a response time estimation for the component they are currently developing by performing a right mouse click on the corresponding class object within the package explorer. In the resulting pop-up menu, an according button is displayed. By clicking this button, the model creation and performance simulation are triggered.

The response time estimations resulting from the simulation are reported back to the developer. If the estimation for a method exceeds a predefined threshold, a notification is displayed in the code editor next to the method declaration. Notifications are displayed as small icons on the left side of the editor, similar to syntax error notifications. Two different thresholds are defined. If the first threshold is exceeded, a yellow icon is displayed. When the second threshold is exceeded, a red icon is displayed. Thresholds are implemented as plugin preferences within the IDE using default values and can be adjusted by the developer.

Before requesting response time estimations, developers are able to supply additional knowledge on the expected runtime behavior of the component operations. The actual control flow of operations at runtime heavily depends on the outcome of *if* statements and the number of iterations loops are passed through. If one outcome of an *if* statement consumes more time than the other, than the probabilities of the outcomes have a high impact on the accuracy of response time estimations. Similar, the more times a loop is passed through, the more time is consumed. The performance awareness approach assumes by default that outcomes of *if* statements have an equal probability and loops are passed through only once. The expected behavior can be specified by developers using annotations within the source code. Annotations can be added either to operation parameters or to local variables. Boolean attributes can be annotated with the probability of having the value *true*. List attributes can be annotated with the expected number of elements. Wherever these attributes are used in the control flow as arguments of an *if* statement or a loop, the generated PCM models are parameterized accordingly.

2.2 Model Creation

During this phase several models which depend on one another are created sequentially. The most important aspects during the creation of each model are described below.

KDM Model Discovery. MoDisco receives a Java project as input and creates a KDM model representation of the discovered artifacts. The scope of the artifacts to be analyzed can be limited using a regular expression. Classes located in standard Java packages are excluded from the analysis by default.

Repository Model Creation. Based on the KDM model extracted from the source code, a repository model is created. The repository model depicts the components of a system, their relationships, as well as resource demands. Relationships between components are represented using interfaces. A component can both provide and require interfaces. In the context of Java EE components may implement multiple interfaces, which in turn inherit operations from the same superinterface. The existing SoMoX implementation was enhanced to support such complex scenarios while avoiding the existence of duplicate component operations. Also private and protected operations were previously not represented within interfaces. When a component calls itself, these operations must, however, be present.

Initially, each class identified in the KDM model is considered as a component candidate. SoMoX attempts to reduce the set of candidates first by merging candidates to a component of higher abstraction level, and then by composing them to a composite component [12]. Classes implementing servlets or enterprise beans are considered in the context of Java EE as implicit components [7]. Our performance awareness approach, therefore, considers each class as an individual component. Merging and composing of component candidates is not performed.

In a next step, SoMoX removes candidates meeting specific criteria from the set. The existing implementation was enhanced to remove candidates representing external components. External components are called by the component which is currently being implemented by the developer. They should not be depicted within the repository model, but are represented by their response time values. However, components sometimes use helper classes, which outsource recurring functions, and cannot be regarded as independent external components. Figure 3 shows an example for this scenario, where *IntComp1* represents the analyzed component and *IntComp2* represents the helper class. Helper classes are often developed simultaneously to the analyzed component and therefore no response time measurements exist for their operations. Helper classes may, in turn, call external components as shown in figure 3a. The approach should therefore consider the control flows of the analyzed component and the one of helper classes together. As shown in figure 3b, candidates representing helper classes are, thus, kept. All classes located within the package of the analyzed component are identified as helper classes. The existing SoMoX implementation removes component candidates representing data objects, e.g., classes having only getter and setter methods. Operations executed on values returned by data objects are then, however, no longer possible. SoMoX was adapted to keep these candidates. The remaining candidates are later converted into actual components.

System Model Creation. The components of the repository model are assembled to a system, as shown in figure 3c. The system model depicts the boundaries and the interfaces of the system. For each component in the repository model SoMoX creates a corresponding instance called assembly context. Assembly connectors are created for each matching pair of provided and required interfaces.

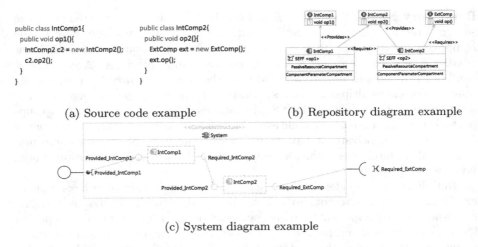

(a) Source code example   (b) Repository diagram example

(c) System diagram example

Fig. 3. Component representation example

The existing SoMoX implementation exposes all interfaces provided by component instances as interfaces provided by the system. The implementation was enhanced so that only interfaces of the analyzed component are exposed as system provided interfaces. Additionally, all required interfaces missing a matching provided interface used to be connected to an artificially created component. A match is missing when external components which are not represented in the repository model are called. Instead of creating an artificial component, the required interfaces are exposed to the system boundary and mapped to response time values for the external component operations.

Resource Environment and Allocation Model Creation. The resource environment model depicts containers where system components are hosted. Containers provide resources such as CPU to components. As our approach ignores resource contention as a factor influencing performance (refer to section 1), the design of the resource environment has no impact on the performance evaluation results. Therefore, the resource environment model consists of a minimum configuration, comprising one server and a single CPU. An allocation model is created for mapping the system to this hardware environment.

Specification of Performance Curves. The response time behavior of external components is specified using performance curves [18]. Performance curves describe the performance of component operations in dependence of their usage or configuration. The SoMoX implementation was extended to iterate through all interfaces required by the system and search for a corresponding performance curve within the performance measurement database.

Usage Model Creation. The existing SoMoX implementation was also enhanced to support the creation of a usage model, which specifies how users interact with the system. The approach identifies all interfaces provided by the system and creates for each one a corresponding transition. Transitions are accessed by users with an evenly distributed probability within a closed workload. Each transition contains a call to an operation of the analyzed component. The number of users and the user think times are specified as plugin preferences within the IDE.

Resource Demanding Service Effect Specifications (RDSEFF) Creation. At the end of the model creation process, for each component operation, a RDSEFF diagram is created. RDSEFF diagrams represent the control flow of operations depicting conditions, loops, internal and external calls. The existing SoMoX implementation was enhanced to support several new functions.

The approach now supports the extraction of annotations from code. Data on the expected parameterization of an operation provided by the developer using annotations is used to parameterize the model (refer to section 2.1). The approach was also enhanced to support the representation of session-level caching. Operations known to cache results within the context of a user session are modeled as branches. When executing the operation for the first time, a transition containing an external call is selected. For subsequent executions, an empty transition is selected. Additionally, the approach was enhanced to support method chaining, calls within return statements and a more fine-grained identification of interfaces providing specific operations. Also, the handling of composed statement containing both internal and external calls was implemented.

2.3   Data Collection and Aggregation

The existence of response time measurements for reused components is a main requirement of the presented approach. We propose the collection of monitoring data from productive environments. Applications are instrumented by Kieker using aspect oriented programming. Manual changes to the application code are, therefore, not necessary. Additionally, the instrumentation can be activated only for certain operations. Kieker was extended with a custom monitoring writer responsible for passing monitoring records to the central database application. Monitoring records consist of response times and the current workload of the application in terms of resource utilization and queue length. Performance measurement can be collected from several application servers for different workloads. It is also possible, that measurements for different versions of the same component are collected. The monitoring writer, therefore, collects information on the deployment where the measurements were performed, including information on the host, the application server instance and meta-data of application binaries. This enables the database application to aggregate performance measurements collected from the same environment and component version. The data model used for storing performance measurements is based on the Software Metrics Meta-Model (SMM) [13]. Records are aggregated to average values, percentiles, and formulas describing the response time of an operation in dependence of the workload. A web-service interface handles requests for aggregated values from clients.

2.4 Performance Simulation

During this phase the workload specified in the usage model is executed on the RDSEFFs representing the component operations. SimuCom first transforms the provided PCM models to Java code and then executes this code. The interaction of users with the system is performed by a corresponding thread. For each component operation, a response time sensor is created. Sensors measure and store response times during the execution of the generated code. Values used to parameterize performance curves, e.g. the current queue length, are determined by the simulation engine based on the current state of the simulation runtime environment. The simulation is executed until a specified time limit is reached or a certain amount of measurements has been collected. At the end of the simulation run, the sensors need to be matched to method declarations within the source code. Sensors representing operations of the analyzed component are matched to method declarations based on their signatures. Response times are then extracted from the identified sensors. Before reporting the results to the developer, average values are calculated.

3 Evaluation

Several assumptions and techniques underlie the performance awareness approach. The feasibility of the approach is evaluated in an experimental setup which aims at investigating the following aspects:

- How accurate are response time estimations for component operations while considering only the component implementation, reused services, and the usage profile?
- What impact does the type of average value used to represent response times of reused services have on the accuracy of estimations?
- What impact does the existence of annotations for boolean and list attributes have on the accuracy of estimations?

A prerequisite for applying the approach is the existence of Java EE components which reuse each other's services. Response time measurements for the reused components are also a requirement. For investigating the accuracy of the response time estimation, measurements for the calling component are needed. For meeting these requirements we employ the SPECjEnterprise2010[2] industry standard benchmark. The setup and the results of the experiment are described in the following sections.

---

[2] SPECjEnterprise is a trademark of the Standard Performance Evaluation Corp. (SPEC). The SPECjEnterprise2010 results or findings in this publication have not been reviewed or accepted by SPEC, therefore no comparison nor performance inference can be made against any published SPEC result. The official web site for SPECjEnterprise2010 is located at http://www.spec.org/jEnterprise2010.

3.1   Experiment Setup

The SPECjEnterprise2010 benchmark consists of a benchmark driver emulating workload of a car manufacturing company and a Java EE application [15]. The Java EE application itself is divided into three domains supporting business processes for customer relationship management, manufacturing and supply chain management. The Orders domain, supporting the supply chain management scenario, fulfills the previously stated requirements and is used as basis for this experiment.

Several types of Java EE components are implemented in the Orders domain, as shown in figure 4. Servlets receive requests from the benchmark driver. The business logic is realized by stateless EJBs which use Java Persistence API (JPA) entities for storing information. EJBs also use the JPA *EntityManager* and *Query* classes for reading and writing data from or to the database. Components of the lower layers are, thus, reused by components of the layer above. The layer of stateless EJBs is selected as the population for this evaluation. The experiment consists of selecting one component from this layer at a time and predicting the response time of each operation. This selection represents the component which is currently implemented by the developer. The set of reused components consists of every component within the Orders domain, which is called by the investigated component. Yet, not every component operation is suitable for the experiment. A preliminary investigation has revealed that several components share the cache provided by the *EntityManager*. For example, when the *OrderSession* retrieves an entity from the database using the *EntityManager*, the *CustomerSession* will retrieve the same entity directly from the cache. As a result, the response time of the corresponding *CustomerSession* operation will be much lower than expected. Component operations affected by this phenomenon are, thus, not eligible for the evaluation. The remaining operations which employ caching are still considered. Other component operations execute queries on the database. The response time of the operation then highly depends on the query string. These operations are assumed to represent data access services which don't reuse other components. These operations are also excluded from the evaluation. The remaining component operations, which are eligible for the evaluation are listed together with some of their characteristics in table 1. This set includes both very simple operations, consisting of only one source line of code (SLOC), and more complex

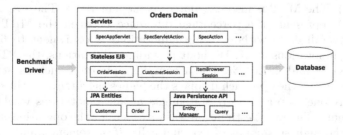

Fig. 4. Layers of the SPECjEnterprise2010 Orders domain

Table 1. Component operations selected for the experiment

Component	Operation	Input Parameters	SLOC	External Operations	If Statements	For Statements
CustomerSession	getCustomer	(Integer)	1	1	0	0
CustomerSession	getInventories	(Integer)	6	2	1	0
CustomerSession	sellInventory	(Integer, long, boolean)	17	5	3	0
CustomerSession	validateCustomer	(Integer)	2	1	0	0
LargeOrderSenderSession	sendOrdersToManufacturing	(List<OrderLine>, int)	26	5	0	1
OrderSession	cancelOrder	(int)	8	2	1	0
OrderSession	getItem	(String)	1	1	0	0
OrderSession	newOrder	(int, ShoppingCart, int, boolean, boolean)	96	22	8	3

ones, containing multiple *if* and *for* statements. The benchmark is executed for collecting response time measurements for the EJB operations and the reused components. While measuring the response times for EJB operations, an eventual overhead created by monitoring the persistence layer should be avoided. Therefore, each layer is instrumented individually and the benchmark is executed several times. Percentile values are used to represent response times of reused components. The evaluation is performed first using median and then using 95-percentile values. Input values for annotations are derived by executing the benchmark and observing the actual runtime behavior.

For measuring response times, the benchmark driver and the Java EE application are deployed on separate hosts in a virtualized environment. The hardware consists of an IBM System X3755 M3 server having 4 AMD Opteron 6172 processors, 12 2.1 GHz cores and 256 GB memory. Storage is provided by an IBM DS3512 SAN. The hardware is virtualized using VMWare ESXi 5.1.0. Virtual machines (VM) run openSUSE 12.3 as operating system. Each VM has 8 virtual processors configured. The VM hosting the Java EE server has 12 GB memory. The Java EE application is deployed on a JBoss 7.1.1 server.

3.2   Experiment Results

For interpreting experiment results, response times estimated by the performance awareness approach are compared to actual measurements. The response time of each component operation was estimated both with and without using annotations and using median values and 95-percentiles for representing response times of reused services. The simulated and the measured response times while using median values for representing response times of reused services are shown in table 2. Measured response times (MRT) are compared to simulated response times (SRT). The MRT is also calculated as median value. The response time error (RTE) represents the percentage difference between the MRT and the SRT. The overall error without using annotations is equivalent to 68 % and while using annotations 42 %. Without providing annotations, the RTE ranges between 9,84 % and 227,01 %. For a complex operation such as *sendOrdersToManufacturing*, the approach displays the highest accuracy. As the operation contains only one loop and no *if* statements at all, annotations would provide limited benefits. The worst estimation was provided for the operation *newOrder*, containing the highest amount of external calls, *if* statements, and loops. The second worst estimation was provided for the operation *getItem* which consists

Table 2. Evaluation results using median values for representing response times of reused services

Component	Operation	MRT	Without Annotations		With Annotations	
			SRT	RTE	SRT	RTE
CustomerSession	getCustomer	0.96 ms	0.71 ms	25.60 %	0.71 ms	25.60 %
	getInventories	0.80 ms	0.71 ms	10.63 %	0.71 ms	10.63 %
	sellInventory	0.91 ms	0.73 ms	19.87 %	0.73 ms	19.87 %
	validateCustomer	1.24 ms	0.71 ms	42.41 %	0.71 ms	42.41 %
LargeOrderSenderSession	sendOrdersToManufacturing	12.35 ms	11.13 ms	9.84 %	11.42 ms	7.52 %
OrderSession	cancelOrder	1.88 ms	0.72 ms	61.57 %	0.73 ms	61.13 %
	getItem	0.28 ms	0.71 ms	152.09 %	0.71 ms	152.09 %
	newOrder	1.83 ms	6 ms	227.01 %	2.30 ms	22.57 %

of only one external call of the *EntityManager* operation *find*. This is because the response time of the operation *find* highly depends on the type of entity which is looked up. In this case, finding an *Item* takes less time than for other entities. While using annotations, the estimation for the operation *newOrder* has dramatically improved. Other operations having less or no *if* and *for* statements have remained relatively constant.

The experiment results while using 95-percentiles for representing response times of reused services are shown in table 3. The overall error without using annotations is equivalent to 38 % and while using annotations 46 %. Regardless of the usage of annotations, the accuracy of most of the estimations has improved. However, in two cases, a significant deterioration can be observed. The estimation for *getItem* is now based on a more pessimistic assumption by using 95-percentiles and differs even more from the measurement. The estimation for *newOrder* is now worse when using annotations. When annotations are provided, the control flow is actually represented more accurately. However, by representing the MRT also as 95-percentile, the reference value of the estimation seems to be too pessimistic.

Table 3. Evaluation results using 95-percentiles for representing response times of reused services

Component	Operation	MRT	Without Annotations		With Annotations	
			SRT	RTE	SRT	RTE
CustomerSession	getCustomer	2.43 ms	2.22 ms	8.63 %	2.22 ms	8.63 %
	getInventories	2.14 ms	2.22 ms	3.92 %	2.22 ms	3.92 %
	sellInventory	2.35 ms	2.28 ms	2.84 %	2.28 ms	2.84 %
	validateCustomer	2.87 ms	2.22 ms	22.71 %	2.22 ms	22.71 %
LargeOrderSenderSession	sendOrdersToManufacturing	19.81 ms	21.92 ms	10.64 %	23.06 ms	16.39 %
OrderSession	cancelOrder	4.83 ms	2.25 ms	53.41 %	2.28 ms	52.83 %
	getItem	0.75 ms	2.22 ms	196.69 %	2.22 ms	196.69 %
	newOrder	14.77 ms	13.34 ms	9.71 %	5.30 ms	64.13 %

The experiment results show that the type of average values used both as input for the estimation and as reference for the estimation results highly impacts the accuracy of the approach. Average values should, therefore, be selected based on the skewness of the probability distribution of the corresponding operation's response time.

4 Related Work

A number of approaches propose automated and integrated means for supporting developers with performance awareness. Existing approaches either provide recorded observations [5,8,10], or performance estimations [17] to developers.

Heger et al. [8] present an approach to integrate performance regression root cause analysis into the development environment. When performance regressions are detected, the approach provides the developer with insights on the causes of the regression. Information on the change and the methods causing the regression is displayed to the developer. The performance measurements used by the approach are collected during the execution of unit tests. The evolution of the performance of affected methods is presented graphically as a function and the methods causing the regression are displayed as a graph.

Bureš et al. [5] propose the integration of performance evaluation and awareness methods into different phases of the development process of autonomic component ensembles. During the design phase, performance goals are formulated. At runtime, performance measurements are collected. The authors propose the presentation of these measurements to the developers within the IDE. Feedback on the measurements is displayed graphically as functions within a pop-up window.

Horký et al. [10] present an approach for enhancing the documentation of software libraries with information on their performance. The performance of libraries is measured using performance unit tests. Tests are executed on demand after the developer looks up a specific method for the first time. Tests can be executed locally or on remote machines. Measurements are then cached and enhanced.

Weiss et al. [17] propose an approach for evaluating the performance of persistence services based on tailored benchmarks during the implementation phase. By applying this approach, developers are able to track the impact of changes on the performance or to compare different design alternatives. The estimated response time is displayed within the IDE using numerical values and graphically as bar charts. The approach is only applicable for Java Persistence API services, however, instructions on how to design and apply benchmark applications to other components are also provided.

The actual performance behavior of applications can only be observed during runtime. At this point it may be, however, unclear whether the observed behavior also reflects the expected one. Bulej et al. [4] propose an approach for supporting the awareness of performance expectations by providing means to formulate, communicate and evaluate these expectations. The approach uses the Stochastic Performance Logic (SPL) to express performance assumptions for specific methods. Assumptions are represented independently from any hardware and are formulated relative to another method. During runtime, measurements are collected and compared to the formulated expectations. Potential violations are reported to the developer.

5   Conclusion and Future Work

In this work, we have presented an approach to introduce performance aware-ness in Java EE development environments by providing automated model-based performance evaluations. The approach performs response time estimations for component operations and provides feedback to the developer within the IDE. Employing the approach requires neither expertise in the performance engineer-ing domain nor additional effort from the developer.

Future research will investigate what improvements in the performance of component operations can be observed by supporting developers with the per-formance awareness approach. The benefits of employing this approach could be quantified by comparing the response times of implementations with same functionality achieved by two groups of developers - one with and one without performance awareness support. Another option could be to measure the amount of cases where the display of performance estimations to developers leads to an optimization of the source code. Further research will also investigate the accu-racy of estimations using more extensive experiment setups. The approach will be extended to support a more extensive interaction with the developer by pro-viding more expressive annotations and indicating expensive external calls using heat maps in the code editor. The accuracy of the approach will be improved by automatically selecting appropriate average values for response times of reused services based on the skewness of their probability distribution. Future work will also investigate if the duration of estimations can be improved by using an analytical approach instead of simulation.

Acknowledgement. This work has been supported by the Research Group of the Standard Performance Evaluation Corporation (SPEC).

References

1. Becker, S.: Coupled Model Transformations for QoS Enabled Component-Based Software Design. Karlsruhe Series on Software Quality, Universitätsverlag Karlsruhe (2008)
2. Brunelière, H., Cabot, J., Dupé, G., Madiot, F.: MoDisco: A model-driven reverse engineering framework. Information and Software Technology **56**(8), 1012–1032 (2014)
3. Brunnert, A., Vögele, C., Danciu, A., Pfaff, M., Mayer, M., Krcmar, H.: Performance management work. Business & Information Systems Engineering **6**(3), 177–179 (2014)
4. Bulej, L., Bureš, T., Keznikl, J., Koubková, A., Podzimek, A., Tůma, P.: Capturing performance assumptions using stochastic performance logic. In: Proceedings of the 3rd ACM/SPEC International Conference on Performance Engineering (ICPE 2012), pp. 311–322. ACM (2012)
5. Bureš, T., Horký, V., Kit, M., Marek, L., Tůma, P.: Towards performance-aware engineering of autonomic component ensembles. In: Margaria, T., Steffen, B. (eds.) ISoLA 2014, Part I. LNCS, vol. 8802, pp. 131–146. Springer, Heidelberg (2014)

6. Danciu, A., Brunnert, A., Krcmar, H.: Towards performance awareness in Java EE development environments. In: Becker, S., Hasselbring, W., van Hoorn, A., Kounev, S., Reussner, R. (eds.) Proceedings of the Symposium on Software Performance: Descartes/Kieker/Palladio Days 2014, pp. 152–159, November 2014
7. DeMichiel, L., Shannon, B.: Java platform, enterprise edition (Java EE) specification, v7 (2013)
8. Heger, C., Happe, J., Farahbod, R.: Automated root cause isolation of performance regressions during software development. In: Proceedings of the 4th ACM/SPEC International Conference on Performance Engineering, pp. 27–38. ICPE 2013. ACM, New York (2013)
9. van Hoorn, A., Waller, J., Hasselbring, W.: Kieker: A framework for application performance monitoring and dynamic software analysis. In: Proceedings of the 3rd ACM/SPEC International Conference on Performance Engineering, pp. 247–248. ICPE 2012. ACM, New York (2012)
10. Horký, V., Libic, P., Marek, L., Steinhauser, A., Tůma, P.: Utilizing performance unit tests to increase performance awareness. In: Proceedings of the 6th International Conference on Performance Engineering (ICPE 2015), pp. 289–300. ACM (2015)
11. Koziolek, H.: Performance evaluation of component-based software systems: A survey. Performance Evaluation 67(8), 634–658 (2010). special Issue on Software and Performance
12. Krogmann, K., Kuperberg, M., Reussner, R.: Using genetic search for reverse engineering of parametric behaviour models for performance prediction. IEEE Transactions on Software Engineering 36(6), 865–877 (2010)
13. OMG: Software metrics meta-model (2012, accessed at 14-10-21)
14. Reussner, R., Becker, S., Happe, J., Koziolek, H., Krogmann, K., Kuperberg, M.: The Palladio Component Model. Universität Karlsruhe (2007)
15. SPEC: SPECjEnterprise2010. http://www.spec.org/jEnterprise2010/ (2012, accessed at 12-04-07)
16. Tůma, P.: Performance awareness: keynote abstract. In: Proceedings of the 5th ACM/SPEC International Conference on Performance Engineering, pp. 135–136. ICPE 2014. ACM, New York (2014)
17. Weiss, C., Westermann, D., Heger, C., Moser, M.: Systematic performance evaluation based on tailored benchmark applications. In: Proceedings of the 4th ACM/SPEC International Conference on Performance Engineering, pp. 411–420. ICPE 2013. ACM, New York (2013)
18. Wert, A., Happe, J., Westermann, D.: Integrating software performance curves with the Palladio Component Model. In: Proceedings of the 3rd ACM/SPEC International Conference on Performance Engineering, pp. 283–286. ICPE 2012. ACM, New York (2012)

Modelling Techniques II

Canonical Form of Order-2 Non-stationary Markov Arrival Processes

András Mészáros[1,2]([⊠]) and Miklós Telek[1,2]

[1] Budapest University of Technology and Economics,
Magyar Tudósok krt. 2, Budapest 1117, Hungary
{meszarosa,telek}@hit.bme.hu
[2] MTA-BME Information Systems Research Group,
Magyar Tudósok krt. 2, Budapest 1117, Hungary

Abstract. Canonical forms of Markovian distributions and processes provide an efficient way of describing these structures by eliminating the redundancy of the general description. Canonical forms of order-2 stationary Markov arrival processes (MAPs) have already been established for both continuous and discrete time. In this paper we prove that the canonical form of continuous time MAPs can be naturally extended to their non-stationary generalisations. We also prove that the equivalence proven for order-2 stationary Markov arrival processes and rational arrival processes also holds for the non-stationary counterparts.

1 Introduction

Markov chain based stochastic models, and, among them, phase type distributions (PHs) and Markov arrival processes (MAPs), are used in a wide array of fields from healthcare [8] to risk theory [3] and, most notably, queueing theory [13,16,19,21]. One of the main benefits of using Markovian structures in queueing models is that they enable the application of the matrix analytic methodology [15], which provides a powerful tool for analysing these systems. When trying to model a real life system these Markovian structures have to be constructed by fitting to empirical data. Several fitting methods have been produced for this purpose. Some use special structures or heuristic fitting methods, e.g. [1,10,11,24], while other methods apply general optimisation techniques such as expectation maximisation [2,6,23], or a mixture of these two, e.g. [9,14]. Using special structures reduces the flexibility of the stochastic model, while using the general structure the efficiency diminishes due to the redundancy in the standard description of the respective stochastic models. This issue can be eliminated by the usage of canonical forms. The canonical form of a Markovian distribution or process is its unique representation that is defined by a minimal number of parameters. This means that every distribution or process has to have a one-to-one correspondence with a canonical form description. In the past years canonical forms for several Markovian structures have been devised. Canonical

This work is partially supported by the OTKA K101150 projects.

© Springer International Publishing Switzerland 2015
M. Beltrán et al. (Eds.): EPEW 2015, LNCS 9272, pp. 163–176, 2015.
DOI: 10.1007/978-3-319-23267-6_11

forms have been established for order-2 phase type distributions (PHs) [18,20] and stationary Markov arrival processes (MAPs) [4,17], and order 3 phase type distributions [12,20] in both continuous and discrete time.

The non-Markovian generalizations of these Markov chain based models, matrix exponential distributions and rational arrival processes, can be efficiently used for overcoming some limitations of the Markovian models. For example Markovian models with low coefficients of variation can be represented far more efficiently with non-Markovian generalizations [5]. In the analysis of these stochastic models it is important to determine whether the Markovian and the non-Markovian class of the same order has the same flexibility or not. In the former case there is no need for the investigation of more complex non-Markovian models. For stationary Markov arrival processes (MAPs) it has been proved that the order-2 Markovian class and the order-2 non-Markovian class are identical [4,17] (both in case of continuous time and discrete time models).

In this paper we focus on the non-stationary extension of (continuous time) order-2 Markov arrival processes and rational arrival processes and investigate their canonical representation and identity. The main contribution of the paper is that we show that the same canonical form is applicable as for the stationary MAPs and that the order-2 Markov arrival processes and rational arrival processes are identical.

The rest of the paper is organized as follows. The next section presents the necessary background of Markov arrival processes and rational arrival processes. Section 3 summarizes the existing results for stationary arrival processes of order 2. Section 4 and 5 present the new results, the canonical representation of non-stationary Markov arrival processes of order 2 and the equivalence of the Markovian and the non-Markovian classes. The paper is concluded in Section 6.

2   Theoretical Background

In this section we present the definitions and some basic characteristics of stationary and non-stationary Markov arrival processes and their non-Markovian generalizations.

Let $\mathcal{X}(t)$ be a point process on $\mathbb{R}^+$ with joint probability density function (joint pdf) of inter-event times $f(x_0, x_1, \ldots, x_k)$ for $k = 1, 2, \ldots$.

Definition 1. $\mathcal{X}(t)$ *is called a stationary rational arrival process if there exists a finite* $(\boldsymbol{H_0}, \boldsymbol{H_1})$ *square matrix pair such that* $(\boldsymbol{H_0} + \boldsymbol{H_1})\mathbb{1} = \boldsymbol{0}$ *(where* $\mathbb{1}$ *and* $\boldsymbol{0}$ *are the column vectors of ones and zeros, respectively, with appropriate size),*

$$\pi(-\boldsymbol{H_0})^{-1}\boldsymbol{H_1} = \pi, \quad \pi\mathbb{1} = \mathbb{1}, \tag{1}$$

has a unique solution, and for $\forall k \geq 0$, $x_0, \ldots, x_k$ *its joint pdf is*

$$f(x_0, x_1, \ldots, x_k) = \pi e^{\boldsymbol{H_0} x_0} \boldsymbol{H_1} e^{\boldsymbol{H_0} x_1} \boldsymbol{H_1} \ldots e^{\boldsymbol{H_0} x_k} \boldsymbol{H_1} \mathbb{1}. \tag{2}$$

In this case we say that $\mathcal{X}(t)$ *is a stationary rational arrival process (RAP) with representation* $(\boldsymbol{H_0}, \boldsymbol{H_1})$, *or shortly,* $RAP(\boldsymbol{H_0}, \boldsymbol{H_1})$.

Definition 2. *If $\mathcal{X}(t)$ is a stationary RAP($\boldsymbol{H_0}, \boldsymbol{H_1}$), where $\boldsymbol{H_0}$ and $\boldsymbol{H_1}$ have the following properties:*

- *$\boldsymbol{H_1}$ has only non-negative elements*
- *$\boldsymbol{H_{0ii}} < 0$, $\boldsymbol{H_{0ij}} \geq 0$ for $i \neq j$, $\boldsymbol{H_0}\mathbb{1} \leq 0$,*

then we say that $\mathcal{X}(t)$ is a stationary Markov arrival process (MAP) with representation $(\boldsymbol{H_0}, \boldsymbol{H_1})$, or shortly, MAP($\boldsymbol{H_0}, \boldsymbol{H_1}$).

The importance of the MAP class comes from the associated stochastic interpretation. Every MAP representation can be mapped to a continuous time Markov chain with generator $\boldsymbol{H} = \boldsymbol{H_0} + \boldsymbol{H_1}$ where $\boldsymbol{H_1}$ contains transition rates with arrivals and $\boldsymbol{H_0}$ contains transition rates without arrivals and the Markov chain starts from initial distribution π which is the stationary probability vector embedded at arrivals. In such a Markov chain (2) is the joint pdf of the inter-arrival times. We note here that an arbitrary $(\boldsymbol{H_0}, \boldsymbol{H_1})$ square matrix pair satisfying (1) does not necessarily define a valid RAP as (2) may still give negative values for some $x_0, \ldots, x_k$. If an $(\boldsymbol{H_0}, \boldsymbol{H_1})$ matrix pair fulfils the additional sign constraints of MAPs in Definition 2, however, then (2) is guaranteed to be positive for arbitrary $x_0, \ldots, x_k$ as can be seen from the mapping to Markov chains. One of the major advantages of MAPs to RAPs is exactly this difference.

RAPs (MAPs) have infinite different representations (as it is demonstrated below for the non-stationary case), i.e. matrix pair sets that give the same $f(x_0, x_1, \ldots, x_k)$ joint probability density function. The different representations might have different sizes [7]. The size of the smallest among those representations is referred to as the order of the RAP (MAP). The class of order n RAPs (MAPs) is denoted by RAP(n) (MAP(n)). From Definition 1 and 2 it follows that MAP(n)$\subseteq$RAP(n).

In Definition 1 and 2 the initial vector in (2), π, has to fulfil (1). That is why π is also referred to as the embedded stationary vector. By relaxing this constraint we obtain the class of non-stationary RAPs and MAPs.

Definition 3. *$\mathcal{X}(t)$ is called a non-stationary rational arrival process if there exists a finite $(\pi_0, \boldsymbol{H_0}, \boldsymbol{H_1})$ initial vector and square matrix pair triple such that*

$$\pi_0 \mathbb{1} = 1$$

$$f(x_0, x_1, \ldots, x_k) = \pi_0 e^{\boldsymbol{H_0} x_0} \boldsymbol{H_1} e^{\boldsymbol{H_0} x_1} \boldsymbol{H_1} \ldots e^{\boldsymbol{H_0} x_k} \boldsymbol{H_1} \mathbb{1}. \tag{3}$$

In this case we say that $\mathcal{X}(t)$ is a non-stationary rational arrival process (NRAP) with representation $(\pi_0, \boldsymbol{H_0}, \boldsymbol{H_1})$, or shortly, NRAP($\pi_0, \boldsymbol{H_0}, \boldsymbol{H_1}$).

Definition 4. *If $\mathcal{X}(t)$ is a non-stationary RAP($\pi_0, \boldsymbol{H_0}, \boldsymbol{H_1}$), where*

- *π_0 has only non-negative elements*
- *$\boldsymbol{H_1}$ has only non-negative elements*
- *$\boldsymbol{H_{0ii}} < 0$, $\boldsymbol{H_{0ij}} \geq 0$ for $i \neq j$, $\boldsymbol{H_0}\mathbb{1} \leq 0$,*

then we say that $\mathcal{X}(t)$ is a non-stationary Markov arrival process (NMAP) with representation $(\pi_0, \boldsymbol{H_0}, \boldsymbol{H_1})$, or shortly, NMAP($\pi_0, \boldsymbol{H_0}, \boldsymbol{H_1}$).

Similar to the stationary case every NMAP representation can be mapped to a continuous time Markov chain with generator $H = H_0 + H_1$ where the initial distribution is π_0, and every NRAP (NMAP) has infinite different representations, i.e. (π_0, H_0, H_1) sets that give the same $f(x_0, x_1, \ldots, x_k)$ joint probability density function. One way to get a different representation of an NRAP(π_0, H_0, H_1) with the same size is the application of the similarity transformation

$$\pi_0' = \pi_0 T, \quad H_0' = T^{-1} H_0 T, \quad H_1' = T^{-1} H_1 T, \tag{4}$$

where T is a non-singular transformation matrix with $T\mathbb{1} = \mathbb{1}$. The transformed representation gives the same joint pdf as

$$
\begin{aligned}
f(x_0, \ldots, x_k) &= \pi e^{H_0 x_0} H_1 \ldots e^{H_0 x_k} H_1 \mathbb{1} = \\
&= \pi T T^{-1} e^{H_0 x_0} T T^{-1} H_1 T \ldots T^{-1} e^{H_0 x_k} T T^{-1} T H_1 T^{-1} \mathbb{1} = \\
&= \pi e^{H_0' x_0} H_1' \ldots e^{H_0' x_k} H_1' \mathbb{1} = f(x_0, x_1, \ldots, x_k), \tag{5}
\end{aligned}
$$

where we used that $T^{-1}\mathbb{1} = \mathbb{1}$ (from $T\mathbb{1} = \mathbb{1}$).

The order of NRAPs and NMAPs is defined similarly as for RAPs and MAPs. The class of order n NRAPs (NMAPs) is denoted by NRAP(n) (NMAP(n)). From Definition 3 and 4 it follows that NMAP(n)$\subseteq$NRAP(n).

3   Previous Results for MAP(2) and RAP(2) Processes

Before discussing the canonical form of NMAP(2) we summarize the results on the canonical structure of MAP(2) from [4] as these will provide the basis for the subsequent argumentation.

For MAP(2)s one of the eigenvalues of matrix $(-H_0)^{-1} H_1$ is 1, since $\pi(-H_0)^{-1} H_1 = \pi$. The other eigenvalue is denoted by γ, for which we have $-1 \leq \gamma < 1$. Based on the sign of γ the following canonical forms can be applied.

Theorem 1. [4] If the γ parameter of the order-2 RAP(H_0, H_1) is

- *non-negative, then it can be represented in the following Markovian canonical form*

$$D_0 = \begin{bmatrix} -\lambda_1 & (1-a)\lambda_1 \\ 0 & -\lambda_2 \end{bmatrix}, \quad D_1 = \begin{bmatrix} a\lambda_1 & 0 \\ (1-b)\lambda_2 & b\lambda_2 \end{bmatrix}.$$

where $0 < \lambda_1 \leq \lambda_2$, $0 < a, b < 1$, $b \geq a\frac{\lambda_1}{\lambda_2}$, $\gamma = ab$, and the associated embedded stationary vector is $\pi = \begin{bmatrix} \frac{1-b}{1-ab} & \frac{b-ab}{1-ab} \end{bmatrix}$,
- *negative, then it can be represented in the following Markovian canonical form*

$$D_0 = \begin{bmatrix} -\lambda_1 & (1-a)\lambda_1 \\ 0 & -\lambda_2 \end{bmatrix}, \quad D_1 = \begin{bmatrix} 0 & a\lambda_1 \\ b\lambda_2 & (1-b)\lambda_2 \end{bmatrix},$$

where $0 < \lambda_1 \leq \lambda_2$, $0 \leq a \leq 1$, $0 < b \leq 1$, $b \geq a\frac{\lambda_1}{\lambda_2}$, $\gamma = -ab$ *and the associated embedded stationary vector is* $\boldsymbol{\pi} = \left[\frac{b}{1+ab} \ 1 - \frac{b}{1+ab} \right]$.

Theorem 2. *[4] For the MAP(2) and RAP(2) sets of point processes we have*

$$MAP(2) \equiv RAP(2).$$

The aim of this paper is to verify the existence of Theorem 1 and 2 for non-stationary processes, NMAP(2) and NRAP(2).

4   Canonical Form of Order-2 NMAP

In this section we present the canonical form of NMAP(2) and prove that such canonical form is Markovian for any valid NMAP(2).

Theorem 3. *An order-2 NMAP($\pi_0, \boldsymbol{H_0}, \boldsymbol{H_1}$) can be represented in the* $(\delta, \boldsymbol{D_0}, \boldsymbol{D_1}) = (\pi_0 \boldsymbol{T}, \boldsymbol{T}^{-1}\boldsymbol{H_0}\boldsymbol{T}, \boldsymbol{T}^{-1}\boldsymbol{H_1}\boldsymbol{T})$ *canonical form, where* $\boldsymbol{T}$ *is the transformation matrix which transforms* $\boldsymbol{H_0}$ *and* $\boldsymbol{H_1}$ *to the MAP(2) canonical form* $(\boldsymbol{T}^{-1}\boldsymbol{H_0}\boldsymbol{T}, \boldsymbol{T}^{-1}\boldsymbol{H_1}\boldsymbol{T})$.

The previous theorem simply means that the canonical form that was used for MAP(2)s can be used for NMAP(2)s with a natural extension to the initial vector. The Markovity of $\boldsymbol{D_0}$ and $\boldsymbol{D_1}$ is trivially satisfied because of Theorem 1, thus the proposed representation is Markovian if and only if the elements of δ are non-negative, which can be formally described as $\delta \boldsymbol{e}_i \geq 0$ for $i = 1, 2$, where $\delta = \pi_0 \boldsymbol{T}$ and $\boldsymbol{e}_i$ is the ith unit column vector (whose elements equal to zero except the ith one which is one). As π_0 is non-negative, δ will also be non-negative if the elements of $\boldsymbol{T}$ are non-negative. (This is a sufficient, but not a necessary condition.) In the following we show that $\boldsymbol{T}$ is indeed non-negative for any initial Markovian ($\pi_0, \boldsymbol{H_0}, \boldsymbol{H_1}$) representation.

Because every Markovian NMAP(2) representation can be obtained from the canonical forms using similarity transformations, the previous statement can be reversed to get the following equivalent: If ($\delta, \boldsymbol{D_0}, \boldsymbol{D_1}$) is an arbitrary NMAP(2) in canonical form, then its similarity transform

$$(\pi_0, \boldsymbol{H_0}, \boldsymbol{H_1}) = (\delta \boldsymbol{T}^{-1}, \boldsymbol{T}\boldsymbol{D_0}\boldsymbol{T}^{-1}, \boldsymbol{T}\boldsymbol{D_1}\boldsymbol{T}^{-1}) \tag{6}$$

is Markovian only if $\boldsymbol{T}$ is non-negative. In other words we "reverse similarity transform" the canonical form (note that here the transformation matrix is $\boldsymbol{T}^{-1}$ while in (4) it was $\boldsymbol{T}$) and examine what could the original representation be and prove that for every possible original representation that satisfies the MAP representation constraints in Definition 2, matrix $\boldsymbol{T}$ is non-negative. In the following we prove this last version of the theorem.

We have different canonical forms for negative and non-negative γ that we have to examine separately. Let us first consider NMAPs with non-negative γ. In this case the matrices of the canonical form are

$$\boldsymbol{D_0} = \begin{bmatrix} -\lambda_1 & (1-a)\lambda_1 \\ 0 & -\lambda_2 \end{bmatrix}, \quad \boldsymbol{D_1} = \begin{bmatrix} a\lambda_1 & 0 \\ (1-b)\lambda_2 & b\lambda_2 \end{bmatrix}.$$

Let

$$T = \begin{bmatrix} 1 - t_1 & t_1 \\ t_2 & 1 - t_2 \end{bmatrix}. \tag{7}$$

From (6) we get

$$H_0 = \begin{bmatrix} -\frac{(1-t_1)(1-at_2)\lambda_1 - t_1 t_2 \lambda_2}{1-t_1-t_2} & \frac{(1-t_1)((1-a(1-t_1))\lambda_1 - t_1 \lambda_2)}{1-t_1-t_2} \\ \frac{t_2((1-t_2)\lambda_2 + at_2\lambda_1 - \lambda_1)}{1-t_1-t_2} & -\frac{(1-t_1)(1-t_2)\lambda_2 - (1-a-at_1)t_2\lambda_1}{1-t_1-t_2} \end{bmatrix}$$

$$H_1 = \begin{bmatrix} \frac{a(1-t_1)(1-t_2)\lambda_1 + t_1\lambda_2(1-b-t_2)}{1-t_1-t_2} & \frac{t_1(\lambda_2(b-t_1)-a\lambda_1(1-t_1))}{1-t_1-t_2} \\ \frac{(1-t_2)(a\lambda_1 t_2 + \lambda_2(1-b-t_2))}{1-t_1-t_2} & \frac{\lambda_2(1-t_2)(b-t_1)-a\lambda_1 t_1 t_2}{1-t_1-t_2} \end{bmatrix}. \tag{8}$$

We have to prove that if the off-diagonal elements of H_0 and the elements of H_1 are non-negative, then $0 \le t_1 \le 1$ and $0 \le t_2 \le 1$. By using the restrictions on a, b in Theorem 1 we can derive constraints from the elements of H_0 and H_1. From the $(1, 2)$ element of H_0 we obtain

$$t_1 + t_2 < 1 \ \wedge \ \left(t_1 > 1 \ \vee \ t_1 < \frac{(1-a)\lambda_1}{\lambda_2 - a\lambda_1} \right) \tag{9a}$$

or

$$t_1 + t_2 > 1 \ \wedge \ t_1 < 1 \ \wedge \ t_1 > \frac{(1-a)\lambda_1}{\lambda_2 - a\lambda_1} \tag{9b}$$

From the $(2, 1)$ element of H_0 we have

$$t_1 < \frac{(1-a)\lambda_1}{\lambda_2 - a\lambda_1} \ \&\& \ t_1 + t_2 > 1 \ \vee \ \left(0 < t_2 \ \wedge \ t_2 < \frac{\lambda_2 - \lambda_1}{\lambda_2 - a\lambda_1} \right) \tag{10a}$$

or

$$t_1 > \frac{(1-a)\lambda_1}{\lambda_2 - a\lambda_1} \ \&\& \ (t_1 < 1) \ \&\& \ \left(t_2 > \frac{\lambda_2 - \lambda_1}{\lambda_2 - a\lambda_1} \ \vee \ (0 < t_2 \ \wedge \ t_1 + t_2 \le 1) \right) \tag{10b}$$

Combining the constraints for the two elements we get

$$t_1 < \frac{\lambda_1(1-a)}{\lambda_2 - \lambda_1} \ \wedge \ 0 < t_2 < \frac{\lambda_2 - \lambda_1}{\lambda_2 - a\lambda_1} \tag{11a}$$

or

$$\frac{\lambda_1(1-a)}{\lambda_2 - a\lambda_1} < t_1 < 1 \ \wedge \ t_2 > \frac{\lambda_2 - \lambda_1}{\lambda_2 - a\lambda_1}. \tag{11b}$$

The first case corresponds to $t_1 + t_2 < 1$, while the second to $t_1 + t_2 > 1$. From the $(1,1)$ element of $\boldsymbol{H_1}$ we have

$$t_1 + t_2 < 1 \;\wedge\; 0 < t_1 < \frac{b\lambda_2 - a\lambda_1}{\lambda_2 - a\lambda_1} \tag{12a}$$

or

$$t_1 + t_2 > 1 \;\wedge\; \left(t_1 > \frac{b\lambda_2 - a\lambda_1}{\lambda_2 - a\lambda_1} \;\vee\; t_1 < 0 \right) \tag{12b}$$

From (11a) and (12a) we get that $t_1 > 0$ and $t_2 > 0$ for the $t_1 + t_2 < 1$ case, from which $0 \le t_1, t_2 \le 1$ if $t_1 + t_2 < 1$. It remains to show the same for the $t_1 + t_2 > 1$ case. From (11b) we have that $0 \le t_1 \le 1$ and $0 \le t_2$ thus we only have to prove that $t_2 \le 1$ also holds. From the $(2,1)$ element of $\boldsymbol{H_1}$ we get

$$t_1 < 0 \;\wedge\; \left(t_2 < \frac{(1-b)\lambda_2}{\lambda_2 - a\lambda_1} \;\vee\; (t_1 + t_2 < 1 \;\wedge\; t_2 > 1) \right) \tag{13a}$$

or

$$t_1 > 0 \;\wedge\; t_1 < \frac{b\lambda_2 - a\lambda_1}{\lambda_2 - a\lambda_1} \;\wedge\; \left(t_2 < \frac{(1-b)\lambda_2}{\lambda_2 - \lambda_1} \;\vee\; (t_1 + t_2 > 1 \;\wedge\; t_2 < 1) \right) \tag{13b}$$

or

$$t_1 > \frac{b\lambda_2 - a\lambda_1}{\lambda_2 - a\lambda_1} \;\wedge\; \left(t_1 + t_2 < 1 \;\vee\; \left(t_2 < 1 \;\wedge\; t_2 > \frac{(1-b)\lambda_2}{\lambda_2 - a\lambda_1} \right) \right) \tag{13c}$$

In the (13c) subcase $t_2 \le 1$ is explicitly stated if $t_1 + t_2 > 1$. For the other two subcases $t_2 \le 1$ holds if $\frac{(1-b)\lambda_2}{\lambda_2 - a\lambda_1} \le 1$. But this is true because from Theorem 1 we know that $b \ge a\frac{\lambda_1}{\lambda_2}$, consequently we can use the transformation

$$t_2 < \frac{(1-b)\lambda_2}{\lambda_2 - a\lambda_1} \le \frac{\left(1 - a\frac{\lambda_1}{\lambda_2}\right)\lambda_2}{\lambda_2 - a\lambda_1} = 1. \tag{14}$$

Substituting this into (13a) and (13b) we obtain that $t_2 < 1$ if $t_1 + t_2 > 1$ in both subcases, thus $0 \le t_1, t_2 \le 1$ is proven for $\gamma \ge 0$, which means that the proposed canonical form is valid for $\gamma \ge 0$.

Let us now consider NMAPs with $\gamma < 0$. The matrices of the canonical form are

$$\boldsymbol{D_0} = \begin{bmatrix} -\lambda_1 & (1-a)\lambda_1 \\ 0 & -\lambda_2 \end{bmatrix}, \quad \boldsymbol{D_1} = \begin{bmatrix} 0 & a\lambda_1 \\ b\lambda_2 & (1-b)\lambda_2 \end{bmatrix}.$$

Using the (6) similarity transformation we get

$$\boldsymbol{H_0} = \begin{bmatrix} -\frac{(1-t_1)(1-at_2)\lambda_1 - t_1 t_2 \lambda_2}{1 - t_1 - t_2} & \frac{(1-t_1)((1-a(1-t_1))\lambda_1 - t_1 \lambda_2)}{1 - t_1 - t_2} \\ \frac{t_2(1-t_2)\lambda_2 + at_2\lambda_1 - \lambda_1}{1 - t_1 - t_2} & -\frac{(1-t_1)(1-t_2)\lambda_2 - (1-a+at_1)t_2\lambda_1}{1 - t_1 - t_2} \end{bmatrix}$$

$$\boldsymbol{H_1} = \begin{bmatrix} \frac{t_1(1-b-t_2)\lambda_2 - a(1-t_1)t_2\lambda_1}{1 - t_1 - t_2} & \frac{a(1-t_1)^2\lambda_1 + t_1(b-t_1)\lambda_2}{1 - t_1 - t_2} \\ \frac{(1-t_2)(1-b-t_2)\lambda_2 - at_2^2\lambda_1}{1 - t_1 - t_2} & \frac{a(1-t_1)t_2\lambda_1 + (b-t_1)(1-t_2)\lambda_2}{1 - t_1 - t_2} \end{bmatrix}. \tag{15}$$

We apply the same approach as before, i.e., we prove that if the respective elements of $\boldsymbol{H_0}$ and $\boldsymbol{H_1}$ are non-negative then $0 \leq t_1, t_2 \leq 1$ has to hold. The $\boldsymbol{H_0}$ matrix is the same as for $\gamma \geq 0$, therefore we get (11) again. In the following we use substitutions

$$e = \frac{b\lambda_2 - 2a\lambda_1}{\lambda_2 - a\lambda_1}, \quad f = \frac{(2-b)\lambda_2}{\lambda_2 - a\lambda_1}, \quad g = \frac{\sqrt{\lambda_2(4a(1-b) + b^2\lambda_2)}}{\lambda_2 - a\lambda_1}$$

From the $(1,2)$ element of $\boldsymbol{H_1}$

$$t_1 + t_2 > 1 \ \wedge \ t_1 > \frac{1}{2}(e + g) \tag{16a}$$

or

$$t_1 + t_2 > 1 \ \wedge \ t_1 < \frac{1}{2}(e - g) \tag{16b}$$

or

$$t_1 + t_2 < 1 \ \wedge \ t_1 < \frac{1}{2}(e + g) \ \wedge \ t_1 > \frac{1}{2}(e - g). \tag{16c}$$

from the constraints on the $(2,1)$ element of $\boldsymbol{H_1}$ we get

$$t_1 + t_2 < 1 \ \wedge \ t_2 < \frac{1}{2}(f - g) \ \wedge \ t1 < \frac{1}{2}(e + g) \tag{17a}$$

or

$$t_1 + t_2 < 1 \ \wedge \ t1 > \frac{1}{2}(e + g) \tag{17b}$$

or

$$t_1 + t_2 < 1 \ \wedge \ t_2 < \frac{1}{2}(f + g) \ \wedge \ t1 < \frac{1}{2}(e - g) \tag{17c}$$

or

$$t_1 + t_2 > 1 \ \wedge \ t_2 < \frac{1}{2}(f + g) \ \wedge \ \frac{1}{2}(e - g) < t1 < \frac{1}{2}(e + g) \tag{17d}$$

or

$$t_1 + t_2 > 1 \ \wedge \ \frac{1}{2}(f - g) < t_2 < \frac{1}{2}(f + g) \ \wedge \ t1 > \frac{1}{2}(e - g) \tag{17e}$$

If (16a) then only the (17e) subcase is possible. None of the constraints in (17) allow (16b) to be true, while is (16c) only possible for the (17a) subcase. Summarising these we get

$$t_1 + t_2 > 1 \ \wedge \ \frac{1}{2}(f - g) < t_2 < \frac{1}{2}(f + g) \ \wedge \ t1 > \frac{1}{2}(e + g) \tag{18a}$$

or

$$t_1 + t_2 < 1 \ \wedge \ \frac{1}{2}(e - g) < t_1 < \frac{1}{2}(e + g) \ \wedge \ t2 < \frac{1}{2}(f - g) \tag{18b}$$

Using the further substitution

$$h = \frac{(1-b)t_1\lambda_2}{t_1\lambda_2 + a(1-t_1)\lambda_1}$$

from the constraints on the $(1,1)$ element of $\boldsymbol{H_1}$ we obtain

$$t_1 + t_2 < 1 \ \wedge \ t_1 < \frac{1}{2}(e - g) \ \wedge \ t_2 > h \tag{19a}$$

or

$$t_1 + t_2 < 1 \ \wedge \ t_1 > \frac{1}{2}(e + g) \tag{19b}$$

or

$$t_1 + t_2 > 1 \ \wedge \ \frac{1}{2}(e - g) < t_1 < -\frac{a\lambda_1}{\lambda_2 - a\lambda_1} \ \wedge \ t_2 < h \tag{19c}$$

or

$$t_1 + t_2 > 1 \ \wedge \ -\frac{a\lambda_1}{\lambda_2 - a\lambda_1} < t_1 < \frac{1}{2}(e + g) \tag{19d}$$

or

$$-\frac{a\lambda_1}{\lambda_2 - a\lambda_1} < t_1 < \frac{1}{2}(e + g) \ \wedge \ t_2 < h \tag{19e}$$

or

$$t_1 > \frac{1}{2}(e + g) \ \wedge \ t_2 > h \tag{19f}$$

We combine these with the previously obtained expressions for the elements of $\boldsymbol{H_0}$ and $\boldsymbol{H_1}$. Let us examine first the $t_1 + t_2 < 1$ case. As (18b) has to hold, only the (19e) subcase is possible. Because of the constraints on the elements of $\boldsymbol{H_0}$, for $t_1 + t_2 < 1$ (11a) has to hold, i.e., $t_1 < \frac{\lambda_1(1-a)}{\lambda_2 - \lambda_1} \le 1$ and $0 \le t_2 \le \frac{\lambda_2 - \lambda_1}{\lambda_2 - a\lambda_1} \le 1$ are true, thus we only have to show that $0 \le t_1$ is also true. From (19e) we have that

$$t_2 = h = \frac{(1-b)t_1\lambda_2}{t_1\lambda_2 + a(1-t_1)\lambda_1}. \tag{20}$$

From Definition 1 we know that $\lambda_2 \ge \frac{a}{b}\lambda_1$ has to hold. The expression is monotonically increasing in t_1 for $t_1 > -\frac{a\lambda_1}{\lambda_2 - a\lambda_1}$ (which is the other constraint in (19e)) and non-negative only if $t_1 \ge 0$. But we know from (11a), that $t_2 \ge 0$ has to hold, thus, for t_2 to have a valid range $t_1 \ge 0$ also has to be true. This proves that $0 \le t_1, t_2 \le 1$ if $t_1 + t_2 < 1$.

Finally, from (11b) we have that $0 \le \frac{\lambda_2(1-a)}{\lambda_2 - \lambda_1} \le t_1 \le 1$ and $0 \le \frac{\lambda_2 - \lambda_1}{\lambda_2 - a\lambda_1} \le t_2$, thus we only have to show that $t_2 \le 1$. First we note that from (19) only the (19f) subcase is possible due to the $t_1 > \frac{1}{2}(e + g)$ condition in (18). We use the constraints on the $(2,2)$ element of $\boldsymbol{H_1}$ with substitutions

$$i = \frac{b\lambda_2 - a\lambda_1}{\lambda_2 - \lambda_1}, \quad j = \frac{(b - t_1)\lambda_2}{(b - t_1)\lambda_2 - a(1 - t_1)\lambda_1}$$

to get

$$t_1 + t_2 < 1 \ \wedge \ t_1 < \frac{1}{2}(e - g) \tag{21a}$$

or

$$t_1 + t_2 < 1 \ \wedge \ \frac{1}{2}(e - g) < t_1 < i \tag{21b}$$

or

$$t_1 + t_2 < 1 \ \wedge \ i < t_1 < \frac{1}{2}(e + g) \ \wedge \ t2 > j \tag{21c}$$

or

$$t_1 < \frac{1}{2}(e - g) \ \wedge \ t_2 < j \tag{21d}$$

or

$$\frac{1}{2}(e - g) < t_1 < i \ \wedge \ t_2 > j \tag{21e}$$

or

$$t_1 + t_2 > 1 \ \wedge \ t_1 > \frac{1}{2}(e + g) \ \wedge \ t_2 < j \tag{21f}$$

For the (21d)-(21f) subcases $t_1 + t_2 > 1$ holds, but because of the constraint $t_1 > \frac{1}{2}(e + g)$ in (18a), subcase (21d) and (21e) are not possible. For the last remaining (21f) subcase the last needed constraint $t_2 \leq 1$ has to hold, because

$$t_2 < j = \frac{(b - t_1)\lambda_2}{(b - t_1)\lambda_2 - a(1 - t_1)\lambda_1} \leq 1$$

as j is monotonously increasing between $t_1 = b$ and $t_1 = 1$ and $j = 1$ if $t_1 = 1$, but we know from (11) that $t_1 \leq 1$, thus $t_2 \leq 1$ also holds.

5   Equivalence of the NMAP(2) and NRAP(2) Classes

In this section we focus on the relation of the NMAP(2) and the NRAP(2) classes.

Theorem 4. *For the NMAP(2) and NRAP(2) sets of point processes we have*

$$NMAP(2) \equiv NRAP(2).$$

That is, every NRAP(2) process has a Markovian representation. The proof follows a similar pattern as the one that proves the equivalence between MAP(2) and RAP(2) in [4], therefore we reiterate some of the main points from there. The kth inter-arrival time in an NRAP has joint probability density function

$$f(X_k = x_k | X_0 = x_0, X_1 = x_1, \ldots, X_{k-1} = x_{k-1}) =$$

$$= \frac{\pi_0 e^{H_0 x_0} H_1 e^{H_0 x_1} H_1 \ldots e^{H_0 x_{k-1}} H_1}{\pi_0 e^{H_0 x_0} H_1 e^{H_0 x_1} H_1 \ldots e^{H_0 x_{k-1}} H_1 \mathbb{1}} e^{H_0 x_k} H_1 \mathbb{1}. \tag{22}$$

The X_k random variable has to have a valid distribution for $\forall k \geq 0$ and $\forall x_0, \ldots, x_k \geq 0$. Let $\pi_k(x_0, x_1, \ldots, x_{k-1})$ be the initial vector before the kth inter-arrival, X_k. It is given by

$$\pi_k(x_0, x_1, \ldots, x_{k-1}) = \frac{\pi_0 e^{H_0 x_0} H_1 e^{H_0 x_1} H_1 \ldots e^{H_0 x_{k-1}} H_1}{\pi_0 e^{H_0 x_0} H_1 e^{H_0 x_1} H_1 \ldots e^{H_0 x_{k-1}} H_1 \mathbb{1}}. \tag{23}$$

If (π_0, H_0, H_1) in the previous expression is an NRAP(2) representation in NMAP(2) canonical form, then the first element of vector $\pi_k(x_0, x_1, \ldots, x_{k-1})$ has to be in the range of

$$0 \leq \pi_k(x_0, x_1, \ldots, x_{k-1}) e_1 \leq \frac{1}{1 - a\frac{\lambda_1}{\lambda_2}}, \tag{24}$$

otherwise the joint pdf in (22) is not strictly non-negative (see [4] for more details). To prove the equivalence of NRAP(2) and NMAP(2) we show that if the H_0, H_1 matrices of an NRAP(2) are in canonical form, then its initial vector is Markovian (non-negative) as well. Let $u(x, t)$ be the first element of the initial vector after an inter-arrival time of length t if the initial vector after the previous arrival was $[x, 1-x]$. Then $u(x, t)$ can be expressed as

$$u(x, t) = \frac{[x, 1-x] e^{H_0 t} H_1 e_1}{[x, 1-x] e^{H_0 t} H_1 \mathbb{1}}. \tag{25}$$

From (24) it is clear that $\delta e_1 \geq 0$ holds regardless of the value of γ. We have to show that $\delta e_1 \leq 1$ ($\delta e_2 \geq 0$) is also true. We will assume a series of arrivals with negligibly small inter-arrival time ($t \to 0$ and consequently $e^{H_0 t} \to I$) and prove that for x to satisfy the constraints in (24) $\delta e_1 \leq 1$ has to hold. First we examine the $\gamma \geq 0$ case. From (25) after using the respective canonical form in Definition 1 and simplifying the expression we get that

$$u(x, 0) = \frac{[x, 1-x] H_1 e_1}{[x, 1-x] H_1 \mathbb{1}} = \frac{ax\lambda_1 + (1-x)(1-b)\lambda_2}{ax\lambda_1 + (1-x)\lambda_2}. \tag{26}$$

This function is a hyperbola that has one or two fix points (points where $u(x, 0) = x$, see Figure 1 for illustration). These are $\frac{(1-b)\lambda_2}{\lambda_2 - a\lambda_1}$ and 1. (There is only one fix point in $x = 1$ if $a = b = 0$ or $\lambda_2 = \frac{a}{b}\lambda_1$.) Because $\lambda_2 \geq \frac{a}{b}\lambda_1$, we know that $x = 1$ is the higher fix point. The first element of δ cannot be higher than this value, because for $1 < x < \frac{1}{1-a\frac{\lambda_1}{\lambda_2}}$ we have $u(x, 0) > x$, which means that the first coordinate of the initial vector would increase after every arrival and would finally go above the upper limit of $x = \frac{1}{1-a\frac{\lambda_1}{\lambda_2}}$ in (24) (this value is the vertical asymptote of the hyperbola). This means that $0 \leq \delta e_1 \leq 1$ has to hold for $\gamma \geq 0$.

Now let us investigate the $\gamma < 0$ case. As before we examine the $u(x, 0)$ function and substitute the canonical form from Definition 1 corresponding to $\gamma < 0$. Doing so we get

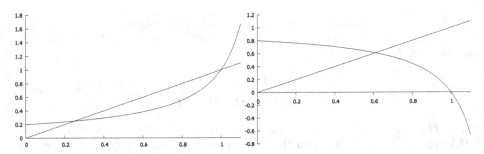

Fig. 1. Behaviour of $u(x,0)$ for positive (left) and negative (correlation) for $\lambda_1 = 1$, $\lambda_2 = 2$, $a = 0.4$, $b = 0.8$

$$u(x,0) = \frac{[x, 1-x]\boldsymbol{H}_1\boldsymbol{e}_1}{[x, 1-x]\boldsymbol{H}_1\mathbb{1}} = \frac{b\lambda_2(1-x)}{(1-x)\lambda_2 + ax\lambda_1}. \tag{27}$$

Again from (24) we know that $\delta e_1 < \frac{1}{1-a\frac{\lambda_1}{\lambda_2}}$, and we have to prove that $\delta e_1 < 1$ The numerator of the expression becomes negative for $x > 1$, while the denominator is negative for $1 < x < \frac{1}{1-a\frac{\lambda_1}{\lambda_2}}$, thus for $\delta e_1 > 1$ the first coordinate of the initial vector would become negative, which is not allowed according to (24). Thus $0 \le \delta e_1 \le 1$ has to hold for $\gamma < 0$ as well.

To summarize, we showed that $0 < \delta e_1 \le 1$ has to hold for any NRAP(2) transformed to the canonical NMAP(2) representation thus we proved that any NRAP(2) can be transformed to a Markovian canonical form, thus NRAP(2)≡NMAP(2).

6   Conclusion

In this paper we proposed a canonical form for NMAP(2)s and proved that this canonical form is Markovian for every NMAP(2). We also showed that the classes of NMAP(2) and NRAP(2) processes are equivalent by proving that every NRAP(2) has a Markovian canonical form. In the course of this work we got informed of a similar effort [22] with partially similar goals. The authors of [22] consider only the first of the two problems investigated in this paper. For that problem they derive the same conclusion as Theorem 3, but with a completely different proof.

References

1. Andersen, A.T., Nielsen, B.F.: A Markovian approach for modeling packet traffic with long-range dependence. IEEE Journal on Selected Areas in Communications **16**(5), 719–732 (1998)
2. Asmussen, S., Nerman, O., Olsson, M.: Fitting phase-type distributions via the EM algorithm. Scandinavian Journal of Statistics 419–441 (1996)

3. Bladt, M.: A review on phase-type distributions and their use in risk theory. ASTIN Bulletin **35**, 145–161 (2005). http://journals.cambridge.org/article_S0515036100014100

4. Bodrog, L., Heindl, A., Horváth, G., Telek, M.: A Markovian canonical form of second-order matrix-exponential processes. European Journal of Operation Research **190**, 459–477 (2008)

5. Buchholz, P., Horvath, A., Telek, M.: Stochastic Petri nets with low variation matrix exponentially distributed firing time. International Journal of Performability Engineering **7**, 441–454 (2011). special issue on Performance and Dependability Modeling of Dynamic Systems

6. Buchholz, P., Panchenko, A.: A two-step EM algorithm for MAP fitting. In: Aykanat, C., Dayar, T., Körpeoğlu, İ. (eds.) ISCIS 2004. LNCS, vol. 3280, pp. 217–227. Springer, Heidelberg (2004)

7. Buchholz, P., Telek, M.: On minimal representation of rational arrival processes. Annals of Operations Research **202**, 35–58 (2013)

8. Fackrell, M.: Modelling healthcare systems with phase-type distributions. Health Care Management Science **12**(1), 11–26 (2009). http://dx.doi.org/10.1007/s10729-008-9070-y

9. Feldmann, A., Whitt, W.: Fitting mixtures of exponentials to long-tail distributions to analyze network performance models. In: Proceedings IEEE INFOCOM 1997. Sixteenth Annual Joint Conference of the IEEE Computer and Communications Societies. Driving the Information Revolution, vol. 3, pp. 1096–1104. IEEE (1997)

10. Fischer, W., Meier-Hellstern, K.: The Markov-modulated Poisson process (MMPP) cookbook. Performance evaluation **18**(2), 149–171 (1993)

11. Horváth, A., Telek, M.: A Markovian point process exhibiting multifractal behaviour and its application to traffic modeling. In: Proceedings of 4th International Conference on Matrix-Analytic Methods in Stochastic models, pp. 183–208 (2002)

12. Horváth, G.A., Telek, M.: A canonical representation of order 3 phase type distributions. In: Wolter, K. (ed.) EPEW 2007. LNCS, vol. 4748, pp. 48–62. Springer, Heidelberg (2007)

13. Houdt, B.V., Alfa, A.S.: Response time in a tandem queue with blocking, Markovian arrivals and phase-type services. Operations Research Letters **33**(4), 373–381 (2005). http://www.sciencedirect.com/science/article/pii/S016763770400 1245

14. Khayari, R.E.A., Sadre, R., Haverkort, B.R.: Fitting world-wide web request traces with the EM-algorithm. Performance Evaluation **52**(2), 175–191 (2003)

15. Latouche, G., Ramaswami, V.: Introduction to Matrix Analytic Methods in Stochastic Modeling. ASA-SIAM Series on Statistics and Applied Probability, Society for Industrial and Applied Mathematics (1999). https://books.google.com/books?id=J31a3P69K8IC

16. Lucantoni, D.M.: New results on the single server queue with a batch Markovian arrival process. Communications in Statistics. Stochastic Models **7**(1), 1–46 (1991). http://dx.doi.org/10.1080/15326349108807174

17. Mészáros, A., Telek, M.: Canonical representation of discrete order 2 MAP and RAP. In: Balsamo, M.S., Knottenbelt, W.J., Marin, A. (eds.) EPEW 2013. LNCS, vol. 8168, pp. 89–103. Springer, Heidelberg (2013)

18. Mitchell, K., van de Liefvoort, A.: Approximation models of feed-forward G/G/1/N queueing networks with correlated arrivals. Performance Evaluation **51**(2), 137–152 (2003)

19. Neuts, M.F.: Models based on the Markovian arrival process. IEICE Transactions on Communications **75**(12), 1255–1265 (1992)
20. Papp, J., Telek, M.: Canonical representation of discrete phase type distributions of order-2 and 3. In: Proc. of UK Performance Evaluation Workshop, UKPEW 2013 (2013)
21. Puhalskii, A.A., Reiman, M.I.: The multiclass GI/PH/N queue in the Halfin-Whitt regime. Advances in Applied Probability **32**(2), 564–595 (2000). http://www.jstor.org/stable/1428204
22. Rodriguez, J., Lillo, R.E., Ramirez-Cobo, P.: A canonical form for the non-stationary two-state Markovian arrival process (2015). submitted for publication
23. Rydn, T.: An EM algorithm for estimation in Markov-modulated Poisson processes. Computational Statistics & Data Analysis **21**(4), 431–447 (1996). http://www.sciencedirect.com/science/article/pii/0167947395000259
24. Yoshihara, T., Kasahara, S., Takahashi, Y.: Practical time-scale fitting of self-similar traffic with Markov-modulated Poisson process. Telecommunication Systems **17**(1–2), 185–211 (2001)

Markov Decision Petri Nets with Uncertainty

Marco Beccuti[1], Elvio G. Amparore[1], Susanna Donatelli[1],
Dimitri Scheftelowitsch[2], Peter Buchholz[2], and Giuliana Franceschinis[3]([✉])

[1] Dipartimento di Informatica, Università di Torino, Torino, Italy
{beccuti,amparore,susi}@di.unito.it
[2] Informatik IV, TU Dortmund, Dortmund, Germany
{dimitri.scheftelowitsch,peter.buchholz}@cs.tu-dortmund.de
[3] DiSIT, Università del Piemonte Orientale, Alessandria, Italy
giuliana.franceschinis@di.unipmn.it

Abstract. Markov Decision Processes (MDPs) are a well known mathematical formalism that combines probabilities with decisions and allows one to compute optimal sequences of decisions, denoted as policies, for fairly large models in many situations. However, the practical application of MDPs is often faced with two problems: the specification of large models in an efficient and understandable way, which has to be combined with algorithms to generate the underlying MDP, and the inherent uncertainty on transition probabilities and rewards, of the resulting MDP. This paper introduces a new graphical formalism, called *Markov Decision Petri Net with Uncertainty* (MDPNU), that extends the *Markov Decision Petri Net* (MDPN) formalism, which has been introduced to define MDPs. MDPNUs allow one to specify MDPs where transition probabilities and rewards are defined by intervals rather than constant values. The resulting process is a Bounded Parameter MDP (BMDP). The paper shows how BMDPs are generated from MDPNUs, how analysis methods can be applied and which results can be derived from the models.

Keywords: Markov Decision Process · Bounded parameter MDP · Markov decision Petri net · Multi-objective optimization

1 Objectives and Contributions

MDPs are a well known mathematical formalism introduced in the 1950s [5] in the context of operations research and dynamic programming. It has been used in a wide range of applications to compute policies (i.e. sequences of decisions) that optimize some reward measure. Indeed an MDP allows a modeler to specify a stochastic control process by means of different states, in which a decision maker may choose any action available in that state. Then, the process responds by randomly moving into a new state according to a specified transition probability, and gives back to the decision maker the corresponding reward (depending on the chosen action and by the source and destination states). An MDP policy defines the choice of actions to be taken in any possible MDP state, so that it

© Springer International Publishing Switzerland 2015
M. Beltrán et al. (Eds.): EPEW 2015, LNCS 9272, pp. 177–192, 2015.
DOI: 10.1007/978-3-319-23267-6_12

is possible to derive the policies that maximize a target function based on the MDP's rewards. Often the expected discounted or expected average reward over an infinite time horizon is used as measure to be optimized.

One limitation of MDPs, like many other modeling formalisms, is that they are based on a complete knowledge of the parameters. Unfortunately, in many real systems parameters are usually estimated and suffer from several limitations or measurement errors. In addition, the knowledge of the system behavior is usually incomplete, hence the real probability distribution of transitions is in most cases an uncertain value in some confidence interval. In this case, the available information defines a set of MDPs rather than a single MDP and the question is what is an optimal policy? It might be the case, that one policy is good for one concrete realization of the parameters whereas it is bad for another realization. A policy is denoted as robust if it shows an acceptable performance for every possible realization of the uncertain parameters and it is of high practical interest to find out robust policies.

BMDPs were introduced in [10]. In a BMDP, action probabilities and rewards may be expressed either as exact values, or as an interval-valued parameter. In this way a BMDP defines a family of MDPs exactly in the way as it is required for situations where parameters are uncertain. In the original paper [10] and successor papers that consider particularly the efficient analysis of BMDPs [8], the computation of policies for the worst or best case (which is the parameter values that minimize resp. maximize the resulting rewards) is described. It is shown that in a BMDP the policies for optimizing the discounted expected reward over infinite horizons for the best and worst cases are stationary, which means that they depend only on the current state. This does not really solve the problem of very pessimistic or optimistic policies because an optimal policy in the worst or best case might yield a bad performance in most other cases.

Recently, the analysis of BMDPs was extended by introducing *multi-objective optimization* [12]. In this setting, worst, average, and best cases are considered together as a single optimization criterion. In some BMDPs it is possible to identify a robust stationary policy (and prove the robustness), which is a single strategy that exhibits optimal or near optimal rewards in all the MDPs of the BMDP. In other words, this strategy shows good performances no matter how the uncertainty is resolved. However, the existence of such a policy is not guaranteed, hence in other cases one has to search for more complex policies or has to be at least aware that the system is very sensitive to parameter changes.

Modeling a system directly at the level of the MDP or BMDP could be a hard task, due to the complexity of most real systems. To cope with this problem, a number of higher-level formalisms have been proposed in the literature (e.g., stochastic transition systems [1], dynamic decision networks [6], probabilistic extensions of reactive modules [2], Markov decision Petri nets and Markov decision well-formed nets [3], Factored Petri Net [9], etc.). Almost all of these approaches have been defined to specify MDPs, parameter uncertainty is not considered. In [7] the uncertainty is introduced to solve factored MDPs with large state space exploiting ϵ-homogeneous partitions of their states. Indeed,

ϵ-homogeneous partition approach are used to induce a family of explicit MDPs corresponding to a BMDP. The derived BMDP has hence a smaller state space, and it can be solved to obtaining a lower and upper bound on the optimal value function of the original MDP.

In this paper we introduce a new extension of Markov decision Petri nets (MDPN), called *MDPN with uncertainties* (MDPNU), where probabilistic distributions and the reward functions are specified with interval-valued parameters, such that the MDPNU semantics is expressed as a BMDP. It is shown how a BMDP is derived from the MDPNU. The resulting models are amenable to the analysis approaches for BMDPs. We show how the new analysis algorithms can be applied and which effort becomes necessary to optimize BMDPs with a varying number of states. The different steps of the approach, from specification, to process generation and solution, ending with the result representation are currently integrated in a new version of the GreatSPN modeling tool which will be part of the next official release of the tool. As an example of application of MDPNU we present the model of the control room of an utility company and provide experimental results computed with the implementation of the new multi-objective BMDP algorithm in [12].

2 Background

Before defining MDPNU as a high level model, we introduce the underlying stochastic models and begin with MDPs in discrete time [11].

Definition 1. *A Markov Decision Process (MDP) is a tuple (S, A, F, R) where*

- *S is a (finite) set of states,*
- *A is a (finite) set of actions and $A_i \subseteq A$ the set of available actions in state $i \in S$,*
- *$F : S \times A \to \mu(S)$ the (partial) transition function which assigns to each possible state action pair a probability distribution over the set of states, and*
- *$R : S \times A \to \mathbb{R}_{\geq 0}$ is the (partial) reward function that assigns to each possible state action pair a non-negative reward value.*

$R(i, a)$ is the reward if action a is chosen in state i and $F(i, a, j)$ is the probability of a transition from state i to j under the condition that action a is chosen in state i. In state i, $|A_i|$ different decisions are possible by choosing one of the available actions. The choice of actions in states is denoted as a policy which is *stationary* if it only depends on the state and not on the history. It is *deterministic*, if a single action is chosen with probability 1. If it is both stationary and deterministic, then it is *pure*. A pure policy can be described by a vector $\mathbf{a}$ of length $|S|$ where $\mathbf{a}(i) \in A_i$ is the action chosen in state i. Each pure policy $\mathbf{a}$ defines a unique transition matrix $\mathcal{P}^{\mathbf{a}}$ and a reward vector $\mathbf{r}^{\mathbf{a}}$. The triple $(S, \mathcal{P}^{\mathbf{a}}, \mathbf{r}^{\mathbf{a}})$ defines a Markov reward process in discrete time. $\mathcal{P}^{\mathbf{a}}$ is a stochastic matrix for every pure policy $\mathbf{a}$.

MDPs are used to compute optimal policies which maximize or minimize gains (= discounted accumulated reward) over infinite time horizons. We consider the maximization of the expected discounted reward for some discount factor $\gamma \in [0, 1)$. It can be shown that the solution of the set of Bellman equations [11]

$$\mathbf{v}^*(i) = \max_{a \in A_i} \left(R(i, a) + \gamma \sum_{j \in S} F(i, a, j) \mathbf{v}^*(j) \right) \text{ for all } i \in S \qquad (1)$$

results in a vector $\mathbf{v}^*$ that contains for all states the maximum of the expected discounted rewards. A corresponding policy is pure and results from the actions that are chosen in the above equations. Since the corresponding actions need not be unique, there can be more than one optimal policy. Other reward measures can be computed similarly [11]. We denote $\mathbf{v}^*$ as the optimal gain vector.

Parameters of an MDP should reflect the behavior of the modeled system and usually have to be estimated based on measured data (resulting in confidence intervals) or, if no data is available, based on expert knowledge or other sources. In general this means that parameters are subject to some uncertainty. The choice of states in MDP modeling may also introduce uncertainty. Often the behavior of the system is not completely memoryless as required, which means that transition probabilities depend slightly on the history, on the sojourn time in a state, or on hidden parameters of the environment that are unavailable at decision time. In all these situations, interval parameters represent a more realistic choice than exact values, which results in a BMDP [10].

Definition 2. *A Bounded Parameter Markov Decision Process (BMDP) is a tuple $(S, A, F_\updownarrow, R_\updownarrow)$ where*

- *S is a (finite) set of states;*
- *A is a (finite) set of actions, $A_i \subseteq A$ the set of actions in state $i \in S$;*
- *$F_\updownarrow = (F_\downarrow, F_\uparrow)$:*
  - *$F_\downarrow : S \times A \to \mu^{low}(S)$, maps a state action pair on a lower bound for a distribution over S, i.e., $\sum_{j \in S} F_\downarrow(i, a, j) \leq 1$ for every $i \in S$ and $a \in A_i$;*
  - *$F_\uparrow : S \times A \to \mu^{up}(S)$, maps a state action pair on an upper bound for a distribution over S, i.e., $\sum_{j \in S} F_\uparrow(i, a, j) \geq 1$ for every $i \in S$ and $a \in A_i$;*
  - *$F_\downarrow(i, a, j) \leq F_\uparrow(i, a, j)$ for every pair $i, j \in S$ and $a \in A_i$;*
- *$R_\updownarrow = R_\uparrow \cup R_\downarrow$ where $R_\uparrow, R_\downarrow : S \times A \to \mathbb{R}_{\geq 0}$ and $R_\downarrow(i, a) \leq R_\uparrow(i, a)$ for all $i \in S$ and $a \in A_i$.*

A BMDP defines a set of MDPs (S, A, F, R), where $F_\downarrow(i, a, j) \leq F(i, a, j) \leq F_\uparrow(i, a, j)$ and $R_\downarrow(i, a) \leq R(i, a) \leq R_\uparrow(i, a)$. Let $\mathcal{BM}(S, A, F_\updownarrow, R_\updownarrow)$ be the set of MDPs that is defined by the bounds. We denote by $(S, A, \overline{F}, \overline{R})$ one specific MDP from the set, the *average case* MDP. If the parameters of the BMDP result from two-sided symmetric confidence intervals, then the average case results from taking the average values between minimum and maximum. If other distributions are assumed, then the average value is computed as the expected value in the interval. For the average case MDP, an optimal policy for the discounted reward can be computed using (1) with $\overline{F}$ and $\overline{R}$ rather than F and R. Let $\overline{\mathbf{v}}$ and $\overline{\mathbf{a}}$ be the value vector and an optimal stationary policy for the average case MDP. The bounds of the gain achieved by any MDP of a BMDP are provided by the following equations:

$$\mathbf{v}_\downarrow(i) = \max_{a \in A_i} \left(\min_{((S,A,F,R) \in \mathcal{BM}(S,A,F_\updownarrow,R_\updownarrow)} \left(R(i,a) + \gamma \sum_{j \in S} F(i,a,j)\mathbf{v}_\downarrow(j) \right) \right) \quad (2)$$

$$\mathbf{v}_\uparrow(i) = \max_{a \in A_i} \left(\max_{((S,A,F,R) \in \mathcal{BM}(S,A,F_\updownarrow,R_\updownarrow)} \left(R(i,a) + \gamma \sum_{j \in S} F(i,a,j)\mathbf{v}_\uparrow(j) \right) \right) \quad (3)$$

for all $i \in S$. Observe that the inner optimization fixes the MDP and the outer maximization determines the policy. It is, of course also possible to formulate similar equations to minimize the gain. The equations can be solved by an extension of value or policy iteration for MDPs [10] which requires more effort than the solution of (1). Nevertheless, as shown in our case study, the vectors can be computed for fairly large BMDPs, with several thousands of states, which is larger than examples on BMDPs available from literature. Obviously, $\mathbf{v}_\downarrow \leq \overline{\mathbf{v}} \leq \mathbf{v}_\uparrow$. Let $\mathbf{a}_\downarrow$, $\overline{\mathbf{a}}$ and $\mathbf{a}_\uparrow$ be policies which are optimal for the worst, average or best case. The policy together with the MDP resulting from the inner optimization defines a transition matrix $\mathcal{P}^{\mathbf{a}}$ where $\mathbf{a}$ equals $\mathbf{a}_\downarrow$ (or $\mathbf{a}_\uparrow$) and a reward vector $\mathbf{r}^{\mathbf{a}}$. If $\mathcal{P}^{\mathbf{a}}$ is ergodic, then the stationary vector $\mathbf{p}^{\mathbf{a}}$ ($\mathbf{p}^{\mathbf{a}}\mathcal{P}^{\mathbf{a}} = \mathbf{p}^{\mathbf{a}}$) can be computed such that $\mathbf{p}^{\mathbf{a}_\downarrow}\mathbf{v}_\downarrow$ and $\mathbf{p}^{\mathbf{a}_\uparrow}\mathbf{v}_\uparrow$ are the worst and best case stationary gain, respectively.

For a given policy $\mathbf{a}$ let $\mathbf{v}^{\mathbf{a}} = (\mathbf{v}_\downarrow^{\mathbf{a}}, \overline{\mathbf{v}}^{\mathbf{a}}, \mathbf{v}_\uparrow^{\mathbf{a}})$ be the vector of worst, average and best case rewards; they can be computed by fixing the action in Equations (2), (1), and (3). We denote a policy $\mathbf{a}$ as robust if $\|\mathbf{v}_\downarrow^{\mathbf{a}_\downarrow} - \mathbf{v}_\downarrow^{\mathbf{a}}\| < \epsilon_\downarrow$, $\|\overline{\mathbf{v}}^{\overline{\mathbf{a}}} - \overline{\mathbf{v}}^{\mathbf{a}}\| < \overline{\epsilon}$ and $\|\mathbf{v}_\uparrow^{\mathbf{a}_\uparrow} - \mathbf{v}_\uparrow^{\mathbf{a}}\| < \epsilon_\uparrow$ for some appropriate thresholds $(\epsilon_\downarrow, \overline{\epsilon}, \epsilon_\uparrow)$. The goal is to compute robust policies which show an acceptable behavior in all cases covered by the parameter uncertainty. It is easy to find a robust policy if $\mathbf{v}_\downarrow = \overline{\mathbf{v}} = \mathbf{v}_\uparrow$ which implies that there is a single optimal gain vector, and all optimal policies are indistinguishable with respect to the chosen optimality criterion. If this is not the case, then there can be many policies which are incomparable. Let $\mathcal{A}$ be the set of all policies, i.e., the set of all vectors $\mathbf{a}$ with $\mathbf{a}(i) \in A_i$. The set of Pareto optimal policies $\mathcal{A}_{opt} \subseteq \mathcal{A}$ is defined as

$$\mathcal{A}_{opt} = \left\{ \mathbf{a} | \mathbf{a} \in \mathcal{A} \wedge \neg \exists \mathbf{a}' \in \mathcal{A} : \mathbf{v}^{\mathbf{a}'} \geq \mathbf{v}^{\mathbf{a}} \wedge \mathbf{v}^{\mathbf{a}'} \neq \mathbf{v}^{\mathbf{a}} \right\}. \quad (4)$$

If the worst, average or best case policies are not adequate, the set $\mathcal{A}_{opt}$ has to be computed or approximated. A method that solves simultaneously (1-3) is described in [12]. Such a method requires much more effort than the analysis of single policies. Sometimes even $\mathcal{A}_{opt}$ does not contain an adequate policy. In this case non-stationary policies that depend on the history might be a solution but these policies are even harder to compute and to implement in a real system.

3   Markov Decision Petri Net with Uncertainty

The Markov Decision Petri Net (MDPN) formalism was first introduced in [3] as a high level formalism to specify MDPs; an example of its application in the context of Fault Tree is shown in [4]. As in MDPs, state changes are the result of a non deterministic choice of an action and a subsequent probabilistic choice: this is realized in MDPNs through the alternation of the executions between two submodels: a non deterministic subnet, denoted PN^{nd}, and a probabilistic one, denoted PN^{pr}. Both subnets are specified using the Petri net formalism, where a single Petri net is defined by a tuple $\langle P, T, I, O, H, prio, m_0 \rangle$ where: P is the finite set of *places*; T is the finite set of *transitions*; $I : T \times P \to \mathbb{N}$, $O : T \times P \to \mathbb{N}$, $H : T \times P \to \mathbb{N}$ are the sets of *input, output* and *inhibitor* arcs; $prio : T \to \mathbb{N}^+$ assigns priorities to transitions; and $m_0 : P \to \mathbb{N}$ is the *initial marking*. Subnets in an MDPN share the same place set P, and have disjoint transition sets: the set of non deterministic transitions T^{nd} and the set of probabilistic transitions T^{pr}. In both subnets the transitions are partitioned into *run* and *stop* subsets, where the firing of *stop* transitions is the base for the alternation between probabilistic and non deterministic behavior. The definition of PN^{pr} includes also a notion of "weight" associated with each transition. The weight of conflicting transitions are then normalized to obtain a probability distribution out of each state of PN^{pr}. The PN^{nd} is often referred to as the *controllable component* since it represents part of the system that is subject to non deterministic choices, that can be interpreted as an external control.

The original definition of MDPN also includes a notion of "components" at system level, so each transition of the two subnets PN^{nd} and PN^{pr} can be mapped onto one or more components. Although components play an important role in modeling, they are not particularly relevant for the extension of MDPN to consider uncertainty, and, for sake of notational simplicity, we shall not include them in the definition of MDPNU given in this section. Components shall be illustrated in the case study.

The dynamics of an MDPN is defined in terms of an alternation between firing sequences of transitions in PN^{nd} and firing sequences of transitions in PN^{pr}. By construction, each (maximal) firing sequence of non deterministic transitions includes zero or more firings of run transitions and exactly one stop transition firing, and a similar situation holds for the probabilistic side. Maximal sequences of non deterministic transitions correspond to actions of the modeled system (and of the underlying MDP). Maximal probabilistic firing sequences are mapped into single probabilistic transitions in the MDP. The probability of a sequence is computed by using the *weight* attributes associated with all probabilistic transitions appearing in the sequence.

An MDPN model includes a reward/cost function specified in terms of state reward/cost, $rs()$, and of non deterministic transition reward/cost, $rt()$; the reward $rg()$ for a state-action pair is then obtained by composing $rs()$ and $rt()$.

MDPNU extends MDPN by introducing uncertainty in the weights of the PN^{pr} transitions and in the reward functions, which leads to an underlying process which is a BDMP.

Definition 3. *A Markov Decision Petri Net with uncertainty (MDPNU) is a tuple* $\mathcal{MN} = \langle N^{pr}, N^{nd}, W \rangle$ *where:*

- N^{pr} *is a Petri net* $\langle P, T^{pr}, I^{pr}, O^{pr}, H^{pr}, prio^{pr}, m_0 \rangle$;
- N^{nd} *is a Petri net* $\langle P, T^{nd}, I^{nd}, O^{nd}, H^{nd}, prio^{nd}, m_0 \rangle$;
- $T^{pr} = Trun^{pr} \uplus Tstop^{pr}$ *and* $T^{nd} = Trun^{nd} \uplus Tstop^{nd}$;
- $T^{pr} \cap T^{nd} = \emptyset$: *a transition is either non deterministic or probabilistic.*
- $W \colon T^{pr} \to (\mathbb{R})^3$ *assigns to each transition* $t \in T^{pr}$ *a triple* $\langle w, l, u \rangle$ *that defines an interval in which the weight of transition* t *may vary (uncertainty on the transition weight). Value* w *represents the reference weight, while* l *and* u *are the lower and upper bounds of uncertainty, so that the weight can vary between* $(w - l)$ *and* $(w + u)$. *We assume that* $l \leq w$;

MDPN rewards are specified with three functions: rs and rt, that associate a reward to any MDPN state and non deterministic transitions respectively, and a function rg that combines them into a single reward (as it is typically expected by MDP solvers). All three functions evaluate to $\mathbb{R}$. Rewards in an MDPNU can have an associated uncertainty, therefore the definition is extended for rs, rt and rg to map on the triplets: reference value, lower and upper bound.

Definition 4 (MDPN reward functions). *Let* $\mathcal{MN}$ *be an MDPN with uncertainty. Then its reward specification is given by:*

- $rs \colon \mathbb{N}^P \to \mathbb{R}^3$ *which defines for every marking an interval reward value.*
- $rt \colon T^{nd} \to \mathbb{R}^3$ *which defines for every transition its interval reward value.*
- $rg \colon rs \oplus rt$, *where* $\oplus$ *is a function with values over triplets of values in* $\mathbb{R}$.

From MDPNU to BMDP. The construction of the BMDP defined by an MDPNU follows the analogous construction for the MDP defined by an MDPN: the probabilistic and the non deterministic subnets are composed to obtain a single "global" Petri net N^g and then the MDP is built based on the reachability graph (RG) of N^g. In the following we shall recall the composition, and explain how uncertainty is combined to build a BMDP.

Figure 1 shows the construction of N^g: the two subnets PN^{nd} and PN^{pr} are abstracted as two boxes in which only the transitions are shown, together with their classification as *Run* and *Stop* transitions. Four places are introduced (Stop^{pr}, Run^{pr}, Stopnd and Runnd) and two transitions ($Pr{\to}Nd$ and $Nd{\to}Pr$) to regulate the alternation between the probabilistic and the non deterministic phases. Each *Run* transition in PN^{nd} (PN^{pr}) is connected with a test arc to the corresponding Run place Runnd (Run^{pr}). Run places in PN^{nd} (PN^{pr}) are emptied by the firing of a stop transition of PN^{nd} (PN^{pr}) that puts a token in place Stopnd (Stop^{pr}), thus activating the switch of control to the other subnet through the firing of the $Pr{\to}Nd$ and $Nd{\to}Pr$ transitions. This construction ensures that the probabilistic behavior is completed/stopped before the decision maker starts the non deterministic phase, and takes decisions before starting the next probabilistic phase. When multiple components are introduced, run and stop transitions are labelled with components identifiers, and places Stop^{pr},

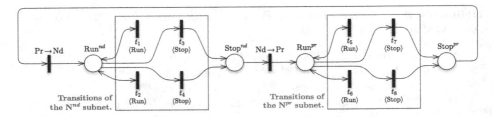

Fig. 1. The construction of the global net N^g from an MDPN.

Run^{pr}, Stop^{nd} and Run^{nd} are replicated as many times as the number of components. The enabling of $Pr{\to}Nd$ ($Nd{\to}Pr$ requires the presence of one token in all Stop^{pr} (Stop^{nd}) places.

More precisely, $N^g = \langle P^g, T^g, \mathcal{I}^g, \mathcal{O}^g, \mathcal{H}^g, prio^g, W^g, m_0 \rangle$ where:

- $P^g = P \cup \{\text{Run}^{pr}, \text{Stop}^{pr}\} \cup \{\text{Run}^{nd}, \text{Stop}^{nd}\}$
- $T^g = T^{pr} \cup T^{nd} \cup \{Pr{\to}Nd, Nd{\to}Pr\}$
- $\mathcal{I}^g, \mathcal{O}^g, \mathcal{H}^g$
  - $\forall p \in P, t \in T^{nd} : \mathcal{I}^g(t,p) = \mathcal{I}^{nd}(t,p), \mathcal{O}^g(t,p) = \mathcal{O}^{nd}(t,p), \mathcal{H}^g(t,p) = \mathcal{H}^{nd}(t,p)$
  - $\forall p \in P, t \in T^{pr} : \mathcal{I}^g(t,p) = \mathcal{I}^{pr}(t,p), \mathcal{O}^g(t,p) = \mathcal{O}^{pr}(t,p), \mathcal{H}^g(t,p) = \mathcal{H}^{pr}(t,p)$
  - $\forall t \in T^{pr} : \mathcal{I}^g(t, \text{Run}^{pr}) = 1$ $\quad$ $\mathcal{I}^g(Pr{\to}Nd, \text{Stop}^{pr}) = 1$
  - $\forall t \in Tstop^{pr} : \mathcal{O}^g(t, \text{Stop}^{pr}) = 1$ $\quad$ $\mathcal{O}^g(Pr{\to}Nd, \text{Run}^{nd}) = 1$
  - $\forall t \in Trun^{pr} : \mathcal{O}^g(t, \text{Run}^{pr}) = 1$ $\quad$ $\mathcal{I}^g(Nd{\to}Pr, \text{Stop}^{nd}) = 1$
  - $\forall t \in T^{nd} : \mathcal{I}^g(t, \text{Run}^{nd}) = 1$ $\quad$ $\mathcal{O}^g(Nd{\to}Pr, \text{Run}^{pr}) = 1$
  - $\forall t \in Tstop^{nd} : \mathcal{O}^g(t, \text{Stop}^{nd}) = 1$
  - $\forall t \in Trun^{nd} : \mathcal{O}^g(t, \text{Run}^{nd}) = 1$
  - for all the other pairs t, p, $\mathcal{I}^g(t,p) = 0, \mathcal{O}^g(t,p) = 0, \mathcal{H}^g(t,p) = 0$;
- $\forall t \in T^{nd}, prio^g(t) = prio^{nd}(t)$, $\quad \forall t \in T^{pr}, prio^g(t) = prio^{pr}(t)$ and $prio^g(Pr{\to}Nd) = prio^g(Nd{\to}Pr) = 0$ (lowest priority).
- $\forall t \in T^{pr}, W^g(t) = W(t)$; for all other $t \in T^g$, $W^g(t)$ is not defined;
- the initial marking m_0 is equal to that of PN^{nd} (which is the same as that of PN^{pr}) for the set P of places, while for the added places we have: $m_0(\text{Run}^{nd}) = 1$ (system starts in a non deterministic state); $m_0(\text{Stop}^{nd}) = 0$; and $m_0(\text{Run}^{pr}) = 0$ and $m_0(\text{Stop}^{pr}) = 0$;

By construction it is not possible to have both non deterministic and probabilistic transitions enabled in any given state. Therefore, the Reachability Set (RS) of N^g can be partitioned into two subsets RS^{pr} and RS^{nd}, such that:

- RS^{pr} contains the probabilistic states (where place Run^{pr} of N^g is marked);
- RS^{nd} has the non-deterministic states (place Run^{nd} of N^g is marked).

Considering this partition of the system states it is possible to identify on the Reachability Graph (RG) the (maximal) sub-paths traversing only states of the same type, so that the RG paths can be described as an alternating sequence of non deterministic and probabilistic sub-paths (we shall consider the particular case of paths with loops in a moment). Then, each probabilistic sub-path can be substituted by a single probabilistic multi-step and assigned an interval probability based on the weight intervals of the transitions firing along the path, as explained in the last part of this section (from Def.6 on). In the same way, a non deterministic maximal sequence can be substituted by a single "long" action, and only the states that start a (maximal) non deterministic sequence σ will appear as states in the BMDP, while the other states in RS^{nd} will be eliminated. Observe that each maximal sequence of probabilistic transitions starts in a state from RS^{pr} and ends in a state from RS^{nd} and each maximal sequence of non deterministic transitions starts in a state from RS^{nd} and ends in a state from RS^{pr}. The reward function has to be extended as follows:

Definition 5. *The transition reward for a non deterministic transition sequence σ is defined by:*

$$rt(\sigma) = \sum_{t \in T^{nd}} rt(t) \, |\sigma|_t$$

where $|\sigma|_t$ is the number of times the non deterministic transition t occurs in σ.

The above definition of $rt(\sigma)$ assumes that reward values are independent from the firing order in the sequence σ.

A BMDP can be generated from an MDPNU if the RG of the MDPNU satisfies the following properties: (1) there are no terminal strongly connected components involving only probabilistic/non deterministic states (and hence there are no deadlock states); (2) there are no loops of non deterministic transitions; (3) loops of probabilistic transitions can be handled under some condition, as discussed next.

The BMDP can be derived from the RG of MDPNU model in two steps:
1) build from the RG a Reduced RG (RRG) in which any non deterministic (maximal) path $\sigma = nd_1 \xrightarrow{t_1} nd_2 \ldots nd_h \xrightarrow{t_h} pr_1$ is replaced by a path $\sigma' = nd_1 \xrightarrow{\{t_1,\ldots,t_h\}} pr_1$ where the first non deterministic state $nd_1 \in \sigma$ is directly connected to the probabilistic state pr_1 through a new arc labeled with the (multi)set of all the transitions in σ, corresponding to the possible actions from nd_1 in the BMDP. For the corresponding transitions we define a function $G(nd, \sigma', pr) \to \{0, 1\}$ where $G(nd, \sigma', pr) = 1$ if $\sigma' = nd \xrightarrow{\{t_1,\ldots,t_h\}} pr$ and 0 otherwise. The reward associated with σ' is computed by $rt(\sigma')$.
2) build from the RRG the BMDP by reducing all the probabilistic paths. Every path starting with a non deterministic state followed by a "composite action" σ and by a (maximal) sequence of probabilistic states ending with a non deterministic state, is substituted in the BMDP by a path directly connecting the two non deterministic states and labeled with action σ and a probability obtained from the probabilistic path.

State space S of the BMDP is given by all $nd \in RS^{nd}$ such that $G(nd, \sigma, pr) = 1$ for some path σ and $pr \in RS^{pr}$ and $A = \{\sigma | \exists pr \in RS^{pr} : G(nd, \sigma, pr) = 1\}$. The transition function F of the BMDP can be computed by combining non deterministic and probabilistic transitions.

Definition 6 (Transition probability matrix $\mathcal{R}_\uparrow$). *The transition probability matrix $\mathcal{R}_\uparrow$ is defined by three matrices $\mathcal{R}_\downarrow$, $\overline{\mathcal{R}}$ and $\mathcal{R}_\uparrow$. Each matrix $\mathcal{R}_* \in \{\mathcal{R}_\downarrow, \overline{\mathcal{R}}, \mathcal{R}_\uparrow\}$ can be computed from*

$$\mathcal{R}_* = \left(\sum_{n=0}^{\infty} \left(\mathcal{R}_*^{(pr,pr)} \right)^n \cdot \mathcal{R}_*^{(pr,nd)} \right). \tag{5}$$

$\mathcal{R}_*^{(pr,pr)}$ *is the matrix encoding the probability of going from state pr to state pr' without hitting any intermediate non deterministic state, and $\mathcal{R}_*^{(pr,nd)}$ encodes the matrix of the transition probabilities of moving from a probabilistic state pr to a non deterministic state nd.*
Moreover $\mathcal{R}_*(i,j)$ is computed as follows:

$$\mathcal{R}_\downarrow(i,j) = \frac{\sum_{i \overset{t}{\to} j}(w_t - l_t)}{\sum_{i \overset{t'}{\to}} w_{t'}}, \ \ \overline{\mathcal{R}}(i,j) = \frac{\sum_{i \overset{t}{\to} j} w_t}{\sum_{i \overset{t'}{\to}} w_{t'}}, \ \ \mathcal{R}_\uparrow(i,j) = \frac{\sum_{i \overset{t}{\to} j}(w_t + u_t)}{\sum_{i \overset{t'}{\to}} w_{t'}} \tag{6}$$

where $\langle w_t, l_t, u_t \rangle = W^g(t)$, $i \overset{t}{\to}$ stands for "all transitions t enabled in state i".

Two possibilities must be considered when computing $\mathcal{R}_*^{(pr,pr)}$. The first corresponds to the situation in which there are no loops involving only probabilistic states. This means that for any probabilistic state $pr_i \in RS_{pr}$ there is a value $n_{0,i}$ such that any sequence of transition firings of length $n \geq n_{0,i}$ starting from such state must reach a non deterministic state $nd_j \in RS_{nd}$. In this case:

$$\exists n_0 : \sum_{k=0}^{\infty} \left(\mathcal{R}_*^{(pr,pr)} \right)^k = \sum_{k=0}^{n_0} \left(\mathcal{R}_*^{(pr,pr)} \right)^k$$

The second corresponds to the situation in which there are possibilities of loops involving only probabilistic states, so that there is a possibility to be "trapped" within a set of probabilistic states. In this case if there is at least one path that allows to exit from the loop arriving in a non deterministic state, then:

$$\sum_{n=0}^{\infty} \left(\mathcal{R}_*^{(pr,pr)} \right)^n = \left[I - \mathcal{R}_*^{(pr,pr)} \right]^{-1}$$

To ensure the convergence of the infinite summation on the left, the dominant eigenvalue of matrix $\mathcal{R}_*^{(pr,pr)}$ must be < 1. This is surely true for $\overline{\mathcal{R}}$ if from each probabilistic state pr_i there is a reachable probabilistic state pr_j such that in pr_j there is a transition with non-zero probability that leads to a non deterministic state nd_k. The condition is necessary and sufficient. The same condition applied to $\mathcal{R}_\downarrow$ is sufficient but not necessary and if it is applied to $\mathcal{R}_\uparrow$ it is necessary but not sufficient. In conclusion $\mathcal{R}_*$ can be rewritten in this way:

$$\mathcal{R}_* = \begin{cases} \left(\sum_{k=0}^{n_0} \left(\mathcal{R}_*^{(pr,pr)} \right)^k \right) \cdot \mathcal{R}_*^{(pr,nd)} & \text{if there are no loops} \\ \left(\left[I - \mathcal{R}_*^{(pr,pr)} \right]^{-1} \right) \cdot \mathcal{R}_*^{(pr,nd)} & \text{if there are loops} \end{cases}$$

Function F of the BMDP is then given by $F_*(i,a,j) = \sum_{k \in RS^{pr}} G(i,a,k)$ $\mathcal{R}_*(k,j)$ where $a \in A$.

4   Case Study

The use case presented in this paper[1] describes the operator of a Control Room receiving incoming calls for different types of emergency services with different priorities (i.e. high and low emergency) and two services (gas and electricity).

Clients call by phone the control room operators to notify a malfunction of the utility. The called operator has to classify the request as either HG (gas high priority), LG (gas low priority), HE (electricity high priority), LE (electricity low priority) or inappropriate (redirected to another service center). Phone call ends with the classification. Request of HG, LG, HE and LE requires the intervention of a company technician at the client's homes, plumbers for gas and electricians for electricity. Deciding whether to allocate a technician to a low or high priority call is an optimization problem, as there are costs and penalties associated with the treatment of an emergency. When a technician is sent to the caller's home, he determines if the emergency call was properly classified. If it was not, a new request with (hopefully) the correct classification is inserted in the system.

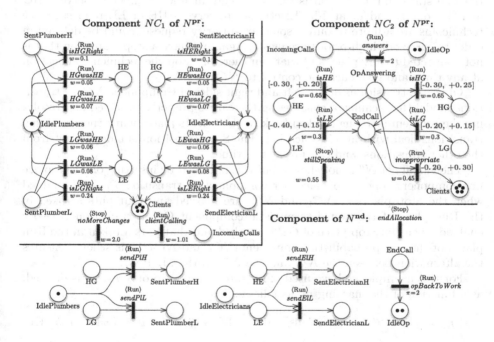

Fig. 2. The three components of the PN^{pr} and PN^{nd} subnets of the MDPNU.

The MDPNU of the system models the allocation part as a set of non deterministic choices (the PN^{nd} subnet) and the operator classification and the technician intervention with possible re-classification of the emergency as probabilistic choices with uncertainty (the PN^{pr} subnet).

---

[1] Taken from a control room model of the Artemis project HoliDes.

The PN^{pr} subnet is actually made of two components: NC_1 and NC_2, (NC stands for "non controllable"). The first models the environment outside the Control room, while the latter models the Control room itself. The two component nets are shown in the upper part of Fig. 2. The reference value w of each triplet $W(t)$ is depicted as $w = \ldots$, while each range of uncertainty is written in brackets as $[-l, +u]$. The PNnd subnet has a single component GC, that decides for the technician allocations (where GC stays for "globally controllable"). Transitions are labeled as either $\langle$Run$\rangle$ or $\langle$Stop$\rangle$. Starting from the NC_2 component of Figure 2, an Incoming Call is taken by an Idle Operator, and the client's call is classified into the four places for HG, LG, HE, LE or as inappropriate and the Call is Ended. Note that *answers* is a transition of priority 2 (denoted π_2), while all the others have priority 1 (the default value), as incoming calls should be answered as soon as possible. The Stop transition has no relation with the rest of the net, it may fire at any time, thus allowing to pass the control to the decision component GC, if also the probabilistic component NC_1 has fired its stop transition. When the control goes to GC, the operator is sent back to the Idle state and a decision is taken to send an Idle Plumber to serve a HG or LG request, and/or an Idle Electrician to serve a HE or LE request. Since technicians are a finite resource, some low-priority requests could be delayed in order to carry out high-priority requests. Again we have a stop transition that is not connected to any place and that can therefore stop the technician allocation at any time, and give back the control to NC_1 and NC_2.

Of the 12 transitions on the left and right sides of Component NC_1, 4 transitions model the technician considering the call as properly classified and resolving the problem (sending a token back to the Clients place),while the remaining transitions deal with misclassification. The Stop transition is independent from the other transitions, and can fire at any time.

The global net N^g is not shown, but it is obtained by inserting one Run place per component, one Stop place per component, a transition $Pr{\rightarrow}Nd$ enabled when the Stop places of NC_1 and NC_2 are marked and that puts a token in the Run place of the non deterministic component GC, a transition $Nd{\rightarrow}Pr$ enabled when the Stop place of GC is marked and that puts a token in the Run places of the two probabilistic componentsNC_1 and NC_2. This schema ensures the alternation between the N^{pr} and the N^{nd} of the MDPNU.

For this example we have used only state rewards, so only $rs(m)$ is defined: as a function of the marking m, as follows:

$$rs(m) = -5 \cdot \#\text{IncomingCalls} - 10 \cdot (\#\text{LG} + \#\text{LE}) - 100 \cdot (\#\text{HG} + \#\text{LE})$$

being $\#P$ the number of tokens of place P in the marking m of the MDPNU.

For the experiments we have considered three variations of the model parameters, as reported in Table 1, that results in BMDPs with an increasing number of states and actions. All experiments are performed on a standard PC with 3.3 GHz processor and 8 GB main memory. The documented times describe the whole computation including the computation of the optimal policies and

Table 1. Model parameters and sizes of the MDPNU RG and its BMDP process.

Model instance	Params				State space			
	Clients	Ops	Electr.	Plumb.	RG States	BMDP States	Actions	Time (sec)
I1	2	2	1	1	220	160	27	2.9
I2	4	2	1	1	2326	1679	35	69.3
I3	6	3	1	1	15420	11868	52	1722.6

Table 2. Experimental results computed for the First Scenario.

Model instance	Min			Avg			Max			Pareto	
	Iter	Stationary gain	Time (sec)	Iter	Stationary gain	Iter	Time (sec)	Stationary gain	Time (sec)	Optim. policies	Time (sec)
I1	7	-1.632e+04	<0.1	4	-1.546e+04	<0.1	9	-1.438e+04	<0.1	1	<0.1
I2	7	-3.632e+04	1.6	6	-3.546e+04	1.2	7	-3.438e+04	1.5	1	5.7
I3	6	-5.632e+04	64.9	5	-5.547e+04	50.8	7	-5.438e+04	59.8	1	421.2

Table 3. Experimental results computed for the Second Scenario.

Model instance	Min			Avg			Max			Pareto	
	Iter	Stationary gain	Time (sec)	Iter	Stationary gain	Iter	Time (sec)	Stationary gain	Time (sec)	Optim. policies	Time (sec)
I1	6	-1.669e+04	<0.1	4	-1.547e+04	<0.1	7	-1.210e+04	<0.1	3597	4587
I2	10	-3.669e+04	1.2	5	-3.547e+04	0.4	10	-1.599e+04	1.4	out-of-memory	
I3	10	-5.669e+04	83.9	5	-5.547e+04	54.7	10	-1.270e+04	96.2	out-of-memory	

the stationary distribution, as well as the input and output operations to read the matrices and write the gain vectors.

First Scenario: in this case, the MDPNU is the one shown in Figure 2, where the uncertainty is on the operator's interpretations of the customer's problem description. The Min, Avg and Max columns report the number of policy iterations to achieve convergence, the stationary gain (computed with discount $\gamma = 0.999$) and the solution time for the computation of the formula in Eq. (2), (1), and (3). The last two columns report the number of Pareto optimal policies, and the time needed to compute them. The main result that emerges from this scenario is that there is just one optimal policy, which is the same for the min, avg and max cases. This means that the optimal strategy remains stable despite the uncertain classification, providing a robust optimal strategy for the entire class of MDPs underlying this BMDP.

Second Scenario: in this case we add an uncertainty of $[-0.7, +2.5]$ to the *client-Calling* transition, simulating the fact that usually, when a problem happens on the utility infrastructure, a lot of customers call the control room simultaneously to signal the problem, while on other days there are few or no customers calling at all. Taking into account an uncertainty on the probability of the clients

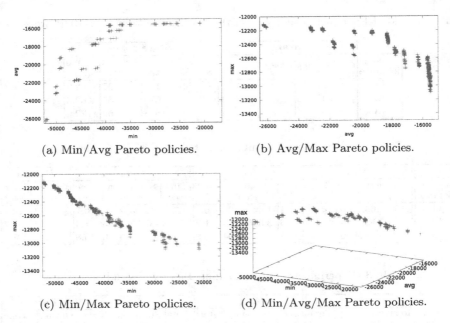

(a) Min/Avg Pareto policies.   (b) Avg/Max Pareto policies.

(c) Min/Max Pareto policies.   (d) Min/Avg/Max Pareto policies.

Fig. 3. Distribution of the optimal Pareto policies in the second scenario.

calling the Control Room results in an BMDP with different parameters, whose solution is reported in Table 3. The main difference with regard to the First Scenario is that now we have many Pareto optimal policies. For the smaller model instance I1, there are 3597 Pareto optimal policies (out of 21950 policies checked in the algorithm). The computation of the Pareto set is very time consuming and requires almost 1.5 hours whereas policies that maximize only one goal (min, avg or max) can be computed for the small configuration in a negligible time. Fig. 3 shows the stationary gain of these policies in the min/avg/max case for all these policies in three dimensions, as seen from the min/avg plane, avg/max plane, min/max plane and in perspective. Each red mark represents an optimal policy, and its coordinate in the min/max/avg axes are the stationary gains in the min/max/avg cases. Interestingly, there is a single policy that minimizes the costs for both the minimum and average case (in the upper-left corner of Fig. 3(a)), which intuitively represents the most conservative choice (performing not that bad in the worst case, and still good in the average case), which is the goal of the Pareto optimization. However, it can also be seen from the results that according to best case, the conservative strategy does not behave well, other more optimistic policies yield much higher rewards in this case.

5   Conclusions

This paper introduces the high-level formalism of MDPNU that translates directly into a BMDP process. This high level formalism allows one to model complex Markov decision processes with uncertainty on the probabilistic distributions and on the reward function, and can be applied to verify the robustness of optimal strategies on the continuous family of MDPs underlying the BMDP.

The MDPNU formalism has been used to model the problem of personnel allocation in a utility company, under different scenarios of uncertainty of the process parameters, that model real system variations. Experimental results show that in some scenarios it is possible to identify a robust strategy (and prove the robustness) that yields optimal results under all possibly resolutions of uncertainty in the parameters, while in other scenarios a single strategy that exhibits a good behavior in all situations does not exist. In such cases, Pareto multi objective optimization often allows one to find strategies with good performance for all possible realizations of uncertainty, i.e. which are less sensitive to the parameter variations of the BMDP.

Results in the paper indicate that the computation of optimal strategies according to a single MDP or even with respect to one of the extreme MDPs defined by a BMDP can be done efficiently for BMDPs with several thousands of states. In the same way the performance of a given strategy can be easily analyzed. Pareto optimal policies computation requires more effort, because the number can be huge and many policies have to be checked and analyzed for Pareto optimality. Thus, there is a need for more efficient algorithms that compute good compromise solutions without analyzing the whole set of Pareto optimal policies. This is, however, a subject for future research.

References

1. de Alfaro, L.: Stochastic transition systems. In: Sangiorgi, D., de Simone, R. (eds.) CONCUR 1998. LNCS, vol. 1466, pp. 423–438. Springer, Heidelberg (1998)
2. Alur, R., Henzinger, T.: Reactive modules. Formal Methods in System Design 15(1), 7–48 (1999)
3. Beccuti, M., Franceschinis, G., Haddad, S.: Markov decision Petri net and Markov decision well-formed net formalisms. In: Kleijn, J., Yakovlev, A. (eds.) ICATPN 2007. LNCS, vol. 4546, pp. 43–62. Springer, Heidelberg (2007)
4. Beccuti, M., Franceschinis, G., Codetta-Raiteri, D., Haddad, S.: Computing optimal repair strategies by means of NdRFT modeling and analysis. The Computer Journal 57(12), 1870–1892 (2014)
5. Bellman, R.: A Markovian Decision Process. Indiana Univ. Math. Journal 6 (1957)
6. Dean, T., Wellman, M.P.: Planning and Control. Morgan Kaufmann Pub, San Francisco (1991)
7. Dean, T., Givan, R., Leach, S.: Model reduction techniques for computing approximately optimal solutions for Markov decision processes. In: 13th Conference on Uncertainty in Artificial Intelligence, pp. 124–131. Morgan Kaufmann (1997)
8. Delgado, K.V., Sanner, S., de Barros, L.N.: Efficient solutions to factored MDPs with imprecise transition probabilities. Artif. Intell. 175(9–10), 1498–1527 (2011)

9. Eboli, M.G., Cozman, F.G.: Markov decision processes from colored Petri nets. In: da Rocha Costa, A.C., Vicari, R.M., Tonidandel, F. (eds.) SBIA 2010. LNCS, vol. 6404, pp. 72–81. Springer, Heidelberg (2010)
10. Givan, R., Leach, S.M., Dean, T.L.: Bounded-parameter Markov decision processes. Artif. Intell. **122**(1–2), 71–109 (2000)
11. Puterman, M.L.: Markov Decision Processes: Discrete Stochastic Dynamic Programming, 1st edn. John Wiley & Sons Inc, New York (1994)
12. Scheftelowitsch, D., Buchholz, P.: Multi-criteria approaches to Markov decision processes with uncertain transition parameters. Tech. rep., TU Dortmund (2015). http://ls4-www.cs.uni-dortmund.de/download/buchholz/pareto.pdf

On-the-fly Fluid Model Checking via Discrete Time Population Models

Diego Latella[1], Michele Loreti[2,3], and Mieke Massink[1](✉)

[1] Istituto di Scienza e Tecnologie dell'Informazione 'A. Faedo', CNR, Pisa, Italy
[2] Dip. di Statistica, Informatica, Applicazioni 'G. Parenti',
Università di Firenze, Firenze, Italy
[3] IMT Advanced Studies Lucca, Lucca, Italy
mieke.massink@isti.cnr.it

Abstract. We show that, under suitable convergence and scaling conditions, fluid model checking bounded CSL formulas on selected individuals in a continuous *large* population model can be approximated by checking equivalent bounded PCTL formulas on corresponding objects in a discrete time, time synchronous Markov population model, using an *on-the-fly* mean field approach. The proposed technique is applied to a benchmark epidemic model and a client-server case study showing promising results also for the challenging case of nested formulas with time dependent truth values. The on-the-fly results are compared to those obtained via global fluid model checking and statistical model-checking.

1 Introduction

Model checking has been widely recognised as a powerful approach to the automatic verification of concurrent and distributed systems, including aspects of their performance. It consists of an efficient procedure that, given an abstract model $\mathcal{M}$ of the system, decides whether $\mathcal{M}$ satisfies a logical formula Φ, typically drawn from a temporal logic. Recently, the integration of mean field and fluid approximation techniques, that originate in statistical physics, with formal modelling techniques has received increasing attention as a way to obtain *highly scalable* approximate model checking techniques, such as fluid model checking [3,4,12] and mean field model checking [16]. These approaches are *independent* of the population size, as long as this is large enough. Such extreme scalability is a prerequisite for the verification of large scale collective adaptive systems, of which performance aspects and emerging behaviour are an essential feature.

Traditional model checking approaches do not scale up to such large systems due to the well-known state space explosion problem. Statistical model-checking [21] is in general performing much better in this respect. It avoids the generation of the state space and approximates the results by a statistical analysis of a number of randomly generated finite executions of the model. This leads to

This work is partially supported by the EU project QUANTICOL (nr. 600708), and the IT MIUR project CINA.

© Springer International Publishing Switzerland 2015
M. Beltrán et al. (Eds.): EPEW 2015, LNCS 9272, pp. 193–207, 2015.
DOI: 10.1007/978-3-319-23267-6_13

better scalability, but the complexity is still linear in the number of objects that the system is composed of. Furthermore, depending on the accuracy required and the particular property of interest, it may be necessary to take a large number of samples into consideration.

Fluid model checking [3,4,12] relies on a global model checking approach for time-inhomogeneous Continuous Time Markov Chains (ICTMC) representing a typical individual object in the context of a large CTMC population model. The rates of the individual may depend on the fraction of the population that is in a particular state. The algorithm relies on the deterministic approximation of the average stochastic behaviour of the system in continuous time, i.e. a fluid approximation. An alternative approach is the one we refer to in this paper by *on-the-fly mean field model checking* [16]. In this approach only as much of the state space as strictly needed to verify the given formula is generated from a high-level specification of the individual behaviour and the population. The algorithm relies on the deterministic approximation of the probabilistic behaviour of the population in discrete time and can be used to verify formulas of the bounded Probabilistic Computation Tree Logic (PCTL)[9].

The main contribution of the present paper is to show that, under suitable convergence and scaling assumptions[1], and for models that are not too stiff[2], fluid model checking can be performed exploiting on-the-fly mean field model checking techniques [16] applied on a time-inhomogeneous *Discrete Time* Markov Chain (IDTMC) model and bounded PCTL formula that are derived from a corresponding ICTMC model and bounded Continuous Stochastic Logic (CSL) [1] formula via a transformation presented in Sect. 2.3. This approach is interesting and differs from other approaches in several respects: 1) the mean field model checking algorithm is implemented as a particular instantiation of an *on-the-fly probabilistic model-checking algorithm* [16]; 2) the latter is *parametric w.r.t. the semantics interpretation* of the model specification language and in this case we instantiate it on the mean-field approximation of a simple probabilistic population description language; 3) the transformation presented in Sect. 2.3 allows one to reuse the implementation once more for a class of CTMC population models; 4) the global fluid model checking algorithm in [3,4] requires the *a priori* calculation of discontinuity points, i.e. points in time in which the truth values of time-dependent (sub)-formulas of an until formula change. This is a non-trivial task and consists in finding all zeros of an analytic function. In the on-the-fly setting such points are detected automatically during the computation of the probabilities, upto a difference that is in the order of a small discrete step size; 5) on-the-fly approaches are particularly efficient when verifying conditional reachability properties because in that case much fewer states need to be generated. Ultimately, however, the on-the-fly mean field algorithm is based on an Euler method to solve differential equations. This poses certain limitations on the continuous time models that can be analysed efficiently this way, in particular they should not be too stiff. For non stiff models the results are promising as shown

---

[1] See Theorem 5 of [4].

[2] Stiff models are those whose rates differ several orders of magnitude.

by the available benchmark models for which also some results for global fluid model checking and statistical model checking are available in the literature.

The outline of the paper is as follows. Sect. 2 introduces discrete and continuous time Markov population models. The relevant temporal logics are recalled in Sect. 3. Sect. 4 presents the model and logic transformation functions and the correctness results w.r.t. fluid model-checking. Sect. 5 provides a comparison with benchmark examples from the literature. Related work and conclusions are presented in Sect. 6 and Sect. 7, respectively. Basic knowledge on Markov chains and related model checking algorithms is assumed.

2   Population Models

We consider two types of Markov population models: continuous time models and discrete time models. In both models we assume that the number of objects in the population is N and that this size remains constant during execution.

2.1   Continuous Time Population Models

For CTMC population models we adopt the notation following [3]. Let $Y_i^{(N)}(t) \in \mathcal{S}$ be the random variable representing the state of object i at time t, where $\mathcal{S} = \{1, 2, \ldots, n\}$ represents the local state space of each object. Multiple classes of agents are represented by partitioning $\mathcal{S}$ into disjoint subsets and allowing state changes only within a single class. Let $\mathbf{Q}^{(N)}(\boldsymbol{x})$ denote the $n \times n$ infinitesimal generator matrix that depends on the fraction of objects $\boldsymbol{x} \in [0,1]^n$ that are in each state. The latter quantity can be computed from $Y_i^{(N)}$ as $\hat{X}_i^{(N)}(t) = \frac{1}{N} \sum_{j=1}^{N} \mathbf{1}\{Y_j^{(N)}(t) = i\}$[3]. $\langle \hat{X}_1^{(N)}(t), \hat{X}_2^{(N)}(t), \ldots, \hat{X}_n^{(N)}(t) \rangle$ is a CTMC $\hat{\mathcal{X}}^{(N)}$ [2] on the state space $[0,1]^n$, also called the *occupancy measure*, with initial state $\boldsymbol{x}_0^{(N)} \in [0,1]^n$. The average infinitesimal variation of $\hat{\mathcal{X}}^{(N)}$, given that it is in state $\boldsymbol{x}$ is $F^{(N)}(\boldsymbol{x}) = \boldsymbol{x}^T \mathbf{Q}^{(N)}(\boldsymbol{x})$, also called the drift[4]. If, for $N \to \infty$, $\mathbf{Q}^{(N)}(\boldsymbol{x})$ converges uniformly to the Lipschitz continuous generator matrix $\mathbf{Q}(\boldsymbol{x})$, and $\boldsymbol{x}_0^{(N)}$ to $\boldsymbol{x}_0$, and, furthermore, if $\boldsymbol{x}(t)$ is the solution of the ODE $\frac{d\boldsymbol{x}}{dt} = F(\boldsymbol{x}) = \boldsymbol{x}^T \mathbf{Q}(\boldsymbol{x})$ for initial condition $\boldsymbol{x}(0) = \boldsymbol{x}_0$, then in the limit the two processes behave almost surely the same for a finite time horizon T [7,13][5].

It is possible to decouple the analysis of a single object from the analysis of the global system by letting the behaviour of the single object depend on the other objects only through the solution of the fluid ODE. This result is known as *fast simulation* [7,18]. The stochastic behaviour of a single object can be defined as $Z^{(N)} = Y_1^{(N)}$ on state space $\mathcal{S}$, assuming, without loss of generality, we are interested in the behaviour of the first object. Note that $Z^{(N)}$ is an

---

[3] $\mathbf{1}\{x = y\}$ yields 1 if $x = y$ and 0 otherwise.

[4] $\boldsymbol{x}^T$ denotes the transpose of vector $\boldsymbol{x}$.

[5] The conditions on uniform convergence and Lipschitz continuity automatically hold for PEPA population models because in that case the rate functions are all piecewise linear [20].

ICTMC. Let $z(t)$ be the ICTMC of an individual object with states in $\mathcal{S}$ such that $\Pr\{z(t+dt) = j|z(t) = i\} = q_{i,j}(\boldsymbol{x}(t))dt$, and let $\mathbf{Q}_z(\boldsymbol{x}(t)) = (q_{i,j}(\boldsymbol{x}(t)))$. We then have that for any finite horizon T and $t \leq T$ the behaviour of the single object $Z^{(N)}(t)$ tends to the behaviour of the object that senses the rest of the system only through its limit behaviour given by $\boldsymbol{x}$, i.e. $z(t)$.

Running Example: Consider the simple PEPA specification of processors and resources that synchronise on a common task [11]:

$$Proc0 := (task1, r_1).Proc1 \qquad Res0 := (task1, r_1).Res1$$
$$Proc1 := (task2, r_2).Proc0 \qquad Res1 := (reset, s).Res0$$
$$Proc0[N_p] \underset{task1}{\bowtie} Res0[N_q]$$

where $Proc0[N_p]$ is a shorthand notation for N_p instances of process $Proc0$ in parallel, and $Res0[N_q]$ denotes N_q instances of process $Res0$ in parallel. Such population oriented PEPA specifications have been given a formal semantics based on ODE by Hillston in [11] and by Tribastone et al. in [20]. In particular, the ODE associated to the example specification can be given as:

$$\frac{d\,proc0(t)}{dt} = -r_1.min(proc0(t), res0(t)) + r_2.proc1(t)$$
$$\frac{d\,proc1(t)}{dt} = -r_2.proc1(t) + r_1.min(proc00(t), res0(t))$$
$$\frac{d\,res0(t)}{dt} = -r_1.min(proc0(t), res_0(t)) + s.res1(t)$$
$$\frac{d\,res1(t)}{dt} = -s.res_1(t) + r_1.min(proc0(t), res0(t))$$

where $proc0(t)$, $proc1(t)$, $res0(t)$ and $res1(t)$ denote the limit occupancy measure at time t for each local state respectively. The function min denotes the minimum function and originates from the specific definition of action synchronisation of the semantics of PEPA [11].

The infinitesimal $\mathbf{Q}$-matrix of an individual object that depends on the behaviour of the global system via its limit occupancy measure can be retrieved as follows (see [3]). From the PEPA semantics of the synchronisation (cooperation) operator we know that the total rate of a shared $task1$ action is given by $min(r_1.proc_0(t), r_1.res_0(t))$. The rate of an *individual process* performing a $task1$ action is then this global rate divided by the fraction of objects present in the system at time t, i.e. $proc_0(t)$. The rate of an individual process performing a $task_2$ action is simply r_2 because this action does not depend on the limit occupancy measure $\boldsymbol{x}$, where $\boldsymbol{x}^T(t) = (proc_0(t), proc_1(t), res_0(t), res_1(t))$. Similar reasoning applies to the rates of a resource object. So, we obtain the following rate functions for the $\mathbf{Q}$-matrix:

$$\mathbf{Q}_{proc0,proc1}(\boldsymbol{x}(t)) = r_1.min(proc0(t), res0(t))/proc0(t)$$
$$\mathbf{Q}_{proc1,proc0}(\boldsymbol{x}(t)) = r_2$$

$$\mathbf{Q}_{res0,res1}(\boldsymbol{x}(t)) = r_1.min(proc0(t), res0(t))/res0(t)$$
$$\mathbf{Q}_{res1,res0}(\boldsymbol{x}(t)) = s$$

The rate functions used in the $\mathbf{Q}$-matrix are all continuous and bounded, at least as long as we do not divide by zero.

2.2   Discrete Time Population Models

For DTMC population models we consider again a system of N interacting objects. Let $W_i^{(N)}(k) \in \mathcal{S}$ be the random variable representing the state of object i at step k, where $\mathcal{S} = \{1, 2, \ldots, n\}$ represents the local state space of each object. Let $\mathbf{K}^{(N)}(\boldsymbol{m})$ denote the $n \times n$ one step transition probability matrix that depends on the fraction of objects $\boldsymbol{m} \in [0,1]^n$ that are in each state of $\mathcal{S}$. This fraction can be computed as $\hat{M}_i^{(N)}(k) = \frac{1}{N} \sum_{j=1}^{N} \mathbf{1}\{W_j^{(N)}(k) = i\}$. It is easy to see that the process $\langle \hat{M}_1^{(N)}(k), \hat{M}_2^{(N)}(k), \cdots, \hat{M}_n^{(N)}(k) \rangle$ is a DTMC $\hat{\mathcal{M}}^{(N)}$ on the state space $[0,1]^n$, with initial state $\boldsymbol{m}_0^{(N)} \in [0,1]^n$.

The average variation of $\hat{\mathcal{M}}^{(N)}$, given that it is in state $\boldsymbol{m}$ is $F^{(N)}(\boldsymbol{m}) = \boldsymbol{m}^T \mathbf{K}^{(N)}(\boldsymbol{m})$. Suppose that, for all i, j and for $N \to \infty$, the elements $\mathbf{K}_{i,j}^{(N)}(\boldsymbol{m})$ converge uniformly in $\boldsymbol{m}$ to some $\mathbf{K}_{i,j}(\boldsymbol{m})$, which is a continuous function of $\boldsymbol{m}$, and $\boldsymbol{m}_0^{(N)}$ converges almost surely to $\boldsymbol{m}_0$, and furthermore define $\boldsymbol{m}(k)$ as follows: $\boldsymbol{m}(0) = \boldsymbol{m}_0$ and $\boldsymbol{m}(k+1) = \boldsymbol{m}(k)^T \cdot \mathbf{K}(\boldsymbol{m}(k))$; then, for any fixed step t, almost surely $\hat{\mathcal{M}}^{(N)}$ converges to function $\boldsymbol{m}(k)$ [18]. As for CTMC population models, it is possible to decouple the analysis of the single object from the analysis of the global system using a fast simulation approach involving the solution of a difference equation rather than an ODE.

Example: Taking probabilities α_i for the rates r_i in the processes and resources example, we obtain the following difference equations for $\boldsymbol{m}^T(k) = (m_{p0}(k), m_{p1}(k), m_{r0}(k), m_{r1}(k))$:

$$m_{p0}(k+1) = m_{p0}(k) - \alpha_1.min(m_{p0}(k), m_{r0}(k)) + \alpha_2.m_{p1}(k)$$
$$m_{p1}(k+1) = m_{p1}(k) + \alpha_1.min(m_{p0}(k), m_{r0}(k)) - \alpha_2.m_{p1}(k)$$
$$m_{r0}(k+1) = m_{r0}(k) - \alpha_1.min(m_{p0}(k), m_{r0}(k)) + \alpha_s.m_{r1}(k)$$
$$m_{r1}(k+1) = m_{r1}(k) + \alpha_1.min(m_{p0}(k), m_{r0}(k)) - \alpha_s.m_{r1}(k)$$

where $m_{pj}(k)$ and $m_{rj}(k)$ denote the limit occupancy measure at step k for processes and resources. We can also retrieve the one step probability matrix for each individual process and resource object using a similar reasoning as in the CTMC case:

$$\mathbf{K}_{p0,p1}(\boldsymbol{m}(k)) = \alpha_1.min(m_{p0}(k), m_{r0}(k))/m_{p0}(k)$$
$$\mathbf{K}_{p1,p0}(\boldsymbol{m}(k)) = \alpha_2$$

$$\mathbf{K}_{r0,r1}(\boldsymbol{m}(k)) = \alpha_1.min(m_{p0}(k), m_{r0}(k))/m_{r0}(k)$$
$$\mathbf{K}_{r1,r0}(\boldsymbol{m}(k)) = \alpha_s$$

The difference equations can be obtained from $\mathbf{K}$ by $\boldsymbol{m}(k+1) = \boldsymbol{m}(k)^T.\mathbf{K}(\boldsymbol{m}(k))$.

2.3   Relationship Between the Models

First note that we can interpret the difference equations obtained from a discrete time population model as an instance of the Euler forward method for approximating the solution of a set of ODEs. The set of ODEs we are interested in solving are those of a corresponding continuous population model. To obtain an acceptable approximation of the solution we need to find a step size for the difference equations such that absolute stability of the method, to avoid that the global error grows exponentially, and a sufficient accuracy [19] are guaranteed. This, in turn, means that we need to derive suitable values for the probabilities α from the rates in the continuous model. What we would be even more interested in is to transform an ICTMC model of an individual (from which the ODEs can be derived) into an IDTMC model, with the same local states and jump structure as the ICTMC, from which we get exactly the set of difference equations that can be used to approximate the solution of the ODEs. We proceed as follows. Using a feature of CTMC uniformisation[6] we can obtain a DTMC with probability matrix $\mathbf{K}$ such that $\mathbf{K} = \mathbf{I} + \frac{1}{q} \cdot \mathbf{Q}$, where $\mathbf{Q}$ is the infinitesimal rate matrix and q the uniformisation rate that is *at least as large* as the maximal exit rate of the states in the original CTMC. This DTMC preserves the local states and the jumps of the original CTMC apart from additional self-loops. Note that in our case the rates in $\mathbf{Q}$ may depend on the occupancy measure $\boldsymbol{m}$. However, $0 \le m_i \le 1$ for all $i \in |\mathcal{S}|$, so assuming rate-functions that include minimum functions and linear combination[7] (but not rational functions) that we derive from PEPA specifications we can easily find a suitable q.

At this point we need to satisfy also the requirements of absolute stability and obtain a satisfactory accuracy following standard procedures [19]. If we are lucky, q is already large enough so that these requirements are fulfilled, otherwise we need to increase q, which is allowed because q is only a lower bound (see above). For linear systems of l differential equations where $u(t) \in \mathbb{R}^l$ and $d\,u(t)/dt = A \cdot u(t)$ where A is an $l \times l$ matrix a necessary condition is that $h\lambda$ is in the stability region of the Euler method for each eigenvalue λ of matrix A and step size h. So, for each eigenvalue λ we need that $-2 \le h\lambda \le 0$ [19]. For non-linear systems we need to determine the range of each eigenvalue and make sure that the step size h is taken small enough so that $h\lambda$ stays within the region of absolute stability for its complete range. Note that $h = 1/q$.

Example: For the running example, with $r_1 = 10, r_2 = 3.0$ and $s = 7.0$, we obtain uniformisation rate $q = 10$ and eigenvalues $\lambda_1 = 0$ or $-r_1 - r_2 \le \lambda_1 \le -r_2$ and $\lambda_2 = 0$ or $-r1 - s \le \lambda_2 \le -s$, showing that all eigenvalues are in a bounded range, with a maximum absolute value of 17. So when taking $h = 1/q$ we get that $0 \le 17 * 1/q \le 2$. This implies that $q = 10$ guarantees stability of the method, but a higher value may be preferred for better accuracy.

3   Properties of Individual Objects

Properties of the behaviour of individuals in the context of a large population model can be expressed as formulas of a suitable temporal logic. For the purpose

---

[6] I.e. transforming a CTMC into one where each state has the same exit rate by adding self-loops where needed, which is an operation that does not alter the transient and steady state properties of the CTMC.

[7] I.e. piecewise linear functions leading to the class of split-free PEPA models [10].

Table 1. Satisfaction relation: CSL fragment.

$$
\begin{aligned}
&s,t \models_C a &&\Leftrightarrow a \in \ell(s) \\
&s,t \models_C \neg\Phi &&\Leftrightarrow \text{not } s,t \models_C \Phi \\
&s,t \models_C \Phi_1 \vee \Phi_2 &&\Leftrightarrow s,t \models_C \Phi_1 \text{ or } s,t \models_C \Phi_2 \\
&s,t \models_C \mathcal{P}_{\bowtie p}(\varphi) &&\Leftrightarrow \Pr\{\sigma \in Paths_C(s,t) \mid \sigma,t \models_C \varphi\} \bowtie p \\
&\sigma,t \models_C \Phi_1 \mathcal{U}^{\leq\tau} \Phi_2 &&\Leftrightarrow \exists \tau_2 \text{ s.t. } 0 \leq \tau_2 \leq \tau, \sigma@\tau_2, t+\tau_2 \models_C \Phi_2 \,\wedge \\
&&&\quad\; \forall 0 \leq \tau_1 < \tau_2, \, \sigma@\tau_1, t+\tau_1 \models_C \Phi_1
\end{aligned}
$$

of this paper, properties of continuous time models are expressed in bounded CSL, and properties of discrete time models are expressed in bounded PCTL. Both are briefly recalled in this section, where we assume set $\mathcal{P}$ of atomic propositions is given and $a \in \mathcal{P}$, $\tau \in \mathbb{Q}_{\geq 0}$, $k \in \mathbb{N}$ and $\bowtie \in \{>, <\}$ and $p \in [0,1] \cap \mathbb{Q}$.

Continuous Stochastic Logic for ICTMC. The syntax of the fragment of bounded CSL we consider is defined below:

$$
\Phi ::= a \mid \neg\Phi \mid \Phi \vee \Phi \mid \mathcal{P}_{\bowtie p}(\varphi) \text{ where } \varphi ::= \Phi \mathcal{U}^{\leq\tau} \Phi.
$$

CSL formulas[8] are interpreted over state labelled ICTMCs $\langle \mathcal{C}, \ell \rangle$, where $\mathcal{C}$ is an ICTMC with state set $\mathcal{S}$ and $\ell : \mathcal{S} \to 2^{\mathcal{P}}$ associates each state with a set of atomic propositions. We define the satisfaction relation on $\mathcal{C}$ and the logic in Table 1. We abbreviate $\langle \mathcal{C}, \ell \rangle$ with $\mathcal{C}$, when no confusion can arise, with $\mathbf{Q}$ its infinitesimal generator matrix. A path σ over $\mathcal{C}$ is a non-empty sequence $s_0 \xrightarrow{t_0} s_1 \xrightarrow{t_1} \cdots$ such that the probability of going from state s_i to s_{i+1} at time $T_i = \sum_{j=0}^{i} t_i$ is positive for all $i \geq 0$. We let $Paths_C(s,t)$ denote the set of all infinite paths over $\mathcal{C}$ starting from state s at time t. We require that all subsets of paths considered are measurable. We let $\sigma@t$ denote the state s_k in σ such that k is the maximum i such that $\sum_{j=0}^{i} t_i \leq t$, i.e. the state reached on path σ at time t. Finally, in the sequel we will consider ICTMCs equipped with an initial state s_0, i.e. the probability mass is initially all in s_0.

Probabilistic Logic for DTMC.[9] The syntax of the fragment of bounded PCTL we consider is defined below:

$$
\Phi ::= a \mid \neg\Phi \mid \Phi \vee \Phi \mid \mathcal{P}_{\bowtie p}(\varphi) \text{ where } \varphi ::= \Phi \mathcal{U}^{\leq k} \Phi.
$$

PCTL formulas are interpreted over *state labelled* DTMCs $\mathcal{D}$ in a similar way as for CTMCs. We assume $\mathbf{P}$ to be the one step probability matrix for $\mathcal{D}$. A path σ over $\mathcal{D}$ is a non-empty sequence of states $s_0, s_1, \cdots$ where $\mathbf{P}_{s_i, s_{i+1}} > 0$ for all $i \geq 0$. We let $Paths_D(s)$ denote the set of all infinite paths over $\mathcal{D}$ starting from

---

[8] For simplicity the time bounds in the formulas are of the form $[0, \tau]$ instead of the more general $[\tau_1, \tau_2]$.

[9] Note that, by making time explicit, the structures used by FlyFast are DTMCs rather than IDTMCs (see Sect. 4).

Table 2. Satisfaction relation: PCTL fragment.

$$s \models_{\mathcal{D}} a \qquad\qquad \Leftrightarrow a \in \ell(s)$$
$$s \models_{\mathcal{D}} \neg\Phi \qquad\qquad \Leftrightarrow \text{not } s \models_{\mathcal{D}} \Phi$$
$$s \models_{\mathcal{D}} \Phi_1 \vee \Phi_2 \quad \Leftrightarrow s \models_{\mathcal{D}} \Phi_1 \text{ or } s \models_{\mathcal{D}} \Phi_2$$
$$s \models_{\mathcal{D}} \mathcal{P}_{\bowtie p}(\varphi) \quad \Leftrightarrow \Pr\{\sigma \in \mathit{Paths}_{\mathcal{D}}(s) \mid \sigma \models_{\mathcal{D}} \varphi\} \bowtie p$$
$$\sigma \models_{\mathcal{D}} \Phi_1 \, \mathcal{U}^{\leq k} \, \Phi_2 \Leftrightarrow \exists 0 \leq h \leq k \text{ s.t. } \sigma[h] \models_{\mathcal{D}} \Phi_2 \,\wedge\, \forall 0 \leq i < h \,.\, \sigma[i] \models_{\mathcal{D}} \Phi_1$$

state s. By $\sigma[i]$ we denote the element s_i of path σ. We will consider DTMCs equipped with an initial state s_0. We define the satisfaction relation on $\mathcal{D}$ and the logic in Table 2.

4   Fluid Model Checking via Discrete Time Models

We first define two transformation functions. Function $\mathcal{T}_M$ takes an ICTMC $z(t)$ with infinitesimal generator matrix $\mathbf{Q}(t)$ and initial state s_0. It takes a step size $d \in \mathbb{Q}$ and a time bound $b > d$. It returns a DTMC with state set $\mathcal{S} \times \{0, \ldots, \lfloor \frac{b}{d} \rfloor\}$, initial state $(s_0, 0)$ and one step transition probability matrix $\mathbf{U}$, as follows:

Definition 1. *For all $0 < d \in \mathbb{Q}, b \in \mathbb{R}$ with $b > d$, and infinitesimal generator matrix $\mathbf{Q}(t)$, $\mathcal{T}_M(\mathbf{Q}(t), d, b)$ is the one step transition probability matrix $\mathbf{U}$:*

$$\mathbf{U}_{(s,i),(s',i')} = \begin{cases} [\mathbf{I} + d \cdot \mathbf{Q}(i \cdot d)]_{s,s'}, & \text{if } i' = i+1, \mathbf{Q}(i \cdot d)_{s,s} \neq 0, \\ 1, & \text{if } i' = i, s' = s, \mathbf{Q}(i \cdot d)_{s,s} = 0, \\ 0, & \text{otherwise} \end{cases}$$

where the indexes of $\mathbf{U}$ are assumed to be ordered as follows:

$$(s_0, 0), ..., (s_n, 0), (s_0, 1), ..., (s_n, 1), ..., (s_0, \lfloor \tfrac{b}{d} \rfloor), ..., (s_n, \lfloor \tfrac{b}{d} \rfloor).$$

Function $\mathcal{T}_F$ below transforms bounded CSL into bounded PCTL formulas:

Definition 2. *For atomic propositions a, bounded CSL formulas Φ, Φ_1 and Φ_2, and $d \in \mathbb{Q}$, function $\mathcal{T}_F$ is defined as follows:*

$$\mathcal{T}_F[\![a]\!]_d = a \qquad\qquad \mathcal{T}_F[\![\Phi_1 \vee \Phi_2]\!]_d = \mathcal{T}_F[\![\Phi_1]\!]_d \vee \mathcal{T}_F[\![\Phi_2]\!]_d$$
$$\mathcal{T}_F[\![\neg\Phi]\!]_d = \neg\mathcal{T}_F[\![\Phi]\!]_d \quad \mathcal{T}_F[\![\mathcal{P}_{\bowtie p}(\Phi_1 \, \mathcal{U}^{\leq\tau} \, \Phi_2)]\!]_d = \mathcal{P}_{\bowtie p}(\mathcal{T}_F[\![\Phi_1]\!]_d \, \mathcal{U}^{\leq\lfloor \frac{\tau}{d} \rfloor} \, \mathcal{T}_F[\![\Phi_2]\!]_d)$$

The definition for the basic state formulas is straightforward. Bounded CSL until formulas translate to bounded PCTL until formulas with the same probability bound and structure, but with a time bound $\frac{\tau}{d}$ where τ was the original time bound in the CSL formula. In the sequel, we let $|\Phi|$ denote the *duration* of Φ, i.e. the length of time to which it refers, as follows:

Definition 3. *For any bounded CSL formula Φ the duration of Φ, $|\Phi|$ is defined as follows:*

$$|a| = 0 \qquad\qquad |\Phi_1 \vee \Phi_2| = \max\{|\Phi_1|, |\Phi_2|\}$$
$$|\neg\Phi| = |\Phi| \qquad\qquad |\mathcal{P}_{\bowtie p}(\Phi_1 \, \mathcal{U}^{\leq\tau} \, \Phi_2)| = \tau + \max\{|\Phi_1|, |\Phi_2|\}$$

Table 3. Comparison of model checking times. Times for Statistical MC (SMC) based on 10000 runs. Data for SMC and Fluid MC from [5]. Time in seconds.

Model	CSL Property	FlyFast ($N > 500$)	Fluid MC ($N > 500$)	SMC (N=100)	SMC (N=1000)
SEIR	$s\,\mathcal{U}^{\leq T}\,r$	~ 0.005 s	~ 0.05 s	~ 5 s	~ 20 s
SEIR	tt $\mathcal{U}^{\leq T}\,\mathcal{P}_{\leq 0.01}(true\ \mathcal{U}^{\leq 10}\ i\,)$	~ 6.3 s	N.A.	N.A.	N.A.
CS	tt $U^{T}(\mathcal{P}_{\leq 0.167}[\text{tt}\ U^{50}\,CR])$	~ 63.9 s	N.A.	N.A.	N.A.

Recall that we assume that time bounds in until formulas are rational numbers. For formula Φ, we let $\boldsymbol{\tau}_\Phi = (\tau_1, \ldots, \tau_l)$, denote the vector of all time bounds occurring in the (until subformulae of) Φ; furthermore we define d_Φ and D_Φ as follows: $d_\Phi = \max\{d \in \mathbb{Q} \mid \frac{\tau_j}{d} \in \mathbb{N}, \text{ for } j = 1 \ldots l\}$ and $D_\Phi = \{d \in \mathbb{Q} \mid$ there exists $w \in \mathbb{N}$ s.t. $d = \frac{d_\Phi}{w}\}$. Note that d_Φ is well defined since $\tau_j \in \mathbb{Q}$, for $j = 1 \ldots l$; actually, letting $\tau_i = \frac{a_i}{b_i}$, s.t. $\gcd(a_i, b_i) = 1$, it is easy to see that $d_\Phi = \frac{1}{\text{lcm}(b_1, \ldots, b_l)}$. We are now ready to state the main Theorem for robust CSL formulas[10]:

Theorem 1. *Let $\mathcal{X}^{(N)}$ be a sequence of CTMC population models, with deterministic fluid limit $x(t)$ for any fixed time $t < T$, under initial condition $x(0) = x_0$, and let $z = z(t)$ be the stochastic process defined from $\mathcal{X}^{(N)}$ as in Sect. 2.1. Let Φ be a robust CSL formula for z. There exists $N_0 \in \mathbb{N}$, such that, for all $d \in D_\Phi$, with $d \leq \frac{1}{q}$ as in Sect. 2.3 and for all $N \geq N_0$ and $b > \lceil \frac{|\Phi|}{d} \rceil$ the following holds:*

$$s, t \models_z \Phi \text{ iff } (s, \lfloor \frac{t}{d} \rfloor) \models_{\mathcal{T}_M(z,d,b)} \mathcal{T}_F[\![\Phi]\!]_d$$

The proof is by induction on the structure of the CSL formula Φ. One is usually interested in the result for $t = 0$. The proof is available in [14]. The result of Theorem 1 shows that it is indeed possible, under suitable scaling and convergence conditions, to use PCTL and a discrete time Markov population model to obtain similar results as by global fluid model checking CSL formulas on ICTMCs. The advantages and limitations of this approach were already outlined in the introduction.

Complexity. For what concerns the complexity of the approach, this depends on the complexity of the underlying on-the-fly probabilistic model-checking algorithm that consists of two phases, an expansion phase and a computation phase, both phases are linear in the number of states and transitions [17] for the time bounded fragment of PCTL. Furthermore it depends on the length and type of the formula, e.g conditional reachability is more likely to lead to the generation of fewer states, the time bounds in the formula and the uniformisation rate needed.

---

[10] We refer to [4] for the definition of formula robustness and to [6,13] for constraints on time horizon T.

5   Benchmark Examples and Comparison

Fluid model-checking is a young field of research and to the best of our knowledge, the global fluid model checking algorithm has not been fully implemented as yet and is not publicly available; consequently we will only use the few benchmark examples available in the literature. More complex examples can be found in [15]. For those we compare results of global fluid model-checking, on-the-fly fluid model checking and statistical model checking for a computer epidemic model and a client-server model [3–5]. Our experiments were conducted with a 1.8 GHz Core i7 Intel processor and 4 GB RAM running Mac OS X 10.7.5. The results are summarised in Table 3.

5.1   A Computer Worm Epidemic Model

The computer worm epidemic model consists of a large number of nodes, each with four local states; susceptible (S), exposed (E), infected (I) and recovered (R) (see Fig. 1). Susceptible nodes can be infected by an external source (inf_e) or by other nodes that are already infected (inf_s) or, rarely, be patched (patch_s). Exposed nodes can be activated (act) and become infected, or they can be patched (patch_e). Infected nodes can be de-activated (de_act) or patched (patch_i), or infect other nodes (inf_s) while remaining infected. Recovered nodes can loose (loss) their protection and become susceptible.

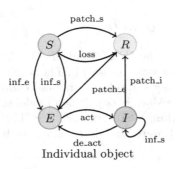
Individual object

Probability functions:

action inf_e :: k_{ext}/q;
action inf_s :: $(k_{inf}/q) * (\mathrm{frc}\, I)$;
action act :: k_{act}/q;
action de_act :: k_{deact}/q;
action patch_i :: k_{high}/q;
action patch_s :: k_{low}/q;
action patch_e :: k_{low}/q;
action loss :: k_{loss}/q;

system worm = $< S[10000] >$;

Fig. 1. Computer Worm Epidemic Process and related rates: $k_{ext} = 0.01$, $k_{inf} = 5$, $k_{act} = k_{deact} = 0.1$, $k_{low} = 0.005$, $k_{high} = 0.1$, $k_{loss} = 0.005$ (left) and derived probability functions using uniformisation rate $q = 10$ (right).

Following the procedure outlined in Sect. 2.3 the model shown in Fig. 1 is transformed into a discrete time model using a suitable rate which guarantees absolute stability and sufficient accuracy of the Euler forward method. The highest exit rate is that of state S, namely 5.2, if we assume that in theory all nodes could be infected at some stage, such that $\mathrm{frc}\, I = 1$. However, to facilitate comparison with results in the literature we take $q = 10$, at the cost of being slightly less efficient. Fig. 1 (right) shows the probabilities for the actions in the discrete time probabilistic model. Note that additional self-loops are added implicitly to the model to make sure that the total outgoing probability for each state is 1.

Fig. 2 shows the correspondence for model checking results (see also [5]) for all three model checking methods for the CSL path formula $s\,\mathcal{U}^{\leq T}\,r$, where s (r, resp.) denote the atomic propositions of being in state S (R, resp.), for T ranging from 0 to 20 showing the probability that a node is patched before being infected within T time steps. This corresponds to an equivalent PCTL formula with T ranging from 0 to 200. Model checking times for all methods for this formula are shown in Table 3. FlyFast generated 601 states. Note that this holds for any large number of nodes. FlyFast is faster than global fluid model-checking in this case. This is likely due to the fact that we are dealing with a *conditional* reachability property, so not all states need to be generated, showing the advantage of an on-the-fly approach. Furthermore, the algorithm uses memoization, meaning that probabilities computed once are preserved for later use. Note that both fluid model-checking approaches are several orders of magnitude faster than statistical model checking for a large population size, providing a scalability compatible with their use for analysing properties of individuals in the context of large scale collective systems, which is the main aim of the current work.

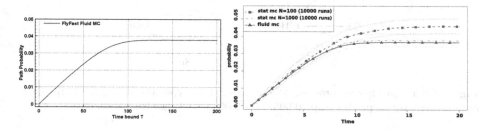

Fig. 2. Results for FlyFast (left) and Stat. MC and global Fluid MC (right).

An example of a nested path formula is $true\ \mathcal{U}^{\leq T}\,\mathcal{P}_{\leq 0.01}(true\ \mathcal{U}^{\leq 10}\ i\ )$, where i denotes to be in state I. It says that eventually, within time $T \in [0, 100]$, a state will be reached in which the probability to get infected within 10 time units is less than 0.01. In PCTL the formula is $true\ \mathcal{U}^{\leq T'}\,\mathcal{P}_{\leq 0.01}(true\ \mathcal{U}^{\leq 100}\ i\ )$, for T' ranging from 0 to 1000, given $q = 10$. Figures are omitted due to space limitations, but a comparison of results for a similar formula are shown in Fig. 4 for a more complex example. The FlyFast model-checking time is approximately 6.3 sec. and the number of states generated is 4000. No data on efficiency is available for the other two techniques for this formula.

5.2   A Client-Server Model

A larger model involving synchronisation is the Client-Server model [3,4]. This model is composed of two populations of processes that synchronise on request and reply actions. A Client process (see Fig. 3) has initial state (CQ) in which it can only perform a request (rq) to the server and then waits (CW) for either a timeout (to) or a reply (rp) from the server to happen. After a timeout it goes to a state to recover (CR), and then returns to the initial state when recovery is completed (rc). After receiving a reply (rp) the Client enters a thinking state (CT) after which it returns to the initial state upon completing thinking (th). The Server process (see Fig. 3) is initially (SQ) ready to receive a request (rq) from a Client. If it receives it, either a timeout (top) may occur or it may process the request (pr) moving to the reply state (SR). From the latter it may either produce a timeout (tor) or deliver a reply (rp) to the client and in both cases the server moves to a log-state (SL) and afterwards returns to the initial state (SQ) upon completion of logging (lg). So, the behaviour of Clients and Servers are synchronising via actions rq and rp, using a PEPA-based interaction paradigm based on a minimum rate principle [3,4]. The various timeouts are occurring independently. The ratio between the number of Clients and Servers is 2 to 1.

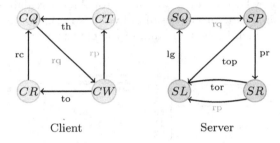

Client   Server

Fig. 3. Client and Server process. *Client* rates: th=1, rp=100, rq=1, rc=1, to=0.01; *Server* rates: rq=1, pr=0.1, top=0.005, tor=0.005, rp=1, lg=10.

As before, we proceed with the transformation of the model shown in Fig. 3 into a discrete time one, by finding a suitable rate. Note that the rate of an action shared by two types of objects can never be higher than the rate of the objects that contribute to the synchronisation, and will also be proportional to the (normalised) population size of the model. Therefore we can choose a rate equal to the maximum total exit rate of any of the states of the objects. In the client-server case this maximum is 100.01 (the sum of rates for Client actions rp and to). This is a large overestimation of the maximal exit rates, since the reply action of the Client is synchronised with that of the much slower Server (with reply rate 1). Therefore $q = 10$ is sufficient, given that the next highest rate is that of action lg which is 10. Table 4 shows the translation of the continuous time model into a discrete time model in the probabilistic input language of FlyFast. Actions of different objects must be distinct which is achieved by appropriate prefixing ($c\_$ for Client actions and $s\_$ for Server ones). Fig. 4 (left) shows results concerning the nested, time-dependent PCTL (path) formula tt $U^T(P_{\leq 0.167}[\text{tt } U^{500} CR])$, concerning a Client timeout; the FlyFast results are given for T varying from 0 to 400 steps. Nested formulas are computationally the most complex to analyse.

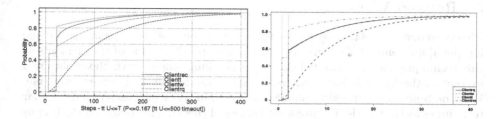

Fig. 4. FlyFast results (left) and global fluid model checking (right) from [3].

Table 4. Client and Server specification in FlyFast with $q = 10$ and the following values for the probabilities: $\alpha_{c\_th} = 1/q$, $\alpha_{c\_rp} = 100/q$, $\alpha_{c\_rq} = 1/q$, $\alpha_{c\_rc} = 1/q$, $\alpha_{c\_to} = 0.01/q$, $\alpha_{s\_rq} = 1/q$, $\alpha_{s\_pr} = 0.1/q$, $\alpha_{s\_top} = 0.005/q$, $\alpha_{s\_tor} = 0.005/q$, $\alpha_{s\_rp} = 1/q$, $\alpha_{s\_lg} = 10/q$.

---

action c_rq : $min(\alpha_{c\_rq} * \text{frc}(CQ), \alpha_{s\_rq} * \text{frc}(SQ))/\text{frc}(CQ)$ state $CQ\ \{c\_rq.CW\}$
action c_to : α_{c_to} state $CW\ \{c\_to.CR + c\_rp.CT\}$
action c_rc : α_{c_rc} state $CR\ \{c\_rc.CQ\}$
action c_th : α_{c_th} state $CT\ \{c\_th.CQ\}$
action c_rp : $min(\alpha_{c\_rp} * \text{frc}(CW), \alpha_{s\_rp} * \text{frc}(SR))/\text{frc}(CW)$

action s_rq : $min(\alpha_{c\_rq} * \text{frc}(CQ), \alpha_{s\_rq} * \text{frc}(SQ))/\text{frc}(SQ)$ state $SQ\ \{s\_rq.SP\}$
action s_pr : α_{s_pr} state $SP\ \{s\_pr.SR + s\_top.SL\}$
action s_tor : α_{s_tor} state $SR\ \{s\_tor.SL + s\_rp.SL\}$
action s_top : α_{s_top} state $SL\ \{s\_lg.SQ\}$
action s_rp : $min(\alpha_{c\_rp} * \text{frc}(CW), \alpha_{s\_rp} * \text{frc}(SR))/\text{frc}(SR)$
action s_lg : α_{s_lg}

system $clientServerSystem =< CQ[1000], SQ[500] >$

The property is the PCTL version of the corresponding CSL nested property analysed using the global fluid model checking approach in [3,4]. Model checking times for FlyFast are shown in Table 3. This time is the cumulative time for the 400 different values for T considered, while it needed to generate 1598 states.

In the original model in [3], the recovery rate of the Client is 100 and not 10. This only affects the curve shown as a solid line in Fig. 4 (left), therefore the other curves continue to show close correspondence to ones on the right of Fig. 4 showing the fluid model checking results for the original model. The latter is a stiff model in which the parameter values differ five orders of magnitude. It would require a uniformisation rate of at least 100 and consequently a time bound of 5000 steps in the PCTL formula. The FlyFast results would still correspond well to the original global fluid model checking results, but model-checking times increase considerably. Global fluid model checking might be a viable alternative in that case because it can exploit existing highly optimised adaptive transient analysis methods, once the points in which the truth values change have been established.

6   Related Work

Closest to our work is that by Bortolussi and Hillston [3, 4] presenting a technique for global fluid model checking. We have briefly recalled some of the elements of this technique and compared their model checking results with those obtained by our on-the-fly technique and to statistical model checking where available. Work on fluid model checking can also be found in [12] which uses in part similar techniques as [4]. On-the-fly probabilistic model checking for bounded PCTL has also been developed by Della Penna et al. [8] but they do not consider its use for on-the-fly *fast mean field* model checking as we did in [16]. Stochastic population models can also be analysed using statistical model checking methods based on simulation (see for example [21]). The computational complexity of the latter increase linearly with the number of objects N, whereas on-the-fly fast mean-field model checking and fluid model checking do not depend on N.

7   Conclusions

We have illustrated an alternative way to perform fluid model checking of bounded CSL properties of individual entities in the context of large CTMC population models. The framework makes use of a prototype implementation of the on-the-fly fast mean field model checker FlyFast to check bounded PCTL formulas of individuals in the context of synchronous, discrete time DTMC population models. We have provided a correctness result and shown promising verification results compared to those available in the literature. Future work will consist in integrating the method and related transformation functions into FlyFast, looking for further optimisations and investigating the possibility of generating error bounds along with the analysis results. We will also consider the extension of the fragment of the logic with intervals and further operators.

References

1. Baier, C., Haverkort, B., Hermanns, H., Katoen, J.P.: Model-Checking Algorithms for Continuous Time Markov Chains. IEEE Transactions on Software Engineering **29**(6), 524–541 (2003). IEEE CS
2. Benaïm, M., Le Boudec, J.: A class of mean field interaction models for computer and communication systems. Performance Evaluation **65**(11–12), 823–838 (2008)
3. Bortolussi, L., Hillston, J.: Fluid model checking. CoRR abs/1203.0920 (2012), version 2, January 2013
4. Bortolussi, L., Hillston, J.: Fluid model checking. In: Koutny, M., Ulidowski, I. (eds.) CONCUR 2012. LNCS, vol. 7454, pp. 333–347. Springer, Heidelberg (2012)
5. Bortolussi, L., Hillston, J.: Checking individual agent behaviours in markov population models by fluid approximation. In: Bernardo, M., de Vink, E., Di Pierro, A., Wiklicky, H. (eds.) SFM 2013. LNCS, vol. 7938, pp. 113–149. Springer, Heidelberg (2013)
6. Bortolussi, L., Hillston, J., Latella, D., Massink, M.: Continuous approximation of collective system behaviour: A tutorial. Performance Evaluation **70**(5), 317–349 (2013)

7. Darling, R., Norris, J.: Differential equation approximations for Markov chains. Probability Surveys **5**, 37–79 (2008)
8. Della Penna, G., Intrigila, B., Melatti, I., Tronci, E., Zilli, M.V.: Bounded probabilistic model checking with the murφ verifier. In: Hu, A.J., Martin, A.K. (eds.) FMCAD 2004. LNCS, vol. 3312, pp. 214–229. Springer, Heidelberg (2004)
9. Hansson, H., Jonsson, B.: A logic for reasoning about time and reliability. Formal Aspects of Computing **6**, 512–535 (1994)
10. Hayden, R.: Scalable Performance Analysis of Massively Parallel Stochastic Systems. Ph.D. thesis, Imperial College London, April 2011. http://pubs.doc.ic.ac.uk/hayden-thesis/
11. Hillston, J.: Fluid flow approximation of PEPA models. In: Proceedings of the Second International Conference on the Quantitative Evaluaiton of Systems (QEST 2005), pp. 33–43 (2005)
12. Kolesnichenko, A., de Boer, P.T., Remke, A., Haverkort, B.R.: A logic for model-checking mean-field models. In: DSN, pp. 1–12. IEEE (2013)
13. Kurtz, T.: Solutions of ordinary differential equations as limits of pure jump Markov processes. Journal of Applied Probability **7**, 49–58 (1970)
14. Latella, D., Loreti, M., Massink, M.: On-the-fly fluid model checking via discrete time population models: Extended version. QUANTICOL TR-QC-08-2014 (2014). www.quanticol.eu
15. Latella, D., Loreti, M., Massink, M.: On-the-fly PCTL Fast Mean-Field Model-Checking for Self-organising Coordination. Science of Computer Programming. Elsevier (2015). http://dx.doi.org/10.1016/j.scico.2015.06.009
16. Latella, D., Loreti, M., Massink, M.: On-the-fly fast mean-field model-checking. In: Abadi, M., Lluch Lafuente, A. (eds.) TGC 2013. LNCS, vol. 8358, pp. 297–314. Springer, Heidelberg (2014)
17. Latella, D., Loreti, M., Massink, M.: On-the-fly probabilistic model-checking. In: Proceedings 7th Interaction and Concurrency Experience ICE 2014. EPTCS, vol. 166 (2014)
18. Le Boudec, J.Y., McDonald, D., Mundinger, J.: A generic mean field convergence result for systems of interacting objects. In: QEST 2007, pp. 3–18. IEEE Computer Society Press (2007). ISBN: 978-0-7695-2883-0
19. LeVeque, R.J.: Finite Difference Methods for Ordinary and Partial Differential Equations. SIAM (2007)
20. Tribastone, M., Gilmore, S., Hillston, J.: Scalable differential analysis of process algebra models. IEEE Trans. Software Eng. **38**(1), 205–219 (2012)
21. å Younes, H.L.S., Kwiatkowska, M., Norman, G., Parker, D.: Numerical vs. statistical probabilistic model checking: an empirical study. In: Jensen, K., Podelski, A. (eds.) TACAS 2004. LNCS, vol. 2988, pp. 46–60. Springer, Heidelberg (2004)

Computing Response Time Distributions Using Iterative Probabilistic Model Checking

Freek van den Berg[1]([✉]), Jozef Hooman[2], Arnd Hartmanns[3],
Boudewijn R. Haverkort[1], and Anne Remke[4]

[1] University of Twente, Enschede, The Netherlands
f.g.b.vandenberg@utwente.nl
[2] Radboud University, Nijmegen & TNO-ESI, Eindhoven, The Netherlands
[3] Saarland University, Saarbrücken, Germany
[4] Westfälische Wihlhems-Universität, Münster, Germany

Abstract. System designers need to have insight in the response times
of service systems to see if they meet performance requirements. We
present a high-level evaluation technique to obtain the distribution of
services completion times. It is based on a high-level domain-specific language that hides the underlying technicalities from the system designer.
Under the hood, probabilistic real-time model checking technology is
used iteratively to obtain precise bounds and probabilities. This allows
reasoning about nondeterministic, probabilistic and real-time aspects in
a single evaluation. To reduce the state spaces for analysis, we use two
sampling methods (for measurements) that simplify the system model: (i)
applying an abstraction on time by increasing the length of a (discrete)
model time unit, and (ii) computing only absolute bounds by replacing
probabilistic choices with non-deterministic ones. We use an industrial
case on image processing of an interventional X-ray system to illustrate
our approach.

1 Introduction

Service-oriented systems are designed for interconnection with other systems and
are commonplace in the domains of business, engineering and operations [12].
Their complexity lies in their capability to handle many service requests in parallel, for multiple kinds of services. To that end, they are equipped with multiple
resources to process services requests, with variable execution times. Service-
oriented systems operate in a real-time manner. When used to perform safety-
critical tasks, they have to respect real-time requirements like bounded response
times. In this way, their safety is determined by their performance.

Performance prediction early in design is difficult [8], especially when the system of concern does not exist yet. Simulation can provide an indication of average
response times, but simulation results tend to be too optimistic. Worst-case execution time analysis [16], on the other hand, leads to absolute bounds. While

This research was supported as part of the Dutch national program COMMIT, and
carried out as part of the **Allegio** project.

© Springer International Publishing Switzerland 2015
M. Beltrán et al. (Eds.): EPEW 2015, LNCS 9272, pp. 208–224, 2015.
DOI: 10.1007/978-3-319-23267-6_14

it can prove that *hard real-time* requirements are met, the computed bounds are often pessimistic, leading to costly, over-dimensioned implementations. For many applications, *soft real-time* guarantees are sufficient, i.e., it is acceptable if deadlines are met with a certain (high) *probability*. In such a scenario, a designer may want to know, e.g., the latency value for which, in the long run, 85 % of the service requests complete. These questions can be answered by probabilistic real-time model checking, in which probabilistic, nondeterministic and timed aspects are combined in one model.

This paper presents an approach that allows computing response time distributions using iterative probabilistic model checking. First a model is specified in a high-level domain specific language, iDSL [2], that abstracts from various under-the-hood technicalities, making it usable for systems engineers. From this model input for the MODEST TOOLSET [7] is automatically generated in the MODEST modelling language [6], which can be used for both simulations and model checking. By calling the model checking procedures iteratively, we are able to efficiently compute response time distributions in an automated fashion, which allows to better compare different designs. This is also the main difference to our previous work [2–4], where this performance evaluation trajectory has first been proposed, however without the ability to compute response time distributions precisely.

We illustrate our approach with a case study on interventional X-ray (iXR) systems as built by our industrial partner Philips Healthcare. These systems provide a continuous stream of X-ray images to a surgeon that operates on a patient. Low latency is necessary for hand-eye coordination [10], i.e., the surgeon must perceive the image stream to be real-time. Low response times of images are thus of vital importance, but a few misses of response deadlines are acceptable.

Related work. The tagged customer approach [5] is a numerical method to compute the response time distribution for open queuing networks, represented as continuous-time Markov chains (CTMCs). It may be used as a fast but approximate measure besides simulation, especially when utilizations are low and service times have high variances. The hierarchical performance evaluation tool (HIT, [1]) supports the model-based evaluation of computing system performance. HIT models are highly structured, based on functional hierarchies and modularization, as with the Y-chart philosophy [11]. HIT models are analysed using various techniques. However, the HIT model at hand determines which techniques can be used, with simulation covering the greatest spectrum. It does not have specific support for response time distributions. Modular Performance Analysis with Real-Time Calculus (MPA, [17]) is based on the Network Calculus and computes hard lower and upper bounds using event streams. Hence, these approaches do not deliver what we do.

Context. The measures that we compute and the type of systems our approach is designed for make it fall right into the field of *performance evaluation* [9]. Typical performance evaluation approaches build on fully stochastic formalisms, such as continuous-time Markov chains or stochastic Petri nets. However, our examples require a mixture of deterministic timing with probabilistic effects and concurrency; we want also to be able to compute hard bounds; and we use abstraction techniques for model simplification that introduce nondeterministic delays. Since they capture exactly these aspects in a compositional fashion, we chose probabilistic timed automata (PTA, [14]) as the semantic basis of our models. The analysis of PTA is supported by a number of tools including PRISM [13] and the MODEST TOOLSET [7]. We use the latter due to its high-level input language and the ability to perform model checking using the included mcsta tool as well as simulation using the modes simulator.

2   Problem Statement and Case Study

We describe service-oriented systems (Section 2.1) and their performance characteristics (Section 2.2), and use the case study on iXR systems as an example.

2.1   Service-Oriented Systems

In service-oriented systems, the system receives a service request after which the system replies with a service response that completes the request. The latency is the elapsed time between service request and response. A service decomposes into a number of atomic tasks that each require access to resources (e.g. a CPU, I/O bus or GPU) for computation or data transfer. These tasks may have variable execution times. A service system can make use of multiple instances of one or more services at the same time, which gives rise to concurrency among service instances. A scheduling policy resolves this concurrency by prescribing an order in which tasks gain access to resources, e.g., first-in, first-out (FIFO). A scenario describes when these service requests arrive; for example, "a service system receives one service request every $100ms$, forever". Service systems can have one or more configurations, each with their own properties.

Example: iXR Systems. provide a continuous stream of images to support a surgeon that operates a patient, i.e., they provide an image processing (IP) service. Service requests are incoming, unprocessed images that arrive with fixed inter-arrival times. The iXR system responds with processed images. The latency should be low enough to enable hand/eye-coordination [10]. The system comprises one resource, the CPU. The service decomposes into a pipeline of twelve image processing steps, all performed on the CPU via a FIFO scheduling policy. Service IP receives 10 images per second. We consider two configurations having image resolutions of 512^2 and 1024^2 pixels, respectively.

2.2   Performance Questions

We define performance questions of service-oriented systems to assess their performance. There are black-box and white-box measures. Black-box measures are

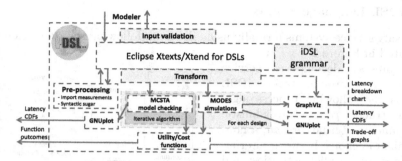

Fig. 1. The iDSL solution chain

observable from the outside of the system, and examples include the latency per service. White-box measures, such as resource utilizations, require knowledge about the inside of the system. We focus on service latencies because they are the prime measure of interest in the case study on iXR systems, viz., latencies should be low enough to enable hand/eye coordination. We show two categories of performance questions that are concerned with service requests of a given service in a given service system, operating in a given scenario, as follows:

Q1. What is the latency for which a given percentage of the service requests completes?

Q2. Which percentage of the service requests has a latency below a given value?

Example: iXR systems have performance questions that are as follows:

Q1a. What is the latency for which 85% of the service requests completes?

Q2a. Which percentage of the service requests has a latency below 55ms?

3 A Formal Model for Service Systems

iDSL comprises a high-level language for modelling service systems, and a toolset to evaluate their performance (see Figure 1 for its solution chain). Each iDSL model leads to the generation of performance artefacts for many so-called designs. iDSL has been developed using the Xtext and Xtend plug-ins of Eclipse for Domain Specific Languages (DSLs). iDSL is thus an Eclipse plug-in with an extensive Integrated development environment (IDE). In a pre-processing phase, measurements can be imported into the model and syntactic sugar is resolved. For each design, performance analysis is done via multiple mcsta (see Section 4) and modes calls of the MODEST TOOLSET [7] for model checking and simulation, respectively. Visualizations are generated with Graphviz and GNUplot.

In Section 3.1, we provide the syntax of the iDSL language by showing its key language constructs. We apply it to iXR systems in Section 3.2 and show three sampling methods for measurements in Section 3.3. In Section 3.4, we define utility and cost functions that answer performance questions using a query on the computed results. In Section 3.5, we define the semantics of iDSL by describing the transformation from iDSL to Modest.

3.1   iDSL Language Syntax

We specify service systems formally using iDSL [2], following the six concepts as illustrated in Figure 2

1. A *process* decomposes service requests into atomic tasks. iDSL provides the following process algebra constructs: *palt*, a probabilistic choice among alternatives; *alt*, a nondeterministic choice between alternatives; *par* for parallel activities; and *seq* for sequential activities. iDSL also offers a *mutex*, a mutual exclusion to run processes uninterruptedly.

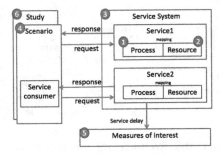

Fig. 2. The iDSL language's concepts.

2. *Resources* are capable of performing one atomic task at a time, in a certain amount of time. A *mapping* assigns atomic tasks to resources.

3. A *service system* consists of one or more services, each implemented using a process, a set of resources and a mapping between processes and resources.

4. A *scenario* comprises a number of invoked service requests over time to observe the performance behaviour of the system in specific circumstances.

5. *Measures of interest* define which performance measures are obtained.

6. A *study* evaluates a selection of systematically chosen systems and scenarios.

3.2   iDSL Model of iXR Systems

Let us now explain the iDSL model of an iXR system:

1. Section **process** (right) contains process "Image-processing" that specifies how images are processed, viz., via two high-level operations "Noise_reduction" and "Refine". They in turn decompose in a sequential pipeline of twelve image operations, each with a load (an amount of work), specified via an abstraction mechanism. These loads are assumed to be independent. Section 3.3 shows how these abstract loads are implemented.

```
Section Process
    ProcessModel Image_Processing seq {
        seq Noise_reduction {
            atom Pre_processing load call preproc
            atom Basic load call basic
            atom Decompose1 load call decomp1
            atom Spatial_noise_red load call spat_nr
            atom Temporal_noise_red load call temp_nr
            atom Compose1 load call comp1
        }
        seq Refine {
            atom Decompose2 load call decomp2
            atom Refine_step1 load call refine1
            atom Compose2 load call comp2
            atom Decompose3 load call decomp3
            atom Refine_step2 load call refine2
            atom Compose3 load call comp3
        }
    }
```

2. Section **resource** comprises resource "Image_processing_PC", that has a CPU with a rate of 1, i.e., it can process 1 unit of load per μs, the time unit of choice. The resource is defined as follows.

```
Section Resource
    ResourceModel Image_Processing_PC decomp { atom CPU rate 1 }
```

3. Section **system** comprises a service named "Image_Processing_Service", which connects process "Image_Processing" to resource "Image_Processing_PC" by defining a mapping, i.e., each of the twelve image operations is performed on the CPU, and a FIFO scheduling policy is used to resolve concurrency.

```
Section System
    Service Image_Processing_Service
        Process Image_Processing
        Resource Image_Processing_PC
        Mapping assign { (Pre_processing,CPU) (Basic,CPU) (Decompose1,CPU)
            (Spatial_noise_red,CPU) (Temporal_noise_red,CPU) (Compose1,CPU)
            (Decompose2,CPU) (Refine_step1,CPU) (Compose2,CPU) (Decompose3,CPU)
            (Refine_step2,CPU) (Compose3,CPU) } scheduling policy { (CPU, FIFO) }
```

4. Section **scenario** comprises scenario "Image_Processing_Run", in which the service is invoked 10 times per second, i.e., once every 100000 s, forever.

```
Section Scenario
    Scenario Image_Processing_run
        ServiceRequest Image_Processing_Service at time 0 us, 100000 us, ...
```

5. Below, section **measure** contains two measures: "ServiceResonseTimes" retrieves average latencies of 100 service requests via simulations, using 3 runs. Simulations provide quick insight into the general behaviour of a system, but are less suitable for showing the extreme behaviour of a system.

Measure "CDF ..." yields a cumulative distribution function (CDF) with latencies, obtained via probabilistic model checking. As usual, a CDF is a function that displays for each latency value l, the percentage of the service requests that has a latency below l, e.g., $cdf(60) = 0.5$ means that half of all service requests have a latency below 60 ms. This measure is obtained via model checking and is thus much slower than simulation, but conveys different insights, e.g., absolute lower and upper bounds. It is explained in detail in Section 4.

```
Section Measure
    Measure CDF of ServiceResponseTimes via PTA model checking
    Measure ServiceResponseTimes using 3 runs of 100 ServiceRequests
```

6. Finally, section **study** allows for design instances to be defined that are each evaluated using the defined measures. We model two iXR systems having image resolutions 512^2 and 1024^2 pixels, respectively.

```
Section Study
    Scenario Image_Processing_run DesignSpace (resolution {"512" "1024" } )
```

iDSL offers two approaches to handle uncertainty: probabilism and nondeterminism. *Probabilism* specifies a range of weighted outcomes, e.g., an image operation completes with probability 0.6 in 45 s, and 0.4 in 46 s. Non-determinism is similar but without probabilities, e.g., an image operation completes in either 45 s, 46 s, or between 45 s and 46 s. *Nondeterminism* can also occur when a system processes multiple service requests, i.e., if a resource is potentially accessed multiple times at the same time, the order of action is undefined.

3.3 Three Sampling Methods for Measurements

In Section 3.2 (process), twelve image operations are defined using abstract loads. In our case, these loads are based on 300 latency measurements each, performed on a real iXR system, to calibrate the model. We show three implementations for "pre-processing", each corresponding to a different way of sampling, viz., *uniform, abstract time*, and *non-deterministic time* sampling. The other image operations are implemented similarly.

Uniform Sampling. Below, the abstract load "preproc" defines the variable load of "Pre_processing", using uniform sampling. In uniform sampling, each measurement has an equal probability to be sampled. The load is defined for two image resolutions, viz., 512^2 and 1024^2 pixels. The *dspace* operator selects the right set of measurements, depending on the resolution of the design instance at hand. *Uniform from file* refers to an external file and a position in that file at which the correct measurements are stored.

```
Abstract load preproc select dspace(resolution) {
    "512":   uniform from file "0512.cdf#Pre_processing"
    "1024":  uniform from file "1024.cdf#Pre_processing"
}
```

iDSL transforms the implementations into basic process algebra constructs, via a so-called model transformation. This leads to one palt-construct per resolution, consisting of measurements. For resolution 512^2, it is as follows:

```
palt  Pre_processing_eCDF {  8    atom Pre_processing load 130 us,  67 atom Pre_processing load 131 us,
                           143 atom Pre_processing load 132 us,  71 atom Pre_processing load 133 us,
                             9 atom Pre_processing load 134 us,   1 atom Pre_processing load 135 us,
                             1 atom Pre_processing load 136 us }
```

E.g., the probability for 130 s is $\frac{8}{300}$, because 8 out of 300 measurements are 130 s, for 131 s it is $\frac{67}{300}$, because 67 out of 300 measurements are 131 s, etc.

Abstract Time Sampling. Next, the abstract load "preproc" defines the variable load of "Pre_processing" using abstract time sampling, as follows.

```
Abstract load preproc select dspace(resolution) {
    "512":  uniform from file "0512.cdf#Pre_processing" time unit 250
    "1024": uniform from file "1024.cdf#Pre_processing" time unit 800
}
```

In abstract time sampling, measurements are divided by a given constant number and rounded to make the model simpler at the price of some precision. We divide by 250 for resolution 512^2, and 800 for 1024^2. The final results will be multiplied by the same constants again. The result of the model transformation for resolution 512^2, is as follows.

```
palt  Pre_processing_eCDF { 300    atom Pre_processing load 1 }
```

Non-deterministic Time Sampling. Finally, "preproc" defines the variable load of "Pre_processing" using non-deterministic time sampling. It is as follows:

```
Abstract load preproc select dspace(resolution) {
    "512":  uniform from file "0512.cdf#Pre_processing" non-deterministic time
    "1024": uniform from file "1024.cdf#Pre_processing" non-deterministic time
}
```

In non-deterministic time sampling, the time of an image operation is defined as the smallest segment that contains all measurements, as follows.

```
palt  Pre_processing_eCDF { 300    atom Pre_processing load 130 to load 136 }
```

Non-deterministic time sampling is typically used to obtain absolute latency bounds. Semantically, the above means that any real value in segment [130 : 136] is a valid sample, but that their individual probabilities are unknown.

3.4   Performance Queries in iDSL

In the following, we add performance queries to the iDSL model. They are specified as so-called *utility* and *cost* functions that specify a query on the performance results, and return a real number. They rely on measures. In the case study, we show two measures that are based on simulations and model checking, resp.

A function is either a cost function when lower values are preferred, e.g., the average latency of a service, or a utility function when higher values are preferred, e.g., the percentage of service requests completed after some time.

The iDSL model comprises both the system model and a scenario in which it operates. We analyze the response times to service requests of a given service S of this system, the latencies. To this end, we introduce four model checking-based functions and two simulation functions, based on Q1 and Q2 (see Section 2.2).

First, we introduce two pairs of model checking functions.

1. Function $Q1a_{lb}$ ($Q1a_{ub}$) returns the minimum (maximum) latency before which P percent of the service requests of service S complete.

2. Function $Q2a_{lb}$ ($Q2a_{ub}$) returns the minimum (maximum) percentage of service requests of service S that has a latency below time T.

Second, we introduce two similar simulation based functions:

1. Function $Q1a_{sim}$ returns the latency before which P percent of the service requests of service S complete, based on R simulation runs of Rq requests each.

2. Function $Q2a_{sim}$ returns the percentage of service requests of service S that has a latency below time T, based on R simulation runs of Rq requests each.

Note that model checking-based functions have two variants, viz., a minimum and maximum one; they return bounds. Simulation has two parameters: runs and requests. The higher these values are, the more accurate the results will be.

Example: iXR Systems. We define two groups of performance queries for iXR systems in iDSL, which are added to the Measure section of iDSL. Each performance question is defined three times, viz., twice for model checking and once for simulation. Simulations are based on 3 runs of 100 requests each.

Questions of type Q1 ask latencies before which a given percentage of service requests completes. They are cost functions, since lower latencies are preferred:

```
Cost Q1a_lb      timeAfterPercentageHasFinished (Service Image_Processing_Service, Percentage 85, tmin)
Cost Q1a_ub      timeAfterPercentageHasFinished (Service Image_Processing_Service, Percentage 85, tmax)
Cost Q1a_sim     percentile 0 of latencies ( Service Image_Processing_Service, Run 3, Request 1..100 )
```

Questions of type Q2 ask percentages of service requests that have latencies below a given time. They are utility functions since higher values are preferred:

```
Utility Q2a_lb   percentageBelowTime ( Service Image_Processing_Service, Time 55 us, pmin )
Utility Q2a_ub   percentageBelowTime ( Service Image_Processing_Service, Time 55 us, pmax )
Utility Q2a_sim  percent below 55 of latencies ( Service Image_Processing_Service , Run 3, Request 1..100)
```

3.5   Translation to Modest

The semantics of the iDSL language is specified via a transformation from iDSL models to MODEST models. MODEST is a high-level modelling language rooted in process algebra with a formal semantics in terms of stochastic hybrid automata (SHA) [6]. Several other popular formalisms such as PTA and discrete-time Markov chains are special cases of SHA. The analysis of MODEST models is supported by the MODEST TOOLSET [7], which in particular includes the tools modes for simulation (or: statistical model checking) and mcsta for model checking of MODEST models conforming to the PTA subset of the language.

iDSL is a high-level language specifically tailored to service systems, yielding, for the iXR case, a model that
is seven times smaller textually than
the autogenerated MODEST code. Additionally, small architectural changes to
the iDSL model can affect the whole
MODEST code, making a MODEST-only
approach hard w.r.t. maintenance.

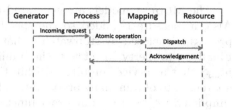

Fig. 3. An interaction diagram.

An iDSL model transforms into one or more MODEST models, each containing
an overarching MODEST process. This process decomposes in a number of parallel
sub-processes of class *generator*, *process*, *mapping* or *resource*. They interact in
the way as shown in Figure 3, viz., a *generator* triggers a *process*, which in turn,
via a mapping, obtains access to a *resource* and receives an acknowledgement.

iDSL processes transform into processes of class *process* in MODEST. Process
algebra constructs in iDSL are thereby translated to their MODEST counterparts. Processes also contain calls to *mapping* processes, for each atomic task.
iDSL resources become *resource* processes in MODEST, containing a queue and a
counter for service time. Mappings lead to *mapping* processes that each connect
an atomic tasks to a resource. iDSL systems and services do not lead to MODEST
code, but merely organize the iDSL model. In the iDSL scenario, service requests
lead to a MODEST generator that sends a trigger to a process periodically and
forever. Measures lead to one or more MODEST models, depending on the measure. In the next section, we show how multiple MODEST models are generated
for model checking purposes. An iDSL study contains design instances that are
evaluated separately. Separate MODEST models are generated for each design.

4   Computing Response Time Distributions

We present a new approach to compute latency response times, based on iterative
probabilistic model checking. It answers performance questions of iDSL models
(of Section 3.4) in five steps: (i) iDSL models are transformed into MODEST
models that are used to retrieve service latencies (Sect. 4.1); (ii) latencies are
aggregated into one overarching latency per service (Sect. 4.2); (iii) mcsta is
applied iteratively to obtain probability bounds (Sect. 4.5); (iv) these bounds
are transformed into a set of possible CDFs (Sect. 4.4); and (v) performance
questions are answered using the set of possible CDFs (Sect. 4.5).

4.1   From iDSL Queries to Modest

We generate a range of MODEST models to answer performance queries, for
each iDSL model i, each service s within that model, and both the minimum
and maximum probability (a flag f). The models have one parameter, $t \in \mathbb{R}_{\geq 0}$,
and return probability p: the probability that a service completes within time t.

The MODEST models are generated using the transformation in Section 3.5. Also, a measure is added to the specific service the model measures, i.e., its process is enclosed by stopwatches that record latencies of its service requests. Finally, a property to retrieve the minimum or maximum probability ($p_{\min}$ or $p_{\max}$) that a service completes within time t is added. Hence, MODEST models are reused to obtain many probabilities, for many values of time t. For the sake of simplicity, we specify an abstract function $\mathcal{M}$ that retrieves such a probability:

$$p = \mathcal{M}(i, s, f, t),$$

where p is either the minimum or maximum probability (depending on flag f) that service s in iDSL model i completes within time t.

$\mathcal{M}$ is implemented using the following three steps: (i) select the MODEST model of iDSL model i, service s and f; (ii) run this model in mcsta with parameter time t; and (iii) return the result of mcsta as probability p.

4.2   Aggregating Latencies of Service Requests

MODEST models have been generated that return the probability that a service request completes within a given time. In iDSL, however, a service leads to an infinite stream of service requests, each with their own latency. Ideally, the average of these latencies is a measure for the performance of the whole service:

$$P_{\Omega}(t) = \lim_{k \to \infty} \frac{1}{k} \sum_{n=1}^{k} P_n(t), \tag{1}$$

where $P_{\Omega}(t)$ is the combined probability, n the service request number, t the latency time, $P_n(t)$ the probability that service request n finishes within time t. However, this infinite sum is not computable. Hence, we show the following two weighted averages of the latencies that can be used to approximate the measure.

First, the *arithmetic mean* considers the first N service requests and weighs them equally, as follows:

$$P_{\Omega}(t) = \frac{1}{N} \sum_{n=1}^{N} P_n(t), \tag{2}$$

where $N \in \mathbb{N}^+$ is the number of service requests considered, e.g., $N = 100$. It is similar to (1) for large values of N. However, even for small values of N, it

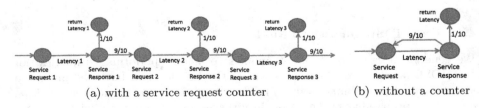

(a) with a service request counter   (b) without a counter

Fig. 4. Binary probabilistic choices induce the geometric distribution

has two drawbacks: (i) it requires a counter to be added to the state in MODEST to keep track of the service request number; and (ii) latencies of the $(N+1)^{th}$ service request and later are neglected.

Second, the *geometric distribution* [15] weighs service requests exponentially decreasing, as follows:

$$P_\Omega(t) = \sum_{n=1}^{\infty} (1-\rho)^{n-1} \rho P_n(t), \tag{3}$$

where $\rho \in (0 : 1)$ is the parameter of the geometric distribution.

It is, again, similar to (1) for ρ close to 0. Lower ρ-values lead to a more complex model but more precise results, and vice versa. Since the geometric distribution considers all service requests and it is capable of finding absolute maximum latencies. In MODEST, it is implemented as a binary probabilistic choice every time a service request completes (as depicted in Figure 4a): either the currently measured latency is returned, with probability ρ, or the next service request is evaluated, with probability $1 - \rho$. Moreover, the geometric distribution is memoryless, i.e., the binary choice does not rely on state information. Consequently, it is possible to omit the service request number from the model, leading to a single reoccurring service request (as in Figure 4b). In the remainder of this paper, we only consider the geometric distribution with $\rho = \frac{1}{10}$, empirically determined.

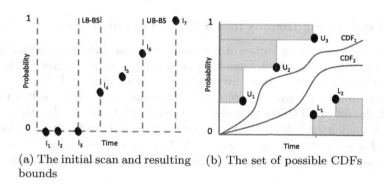

(a) The initial scan and resulting bounds

(b) The set of possible CDFs

Fig. 5. Cumulative Distribution Functions (CDFs) based on function $\mathcal{M}$

4.3   Iterative Model Checking for Probability Bounds

We provide an algorithm to compute function $\mathcal{M}$, for a given iDSL model i, service s in this model and a minimum/maximum bound flag f. $\mathcal{M}(i, s, f, t)$ is iteratively applied for different values t, comprising three stages, viz., an initial scan, a binary lower & upper bound search, and a brute force computation.

Initial Scan. The initial scan gives an idea of the order of magnitude of the time values. We compute $\mathcal{M}(i, s, p_{\min/\max}, t)$ for $t = 1, 2, 4, 8, 16, \ldots, 2^m, 2^{m+1}$, $\ldots$, 2^n, 2^{n+1} until $\mathcal{M}(i, s, p_{\min/\max}, 2^{n+1}) = 1$. The lower bound is then located between 2^m and 2^{m+1} with $\mathcal{M}(i, s, p_{\min/\max}, 2^m) = 0$ and $\mathcal{M}(i, s, p_{\min/\max}, 2^{m+1}) > 0$, and the upper bound between 2^n and 2^{n+1} with $\mathcal{M}(i, s, p_{\min/\max}, 2^n) < 1$ and $\mathcal{M}(i, s, p_{\min/\max}, 2^{n+1}) = 1$. Note that m and n are unique values.

Figure 5a depicts the initial scan graphically. It shows computations i_1, i_2, $\ldots$, i_7, with i_7 having a probability of 1. We observe that the lower bound is located between i_3 and i_4, and the upper bound between i_6 and i_7.

Binary Lower & Upper Bound Search. Next, two binary searches are performed to determine the *exact* lower and upper bound, using the ranges of the initial scan. The binary searches are applied to $[2^m : 2^{m+1}]$ and $[2^n : 2^{n+1}]$ for the lower and upper bound, respectively. They lead to lower and upper bound lb and ub, respectively. By definition, $\mathcal{M}(i, s, p_{\min/\max}, t) = 0$, for $t < lb$, and $\mathcal{M}(i, s, p_{\min/\max}, t) = 1$, for $t > ub$. Thus, only $\mathcal{M}(i, s, p_{\min/\max}, t)$, for $t \in [lb : ub]$, need to be determined yet.

Brute Force Computation. We obtain $\mathcal{M}(i, s, p_{\min/\max}, t)$ for all times $t \in [lb : ub]$. We compute them on c CPU cores by distributing the possible values for t equally to the available CPU cores.

Finally, a cache is used throughout all computations for $\mathcal{M}$ to avoid duplicate computations, which is possible because $\mathcal{M}$ is deterministic.

4.4   Transforming Bounds into a Set of Possible CDFs

By iteratively computing values of function $\mathcal{M}$, lower and upper bound probabilities ($p_{\min}$ and $p_{\max}$) of latencies have been computed, for a given iDSL model i and service s. Figure 5b shows five probabilities (upper bounds U_1, U_2 and U_3, and lower bounds L_1 and L_2) and two CDFs that respect these bounds. We consider the set of all CDFs that respect these bounds, i.e., each CDF is below the upper bounds and above the lower bounds, for all times t. Formally, function $CDF_{all} : I \times S \to 2^{\widehat{CDF}}$ returns, where $\widehat{CDF}$ is the universe of all CDFs, given an iDSL model i and service s, the set of CDFs that respect the bounds in $\mathcal{M}$:

$$CDF_{all}(i, s) = \{ \, cdf \in \widehat{CDF} \mid cdf(0) = 0 \wedge \mathcal{M}(i, s, p_{\min}, t) = p_1 \Rightarrow cdf(t) \geq p_1$$
$$\wedge \mathcal{M}(i, s, p_{\max}, t) = p_2 \Rightarrow cdf(t) \leq p_2 \, \}$$

Constraint $cdf(0)$ requires all the values to be greater than or equal to 0.

4.5   Answering the Performance Queries Using the CDFs

We now use CDF_{all} to answer the performance queries, as follows. Queries of type Q1, the minimum time for which a service request completes with probability p, are determined, as follows:

$$Q1(i, s, p, T_{\min}) = \min \{ \, t \mid (t, p) \in cdf \wedge cdf \in CDF_{all}(i, s) \, \}$$

Queries of type Q2, the minimum probability that a latency is below a given time t, are determined, as follows:

$$Q2(i, s, t, P_{\min}) = \min \{ p \mid (t, p) \in cdf \wedge cdf \in CDF_{all}(i, s) \}$$

The maximum cases of Q1 and Q2 are determined analogously.

5   Case Study Results

We apply the performance analysis approach of Section 4 to the iXR system of the case study. We define three experiments and compare their results with simulations and real measurements, in three steps: (i) we present CDFs with latency times; (ii) we show the execution times and model sizes; and (iii) we show the answers to the performance questions (for Experiment 1).

Three experiments are defined, based on the sampling methods in Section 3.3, respectively. Experiment 0 uses uniform sampling. Running MCSTA leads

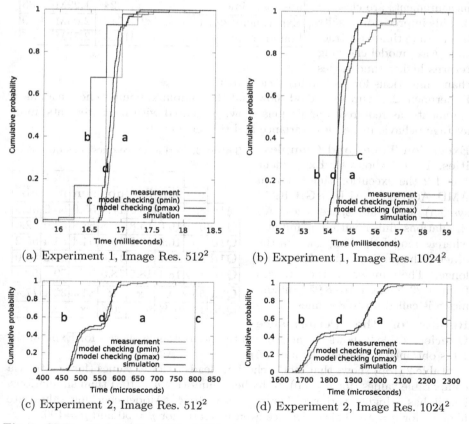

(a) Experiment 1, Image Res. 512^2 (b) Experiment 1, Image Res. 1024^2

(c) Experiment 2, Image Res. 512^2 (d) Experiment 2, Image Res. 1024^2

Fig. 6. CDFs with latencies of IP of iXR systems: measurements on a real iXR system (a), model checking (lower & upper bounds) (b+c) and simulations (d).

to a incremental generation of the state space, but runs out of memory and stalls after having generated 38 million states. Experiment 1 uses abstract time sampling. Experiment 2 uses non-deterministic time sampling, but on a subset of the system to reduce complexity, i.e., only the first 3 image operations are considered, instead of all 12.

CDFs with Latencies. Figure 6 shows latency CDFs, for resolutions 512^2 and 1024^2, and experiment 1 and 2. They show the percentage of service requests that complete, on the Y-axis, within a given latency, on the X-axis.

Experiment 1: Figure 6a and 6b convey that the bounds, obtained by model checking, do not enclose all measurements in both cases, e.g., in Figure 6a the lower bound has probability 0.16 for time 16.5 while measurements are close to 0. This imprecision results from the time abstraction. Also, model checking returns higher time values than simulations for probabilities close to 1.

Table 1. Two experiments: execution times for simulations and model checking, the number of states of the Modest model and the number of MCSTA calls.

Exp	sim. time	MC time	img. res.	states	calls
Experiment 1: abstract time	56"	3:17'28"	512^2	8.05M	88
			1024^2	1.29M	85
Experiment 2: bounds only	44"	5:59'22"	512^2	2.03M	49
			1024^2	2.77M	57

Experiment 2: Figure 6c and 6d show that the computed bounds enclose all measurements, as required. Simulations show, compared with measurements, more average behaviour, i.e., less variance and tighter bounds.

Execution Times and Complexities. Table 1 shows for Experiment 1 and 2 the execution times (on an AMD A6-3400M APU, 8 GB RAM system) and state space sizes. All simulations finish within a minute, whereas model checking takes in the order of hours, i.e., up to 500 times longer. The number of states ranges from 1.29 million to 8.05 million. mcsta is called up to 88 times.

Table 2. Performance questions outcomes

		512^2			1024^2		
	n	sim	lb	ub	sim	lb	ub
Q1a	85	16.9	16.8	17.0	55.0	55.2	56.0
Q1b	0	16.6	15.8	15.8	53.8	52.0	52.0
Q1c	50	16.8	16.5	16.8	54.4	54.4	55.2
Q1d	90	17.0	16.8	17.0	55.1	55.2	56.0
Q1e	100	18.1	18.5	18.5	57.0	59.2	59.2
Q2a	55	x	x	x	84%	32%	77%
Q2b	17	91%	91%	96%	x	x	x

Results of the Performance Queries. Table 2 shows the answers to the performance queries for Experiment 1 (as obtained in Section 4.4 and 4.5).

Table 2 (top) shows that model checking leads to lower values (for $n = 0$), via comparable values (for $n = 50$), to higher values (for $n = 100$) than simulations, i.e., it has a higher variance. For resolution 1024^2, Model checking values are higher , for $n = 85$. This difference even increases for $p = 90$ and $p = 100$.

Table 2 (bottom) shows that, for a given latency, the percentage of service requests that meet a latency deadline can be obtained. E.g., if 90 % of the images need to be in time, then a latency of 17 ms for resolution 512^2 is met.

6 Conclusion

We have introduced a high-level domain specific language to model service systems and retrieve their response time distributions, usable by system designers. Besides the traditionally used simulations, response times are also obtained via iterative probabilistic model checking. Since model checking faces the state space explosion problem, we have introduced sampling methods to reduce the model complexity: (i) increasing the model time unit, and (ii) eliminating probabilism. A case studty on iXR systems shows the feasibility of our approach.

References

1. Beilner, H., Mater, J., Weissenberg, N.: Towards a performance modelling environment: news on HIT. In: Modeling Techniques and Tools for Computer Performance Evaluation, pp. 57–75. Plenum Press (1989)
2. van den Berg, F., Remke, A., Haverkort, B.R.: A domain specific language for performance evaluation of medical imaging systems. In: 5th Workshop on Medical Cyber-Physical Systems, pp. 80–93. Schloss Dagstuhl (2014)
3. van den Berg, F., Remke, A., Haverkort, B.: iDSL: Automated performance prediction and analysis of medical imaging systems. In: Computer Performance Engineering, LNCS, vol. 9272. Springer (2015) (to appear)
4. van den Berg, F., Remke, A., Mooij, A., Haverkort, B.: Performance evaluation for collision prevention based on a domain specific language. In: Balsamo, M.S., Knottenbelt, W.J., Marin, A. (eds.) EPEW 2013. LNCS, vol. 8168, pp. 276–287. Springer, Heidelberg (2013)
5. Grottke, M., Apte, V., Trivedi, K., Woolet, S.: Response time distributions in networks of queues. In: Queueing Networks, pp. 587–641. Springer (2011)
6. Hahn, E., Hartmanns, A., Hermanns, H., Katoen, J.P.: A compositional modelling and analysis framework for stochastic hybrid systems. Formal Methods in System Design **43**(2), 191–232 (2012)
7. Hartmanns, A., Hermanns, H.: The modest toolset: an integrated environment for quantitative modelling and verification. In: Ábrahám, E., Havelund, K. (eds.) TACAS 2014 (ETAPS). LNCS, vol. 8413, pp. 593–598. Springer, Heidelberg (2014)
8. Haveman, S., Bonnema, G., van den Berg, F.: Early insight in systems design through modeling and simulation. Procedia Computer Science **28**, 171–178 (2014)
9. Jain, R.: The Art of Computer Systems Performance Analysis. John Wiley & Sons (1991)
10. Johnson, J.: Designing with the Mind in Mind: Simple Guide to Understanding User Interface Design Rules. Morgan Kaufmann (2010)
11. Kienhuis, B., Deprettere, E.F., van der Wolf, P., Vissers, K.: A methodology to design programmable embedded systems. In: Deprettere, F., Teich, J., Vassiliadis, S. (eds.) SAMOS 2001. LNCS, vol. 2268, pp. 18–37. Springer, Heidelberg (2002)

12. Kontogiannis, K., Lewis, G., Smith, D. and Litoiu, M., Muller, H., Schuster, S., Stroulia, E.: The landscape of service-oriented systems: a research perspective. In: Proceedings of the International Workshop on Systems Development in SOA Environments, p. 1. IEEE Computer Society (2007)

13. Kwiatkowska, M., Norman, G., Parker, D.: PRISM 4.0: verification of probabilistic real-time systems. In: Gopalakrishnan, G., Qadeer, S. (eds.) CAV 2011. LNCS, vol. 6806, pp. 585–591. Springer, Heidelberg (2011)

14. Kwiatkowska, M., Norman, G., Segala, R., Sproston, J.: Automatic verification of real-time systems with discrete probability distributions. Theor. Comput. Sci. **282**(1), 101–150 (2002)

15. Philippou, A., Georghiou, C., Philippou, G.: A generalized geometric distribution and some of its properties. Statistics & Probability Letters **1**(4), 171–175 (1983)

16. Puschner, P., Burns, A.: Guest editorial: A review of worst-case execution-time analysis. Real-Time Systems **18**(2–3), 115–128 (2000)

17. Wandeler, E., Thiele, L., Verhoef, M., Lieverse, P.: System architecture evaluation using modular performance analysis: a case study. International Journal on Software Tools for Technology Transfer **8**(6), 649–667 (2006)

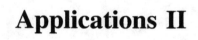

Applications II

Applications I

iDSL: Automated Performance Prediction and Analysis of Medical Imaging Systems

Freek van den Berg[✉], Anne Remke, and Boudewijn R. Haverkort

University of Twente, PO Box 217, 7500 AE Enschede, The Netherlands
{f.g.b.vandenberg,a.k.i.remke,b.r.h.m.haverkort}@utwente.nl

Abstract. iDSL is a language and toolbox for performance prediction of Medical Imaging Systems; It enables system designers to automatically evaluate the performance of their designs, using advanced means of model checking and simulation techniques under the hood, and presents results graphically. In this paper, we present a performance evaluation approach based on iDSL that (i) relies on few measurements; (ii) evaluates many different design alternatives (so-called "designs"); (iii) provides understandable metrics; and (iv) is applicable to real complex systems. Next to that, iDSL supports advanced methods for model calibration as well as ways to aggregate performance results. An extensive case study on interventional X-ray systems shows that iDSL can be used to study the impact of different hardware platforms and concurrency choices on the overall system performance. Model validation conveys that the predicted results closely reflect reality.

1 Introduction

Embedded systems have faced a significant increase in complexity over time and are confronted with stringent costs constraints. They are frequently used to perform safety critical tasks, as with Medical Imaging Systems (MIS). Their safety is significantly determined by their performance. As an example of an important class of MIS, we consider interventional X-ray (iXR) systems, as built and designed by Philips Healthcare.

These systems create images continuously based on X-ray beams, which are observed by a surgeon operating a patient. Images need to be shown quickly for hand-eye coordination [14], viz., the surgeon perceives images to be real-time.

In earlier work, when the ASD method [10] was considered to be used for the design of iXR machines, we have evaluated their performance using simulation models, derived from the design specification by hand.

This paper presents a fully formalised performance evaluation trajectory in which we go from real measurements, via a formal model, to performance predictions, for many different designs, in a fully automated way. Starting point

This research was supported as part of the Dutch national program COMMIT, and carried out as part of the **Allegio** project under the responsibility of the ESI group of TNO, with Philips Medical Systems B.V. as the carrying industrial partner.

© Springer International Publishing Switzerland 2015
M. Beltrán et al. (Eds.): EPEW 2015, LNCS 9272, pp. 227–242, 2015.
DOI: 10.1007/978-3-319-23267-6_15

for this evaluation are models expressed in the IDSL formalism, which has been introduced in [4] and is extended here to fit our new approach. From such models input for the Modest toolset [11] is automatically generated and results are visualized using Graphviz and Gnuplot. This is not only very efficient, it also brings advanced formal performance evaluation techniques, e.g., based on model checking of timed automata and Markov chains, and discrete-event simulation, at the fingertips of system designers, without bothering them with the technical details of these. Furthermore, the approach allows to efficiently predict the performance of a large number of design variants, to compare them, and select the best design given a set of constraints and measures of interest. However, note that in contrast to Design Space Exploration (DSE) [2], in which a few optimal designs are being searched for, we evaluated a large and fixed amount of designs.

Even though the presented approach is fairly general, we illustrate its feasibility on so-called *biplane* iXR systems, which comprise two imaging chains, positioned in perpendicular planes to enable 3D-imaging. They are currently implemented using two separate hardware platforms. However, for various reasons, e.g., costs, physical space, energy consumption and failure rate, it is worth investigating running the software for both image chains on shared (but more powerful) hardware. Hence, we use the above mentioned approach to predict the performance for shared hardware as a case study. Sharing hardware gives potential to concurrency, which may result in increased latency and jitter of images, which, in their turn, affect (perceived) system safety.

We have identified four key objectives that such an integral and fully automated performance evaluation approach should meet, i.e., it should

O1: use as few costly measurements as possible;
O2: be able to evaluate a large number of complex designs;
O3: present its predictions intuitively via understandable (aggregated) metrics;
O4: be applicable to real complex systems.

These objectives are realized through the following four contributions made in this paper. First, the model is calibrated using measurements and measurement predictions to rely on few costly measurements. In contrast, current Design Space Exploration approaches typically require many measurements to be readily available [2,13]. Second, we use iDSL [4], a language and toolbox for automated performance evaluation of service systems, and extend it to support the prediction of unseen empirical cumulative distribution functions (eCDFs). The automation allows us to evaluate many designs, using Modest [11] for simulations, in line with previous work [5,12]. Third, we use a variety of aggregation functions to evaluate designs on different aspects. Fourth, we conduct a case study on a real-life MIS, viz., Image Processing of iXR systems. We validate our model by comparing its predictions with corresponding measurements. Also, the predictions are used to gain insight in the performance of biplane iXR systems with shared hardware.

Paper Outline: This paper is organised as follows: Section 2 provides the methodology of our approach. Section 3 describes how measurements are taken, predicted and applied. Section 4 sketches the iDSL tool chain and model. Section 5 presents the results of the case study. Section 6 concludes the paper.

2 Methodology

We specify our approach as a solution chain as depicted in Figure 1, consisting of three consecutive stages, viz., pre-processing, processing and post-processing. The iDSL toolbox automates these steps and connects them seamlessly.

During **pre-processing**, measurements are performed and execution times derived from them. They are performed for different iXR system configurations and yield large sets of so-called activities for every single design. An activity specifies, for a particular resource and a performed function, the time interval of execution. Activities are visualized automatically in Gantt charts [15]. They are grouped to obtain total execution times per function and in turn aggregated into so-called empirical cumulative distribution functions.

During **processing**, we start with many inverse eCDFs, all based on measurements, that cover all

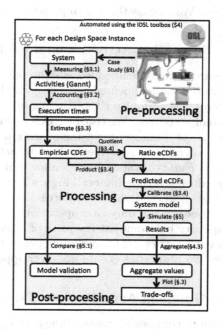

Fig. 1. The solution chain of the approach comprising pre-processing (performing measurements and deriving execution times), processing (predicting eCDFs and simulating) and post-processing (aggregate functions).

possible designs of interest. Many are used for model validation (explained below) and a fraction of them is used to predict new, inverse eCDFs. Hence, one may reason about the performance of many designs, while relying on only few measurements, in line with Objective **O1**.

Next, the iDSL model is executed to obtain performance results, in two steps: (i) iDSL predicts eCDFs for all designs and calibrates the model based on these eCDFs; and (ii) iDSL performs many simulations via the Modest toolset (see Section 4), yielding results for all designs, meeting Objective **O2**.

During **post-processing**, results are processed into aggregated, understandable metrics, facilitating the interpretation of the results (Objective **O3**).

3   Measurements and Emperical CDFs

In this section, measurements performed on design instances are used to predict the performance of other design instances, in four steps: (i) we perform measurements that yield activities; (ii) these activities are grouped into execution times; (iii) these execution times are used to estimate emperical CDFs (eCDFs); and (iv) we predict eCDFs for the complete design space, relying on few estimated eCDFs. We discuss these 4 steps below in more detail.

3.1   Measuring Activities on a Real System

Measurements on embedded systems are typically performed by executing real program code augmented with stopwatches, during a so-called execution run. Stopwatches administer the starting and ending times of functions that run on different resources. We consider iXR systems that loop in cycles and perform a sequence of n image processing operations $(f_1, f_2, ..., f_n)$ on m parallel resources $(r_1, r_2, ..., r_m)$. Measurements lead to activities $Act : Res \times Cycle \times Time \times Time \times Func$ that specify a resource that performs a given function, in a certain cycle, during a time interval. Figure 2 visualizes this in a Gannt-chart.

The system designer requires iXR systems to meet two properties: (i) resources process only one operation for one image at a time, which reduces complexity but comes at the price of a reduced utilization; and (ii) iXR systems adhere to a strict FIFO scheduling policy to preserve the image order. Combined, these two properties ensure non-overlapping functions.

3.2   Grouping Activities into Execution Times

To reduce complexity, we combine activities that perform the same functionality, in the same cycle, but on different resources, into one execution time; formally:

$$E_f(c) = \max\{t_2 \mid (r_i, c, t_1, t_2, f) \in Act\} \ - \ \min\{t_3 \mid (r_j, c, t_3, t_4, f) \in Act\}, \quad (1)$$

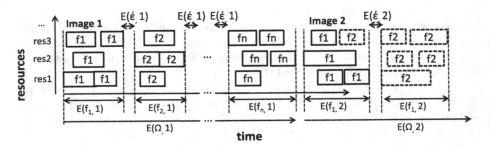

Fig. 2. Activities displayed in a Gannt-chart: The time is on the X-axis and corresponding resources are on the Y-axis. It shows how E activities are grouped into execution times. Activities form rectangles that are labelled with the performed function.

Fig. 3. The empirical distribution function and its inverse, both based on k samples. They are used to determine the probability that a random variable is below a certain value, and for sampling, respectively. It shows the execution time v (X-axis) and corresponding cumulative probability p (Y-axis).

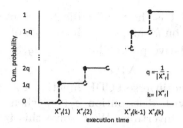

where f is a function, c the cycle, r_i and r_j resources, and t_1, t_2, t_3 and t_4 times. Execution time $E_f(c)$ may include time during which all resources idled. This may result from executing code without stopwatches, or a resource waiting for another resource. Either way, this idle time is attributed to $E_f(c)$ to not underestimate execution times. Finally, $E_\Omega(c)$ represents the overall execution time; formally:

$$E_\Omega(c) = \max\{t_2 \mid (r_i, c, t_1, t_2, f_i) \in Act\} - \min\{t_3 \mid (r_i, c, t_3, t_4, f_j) \in Act\}. \quad (2)$$

3.3   Using Execution Times to Estimate eCDFs

We now *estimate* eCDFs that summarize execution times for different functions. We group the execution times for function f in an array, where we delete the first j samples from $j + k$ measured cycles to eliminate initial transient behaviour. In order to chose a suitable truncation point j, we use the Conway rule [7], and define j as the smallest integer for each function f that is neither the minimum nor the maximum of the remaining samples:

$$\min(E_f(j + 1), ..., E_f(j + k)) \neq E_f(j + 1) \neq \max(E_f(j + 2), ..., E_f(j + 1)).$$

This results in array X_f with $|X_f| = k$ elements, where $X_f(i)$, with $1 \leq i \leq |X_f|$, denotes the ith element of X_f:

$$X_f = (E_f(j + 1), E_f(j + 2), ..., E_f(j + k)). \quad (3)$$

Now, let X_f^* be a numerically-sorted permutation of X_f, such that $X_f^*(i) \leq X_f^*(j)$, for all $i \leq j$. Clearly, $|X_f^*| = |X_f| = k$ and again, $X_f^*(i)$ with $1 \leq i \leq |X_f^*|$ denotes the ith element of X_f^*.

In the following, we define the eCDF function e_f and its inverse e_f^{-1} based on X_f^*, for all functions f. The eCDF function $e_f(v) : \mathbb{R} \to [0 : 1]$ is a discrete, monotonically increasing function that returns the probability that a random variable has a value less than or equal to v. It is defined, for each function f, using the commonly known empirical distribution function [1], as follows:

$$e_f(v) = \frac{1}{k} \sum_{i=1}^{k} \mathbf{1}\{X_f^*(i) \leq v\}, \quad (4)$$

where $\mathbf{1}$ is the usual indicator function. Figure 3 shows an example plot of e_f, based on k values, which consists of $|X_f^*| + 1$ horizontal lines, one for each of the cumulative probabilities $(0, q, 2q, 3q, ..., 1)$. It shows that $e_f(v) = \frac{1}{|X_f^*|} = q$, for $X_f^*(1) \le v < X_f^*(2)$.

The inverse eCDF function $e_f^{-1} : [0 : 1] \rightarrow \mathbb{R}$ is used to draw samples in line with distribution $e_f(v)$, when simulating. Due to the discontinuities, e_f is not invertible. We resolve this by rounding each probability p to the next higher probability p' for which $e_f^{-1}(p')$ is defined (see the vertical dotted lines in Figure 3). Thus, $e_f^{-1}(p)$ returns for each $p \in [0 : 1]$ a value v, as follows:

$$e_f^{-1}(p) = \begin{cases} X_f^*(1), & \text{if } p = 0, \\ X_f^*(\lceil |X_f| \, p \rceil), & \text{if } 0 < p \le 1. \end{cases} \tag{5}$$

This inverse eCDF $e_f^{-1}(p)$ can be used within the inverse transformation method [8]. Due to the above definition, only actual sample are returned.

3.4   Predicting eCDFs for the Complete Design Space

We now *predict* eCDFs for different designs choices. Formally, a Design Space has n dimensions, each comprising a set of designs alternatives $dim_i = \{val_1, val_2, ..., val_{m_i}\}$, for $1 \le i \le n$. The Design Space Model $DSM : dim_1 \times dim_2 \times ... \times dim_n$ is then the n-ary Cartesian product over all dimensions. A Design Space Instance DSI, also called a "design" or "design instance", provides a unique assignment of values to all dimensions: $\overline{x} = (x_1, x_2, ..., x_n)$, where each entry $x_i \in dim_i$ represents the respective design choice for dimension i.

For the sake of simplicity, $Q_{\overline{x}}$ denotes an inverse eCDF e_f^{-1} that is based on a set of measurements of an execution run for design $\overline{x}$. Additionally, $Q_{\overline{x}}(p)$ denotes a sample drawn from $Q_{\overline{x}}$, for probability p.

Clearly the number of designs can grow large, making it costly and infeasible to perform measurements for all possible designs. Hence, we *predict* inverse eCDFs based on other inverse eCDFs without additional measurements, as follows. We carefully select a base design $\overline{b}$ to serve as basis for all eCDF predictions, i.e., $\overline{b}$ is a design that performs well so that its execution times mostly comprise service time and no queueing time. Consequently, set $\hat{Q}$ comprises all inverse eCDFs that need to be acquired through measurements. They correspond to $\overline{b}$ and all neighbours of $\overline{b}$ that differ in exactly one dimension, specified as a union over all dimensions, as follows.

$$\hat{Q} = \cup_{i=1}^{n} \{ Q_{\overline{b}[v_i]_i} \mid v_i \in dim_i \}, \tag{6}$$

where i is the dimension number, and $\overline{b}[v_i]_i = (b_1, b_2, ..., b_{i-1}, v_i, b_{i+1}, ..., b_n)$.

Let $\overline{t}$ be the design for which the inverse eCDF has to be predicted. We assume that all n design dimensions are independent. As we will see below, this assumption does well in the case we have addressed so far.

Fig. 4. A geometric interpretation of a 3D Design Space Model; each spatial dimension relates to a design space dimension. Each point in 3D-space represents a Design Space Instance by assigning a value to each dimension. An arrow depicts a ratio between two Design Space Instances.

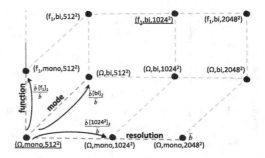

Using only inverse eCDFs in $\hat{Q}$, we specify the product of n ratios that each compensate for the difference between $\bar{b}$ and $\bar{t}$ in exactly one dimension:

$$R(p) = \prod_{i=1}^{n} \frac{Q_{\bar{b}[t_i]_i}(p)}{Q_{\bar{b}}(p)}, \tag{7}$$

where $\bar{t}=(t_1, t_2, ..., t_n)$, p the probability, and n the number of dimensions.

Measuring all eCDFs in a design space with n dimensions and maximally v values per dimension requires $|DSM| = \mathbb{O}(v^n)$ measurements, while the prediction approach only requires $|\hat{Q}| = \mathbb{O}(vn)$ measurements. Predicting eCDFs is particularly efficient for many dimensions, e.g., for 5 dimensions having 5 values each, prediction requires only 25 out of 3125 (0.8%) eCDFs to be measured.

We illustrate eCDF prediction on an iXR machine with three design dimensions: (i) the image processing function, which is f_1, f_2,..., f_n, or Ω (the sum of all functions); (ii) the mode is either $mono$(plane) for one imaging chain, or bi(plane) for two parallel imaging chains; (iii) the resolution is the number of pixels of the images processed, and is either 512^2, 1024^2 or 2048^2 pixels.

Let $\bar{d} = (f_i, m_j, r_k)$ denote design instance $\bar{d}$ with function f_i, mode m_j, and resolution r_k. It is presented conveniently in 3D-space (see Figure 4). Additionally, $Q_{\bar{d}}$ denotes the inverse eCDF of this particular design.

Let $\bar{t} = (f_1, bi, 1024^2)$ be the design, for which we predict an inverse eCDF. Let $\bar{b}=(\Omega, mono, 512^2)$ be the selected base design on which this prediction is based. We then require eCDFs based on measurements for design $\bar{b}$ and for $(f_1, mono, 512^2)$, $(\Omega, bi, 512^2)$ and $(\Omega, mono, 1024^2)$ that each differ from $\bar{b}$ in exactly one dimension and from $\bar{t}$ in all other dimensions. We assume that the three design dimensions are independent. $R(p)$ is then the product of three ratios that each compensate for the difference between design $\hat{b}$ and $\hat{t}$ in one dimension:

$$R(p) = \frac{Q_{f_1,mono,512^2}(p)}{Q_{\bar{b}}(p)} \cdot \frac{Q_{\Omega,bi,512^2}(p)}{Q_{\bar{b}}(p)} \cdot \frac{Q_{\Omega,mono,1024^2}(p)}{Q_{\bar{b}}(p)}. \tag{8}$$

The eCDF of the design $Q_{\bar{t}}$ is then predicted as follows: $Q_{\bar{t}}(p) \approx Q_{\bar{b}}(p) \cdot R(p)$, for probabilities $p \in [0 : 1]$. To validate, we compare $R(p)$ with ratio $Q_{\bar{t}}(p)/Q_{\bar{b}}(p)$ that is obtained when measuring $Q_{\bar{t}}(p)$ in Figure 5, for all probabilities $p \in [0 : 1]$. Figure 5 shows the three ratio terms of (8):

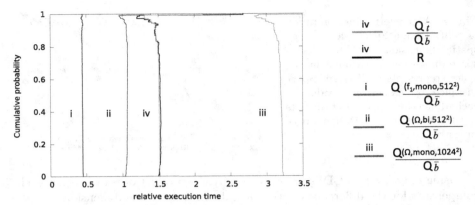

Fig. 5. Inverse eCDFs with relative execution times, which are the quotient of two eCDFs. On the X-axis, it shows relative execution times, and on the Y-axis, cumulative probabilities. Both axes show ratios and are therefore unitless.

(i) $Q_{(f_1,mono,512^2)}$ / $Q_{\bar{b}}$ (dark blue) compares the execution times of function f_1 and Ω. Function f_1 takes about 0.4 of the total execution time;

(ii) $Q_{(\Omega,bi,512^2)}$ / $Q_{\bar{b}}$ (red) compares the performance of a mono and biplane system. Most values are close to 1. Hence, their performance is comparable;

(iii) $Q_{(\Omega,mono,1024^2)}$ / $Q_{\bar{b}}$ (purple) shows the performance effect of a resolution increase from 512^2 to 1024^2 pixels, which is 3.2 for most probabilities p, which is less than the fourfold increase of pixels.

Presumably, image processing comprises a constant and pixel dependent part, leading to relatively faster processing for larger images. We also see that (iv) $R(p)$ matches its measurement-based counterpart $Q_{\bar{t}}(p)/Q_{\bar{b}}(p)$ well.

The shown graphs are fairly constant for most probabilities p, which indicates that design instances are linearly dependent. However, they display smaller values for probabilities p close to 1. This is because of the inverse eCDF $Q_{\bar{b}}$, which has high execution times for probabilities near 1. Since all ratios discussed have $Q_{\bar{b}}$ in their numerator, they consequently display smaller values for the same probabilities. In Section 5, we show the results of predicting the performance of designs, using these ratios.

4   Extending the iDSL Language and Solution Chain

In this section, we explain how we use iDSL [4] to automate the solution chain of Figure 1. For this purpose, we have build on previous work of iDSL in which a language and toolbox for performance evaluation of service systems has been constructed. The language comprises six sections that constitute the conceptual model, i.e., Process, Resource, System, Scenario, Measures and Study (see Figure 7). Figure 6 shows the iDSL solution chain that automates the methodology. To support it, the iDSL toolbox has been extended with functionalities "Resolve eCDF", realizing the concepts of Section 3, and "Compute aggregate"

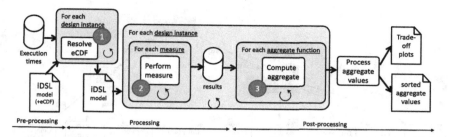

Fig. 6. The fully automated iDSL solution chain. An iDSL model and execution times are used to predict eCDFs, leading to an iDSL model having the predicted eCDFs incorporated in it. For each design, measures are performed and a number of aggregate functions are computed using these measures. Finally, the aggregate values of all design instance are sorted and turned into trade-off plots.

(see Figure 6, component 1 and 3). Below, we discuss the iDSL model of iXR systems (Section 4.1), followed by the two extensions (Section 4.2 and 4.3).

4.1 The iDSL Model of iXR Systems

The iDSL model is defined as follows. *Process* Image Processing (IP) encompasses two high-level functions "Noise reduction" and "Refine", which in turn decompose into a sequence of n atomic functions $f_1, f_2, \ldots, f_n$ (as depicted in Figure 8). IP is enclosed by a **Mutual Ex**clusion to enforce a strict FIFO scheduling policy by processing images in one go. The only *resource*, CPU, is equipped with a FIFO queue. *Service* IP maps all processes to the CPU. The *Scenario* prescribes that images arrive f times per second with fixed inter-arrival times, where f is the frame-rate. The *Measure* simulation yields, in one run, latencies of 50 images and the utilization of the CPU. The *Design space* is the Cartesian product resolution and mode. However, to compute two trade-off graphs (as in Figure 9), we also vary the buffer size and frame-rate.

Fig. 7. The concepts of iDSL [4]: A **service system** provides services to consumers. A **service** is implemented using a **process, resource** and mapping. A **process** decomposes service requests into atomic tasks, each assigned to resources by the mapping. Resources perform one atomic task at a time. A **scenario** comprises invoked service requests over time. A **study** evaluates a set scenarios to derive the system's characteristics. **Measures** of interest are the retrieved measures.

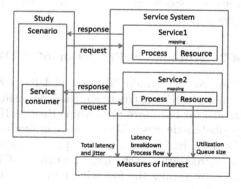

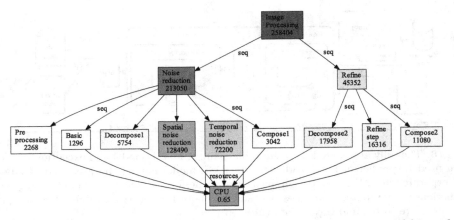

Fig. 8. Service IP contains process IP, which decomposes into a sequential hierarchy of functions. All atomic functions map to resource CPU. In the figure, it shows latency times (in μs) for each function and the utilization for Resource CPU. Latency demanding functions are dark to be easily pinpointed by the system designer. Green Resource CPU has a low utilization. This visual is auto-generated from the iDSL description.

The iDSL model of iXR systems can be depicted as a Discrete-time Markov chain (DTMC) informally, as follows. Its *states* are composed of an image counter, a function counter, the accrued service time of the processed function, the time until the next image arrives, and the queue of Resource CPU. Its key events are the arrival of a new image, which is placed in the queue or discarded when the queue is full, and the current function finishing processing. All states have one or two outgoing *transitions*. The latter case represents a binary probabilistic choice, driven by an eCDF, to decide whether the currently processed function receives further processing, or finishes. Principally, the number of states is infinite due to an ever-increasing image counter. However, omitting this counter from the state, yields a finite-state DTMC that can be analyzed via model checking [4] to retrieve aggregated latencies, e.g., the maximum latency. iDSL generates the DTMC via the Modest language, which in turn transforms into it a Stochastic Timed Automata [11] network. We only use simulations in this paper.

4.2   Automated Prediction of eCDFs for the Complete Design Space

Pre-processing step "Resolve eCDF" performs iDSL model transformations in which eCDF constructs are resolved, using execution times of designs. Our iDSL model (as in Figure 8) contains Process IP with n functions $(f_1, f_2, \ldots, f_n)$ whose probabilistic execution times are individually computed as in (8), for each design. Concretely, "Resolve eCDF" predicts eCDFs using execution times, followed by a discretization step that turns these eCDFs into finite probabilistic choices. It thereby applies the following four steps, for each function, mode and resolution.

First, the required eCDFs, as in the right hand of (8), are obtained by retrieving the corresponding execution times, from which in turn eCDFs are estimated. Second, solving this equation yields the eCDF to be predicted. Third, n samples are taken from this predicted eCDF for probabilities $(\frac{1}{n}, \frac{2}{n}, \cdots, 1)$ to discretize it, i.e., we use $n = 1000$. Fourth, these n samples are combined in a probabilistic choice, a process algebra construct that iDSL supports by default. After this, the resulting probabilistic choices are ordered by design and function within that design, and added to the iDSL model.

4.3 Automated Aggregation of Latencies

The post-processing step "Compute aggregate" applies, for each design, a number of aggregate functions on n obtained latencies from simulations (we use $n = 50$). We selected the *average*, *maximum* and *median* as the functions of interest (for an example, see Table 1). Concretely, "Compute aggregate" executes when a simulation run finishes and computes the specified aggregate functions.

Next, "Process aggregate values" generates trade-off plots [6, 9] that help the system designer with balancing between two system aspects by plotting these aspects of designs in a 2D-plane (for examples, see Figure 9). They visualize how gains on one system aspect pay its toll on another. A design dominates another design when it ranks better on one aspect and is at least a good on the other aspect. Dominated designs are called Pareto suboptimal, others Pareto optimal.

Finally, "Process aggregate values" sorts design instances on each individual aspect. This enables the comparison of designs on a particular system aspect.

5 Results of a Case Study on iXR Systems

In this section, we study the performance results of an iXR system to show the validity and applicability of our work by evaluating a concrete iXR system.

We obtained all results by executing the constructed iDSL model on a PC (AMD A6-3400M, 8Gb RAM) using 32'27" (minutes, seconds). Predicting eCDFs took 1'48" (6%), simulations 30'13" (91%) and aggregate functions 19" (1%).

5.1 The Performance of an iXR System

In the following, we present eCDFs with execution times and corresponding aggregate metrics, a latency break-down graph, and two trade-off graphs.

We assess if the performance of biplane iXR systems on shared hardware is as good as monoplane ones. We use **eCDFs with execution times** in which we compare the measured performance (in Figure 10) of these biplane systems (green) with monoplane ones (red), for resolutions 512^2 (top), 1024^2 (middle) and 2048^2 (bottom). As an effect of sharing hardware, biplane systems perform worse than monoplane ones for image resolutions 512^2 and 1024^2, viz., their average latencies are 6% and 2% higher (as in Table 1), respectively. In contrast,

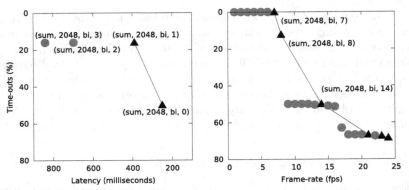

Fig. 9. Two trade-off graphs. Left, designs $(\Omega, 2048, bi, b)$ where b is the buffer size, which affects the relative number of time-outs (y-axis) and the average latency (x-axis). Designs $(\Omega, 2048, bi, 0)$ and $(\Omega, 2048, bi, 1)$ are Pareto optimal (black triangles), opposed to the other designs that are Pareto suboptimal (green circles). Right, designs $(\Omega, 2048, bi, f)$ where f is the frame-rate (x-axis) affecting the time-out ratio (y-axis).

biplane systems with an image resolution of 2048^2 perform 9% better than their monoplane counterparts, due to more powerful hardware biplane systems entail.

Additionally, we assess whether the predictions reflect reality. Therefore, we compare the predicted performance (in Figure 10) of these biplane systems (blue) with the monoplane one (black). They are consistently 6-7% slower, just as the difference in the average and median latency of the **aggregate metrics** (in Table 1). This difference stems from eCDF prediction, i.e., it is the ratio for mode between monoplane and biplane: $Q_{(\Omega, bi, 512^2)}$ / $Q_{\bar{b}}$ (in Figure 5).

Furthermore, iDSL creates **latency breakdown charts** for all designs, including design $(\Omega, 2048, bi)$ (see Figure 8). "Spatial noise reduction" and "Temporal noise reduction" (dark gray) are on average the most time consuming functions of IP. We consider the utilization of CPU of 0.65 "good" (green).

Finally, we show two **trade-off graphs** (in Figure 9). They provide insight in how an increase in one system aspect implies a loss in another one. For illustration purpose, the design space is therefore expanded here with the dimensions buffer size and frame-rate. First, Figure 9, (left) shows how the buffer size influences both the average latency (x-axis) and the time-out ratio (the relative amount of images rejected by the system due to overuse) (y-axis), for designs $(\Omega, 2048, bi, b)$ where $b \geq 0$ is the buffer size. Note that both axis are reversed, so that designs that are on the top-right in the graph are preferable, e.g., design $(\Omega, 2048, bi, 1)$ is preferred to $(\Omega, 2048, bi, 2)$. The design with $n = 0$ yields 50% time-outs and a latency of 119ms, whereas the design with $n = 1$ leads to 16% time-outs, but at the price of a latency of 184ms. All designs with $n \geq 2$ yield 16% time-outs, but with an ever increasing latency due to queuing time as n increases, making them Pareto suboptimal.

Figure 9 shows that the frame-rate and the time-out ratio are positively correlated, for designs $(\Omega, 2048, bi, f)$ where f is the frame-rate. For $f \leq 7$, no time-outs occur, whereas for $f > 7$ the number of time-outs increases steadily.

5.2   The Validity and Applicability of the iDSL Model

We compare the predicted and measured eCDFs of six designs to see if the predicted results reflect reality, using two similarity functions for eCDFs: the Kolmogorov distance Kd and the maximum execution ratio Er. Er is inspired by Kd, but returns the "horizontal distance"; these distances are also indicated in the plot in Figure 10. Er is normalized using the median values of its arguments, making Er symmetric and unitless. They are defined as follows.

$$Kd_{m,n} = \sup_{x \in \mathbb{R}} |F_m(x) - F_n(x)|, \qquad Er_{m,n} = \frac{\sup_{p \in [0:1]} |G_m(p) - G_n(p)|}{\frac{1}{2}G_m(0.5) + \frac{1}{2}G_n(0.5)},$$

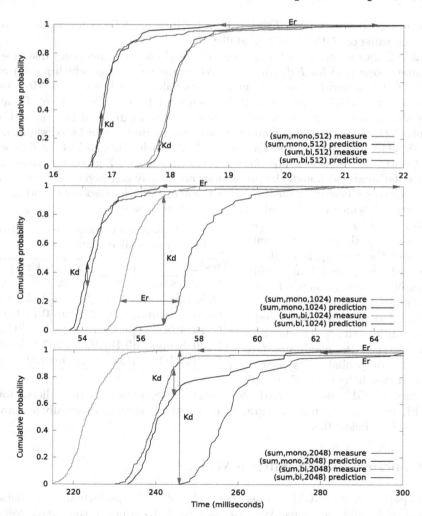

Fig. 10. Measured and predicted execution times eCDFs, for resolution 512 (top), 1024 (middle) and 2048 (bottom), and mode monoplane and biplane. (iDSL auto-generated)

Table 1. For three aggregate functions, the predicted and measured outcomes (in ms) and their difference Δ, based on the first 50 latency values, for six designs.

Design	Average latency			Maximum latency			Median latency		
	Pred.	Meas.	Δ	Pred.	Meas.	Δ	Pred.	Meas.	Δ
$(\Omega, 512, mono)$	9	9	0%	9	12	-22%	9	9	0%
$(\Omega, 512, bi)$	9	9	0%	12	15	-21%	9	9	0%
$(\Omega, 1024, mono)$	27	28	-2%	29	33	-12%	27	27	0%
$(\Omega, 1024, bi)$	29	28	4%	30	29	3%	29	28	4%
$(\Omega, 2048, mono)$	122	123	-1%	149	198	-25%	120	120	0%
$(\Omega, 2048, bi)$	130	112	16%	149	126	19%	128	112	14%

where m and n are eCDFs, $F_i(x)$ the probability of eCDF i for value x, and $G_i(p)$ the value of eCDF i for probability p.

Table 2 shows outcomes for Kd and Er. It shows the maximum distance p and time it occurred for Kd, and the maximum time ratio at which p occurred for Er. Kd is generally low, i.e., most of its values are below 0.16. However, for design $(\Omega, 1024, bi)$ and $(\Omega, 2048, bi)$, Kd is high, 0.86 and 1, resp. Table 1 shows the measured and predicted outcomes of the aggregated functions. Like Kd, predictions for the average and median latency are high for two designs, i.e., for design $(\Omega, 1024, bi)$, 4% and 4%, and for $(\Omega, 2048, bi)$ 16% and 14% difference, resp. Contrarily, the eCDFs for these designs (in Figure 10, green and blue) are not that far apart, although the predictions are clearly conservative. This is due to the relative efficiency gain that occurs when both the resolution and mode are increased, which eCDF prediction does not address.

Note that Kd is high when the execution times do not vary much and the overlap is small (in Figure 10, green and blue), while the graphs are fairly similar. Hence, we propose measure Er, tailored to this domain, comparing relative execution times. In the case study, Er has its maximum (see Table 2), for probabilities near 1, the worst case behaviour. Er is high

Table 2. Comparing measured and predicted eCDFs via two similarity functions.

Design	Kd	x	Er	p
$(\Omega, 512, mono)$	0.13	8.4 ms	0.27	1.00
$(\Omega, 512, bi)$	0.08	8.9 ms	0.35	1.00
$(\Omega, 1024, mono)$	0.22	27.0 ms	0.14	1.00
$(\Omega, 1024, bi)$	0.86	28.4 ms	0.04	0.20
$(\Omega, 2048, mono)$	0.15	122.6 ms	0.50	0.98
$(\Omega, 2048, bi)$	1.00	125.1 ms	0.24	0.98

for designs $(\Omega, 512, mono)$, $(\Omega, 512, bi)$ and $(\Omega, 2048, mono)$ due to outliers. However, Figure 10 shows that their graphs are relatively similar, especially for probability values below 0.8.

6   Conclusions and Future Work

In this paper, we used iDSL, a language and toolbox for performance prediction of medical imaging systems. We extended iDSL to support the prediction of unseen eCDFs based on other measured eCDFs, and aggregate functions.

iDSL provides a performance evaluation approach in which we (i) rely on few costly measurements; (ii) use the iDSL toolset to automatically evaluate many designs and present the results visually; (iii) automatically generate aggregated metrics; and (iv) evaluate the performance of complex iXR systems.

In a case study, we have investigated the performance effect of biplane iXR systems on shared hardware. Measurements indicate that these systems perform as good as monoplane ones, but predictions show more conservative results.

iDSL generates latency breakdown charts for each design that show the system designer the process structure, the time consuming processes and resource utilizations, at one glance. iDSL also generates trade-off graphs, in which designs are plotted on two oppose system aspects. They provide the system designer insight in how an increase on one system aspect implies a loss on another one.

We validated the model by comparing its outcomes with measurements. They mostly reflect reality, but are conservative for high resolution biplane systems. The case study involved a medical imaging system, but we consider the approach applicable to many service-oriented systems.

In parallel work we have extended iDSL with probabilistic model checking to obtain execution time eCDFs [3] using the Modest toolset [11].

Acknowledgements. We would like to thank Arnd Hartmanns of the Modest team at Saarland University for his efforts made during the development of iDSL, and Mathijs Visser at Philips Healthcare for giving us insight in the iXR system case study and for performing measurements.

References

1. Ayer, M., Brunk, D., Ewing, G., Reid, W., Silverman, E.: An empirical distribution function for sampling with incomplete information. The Annals of Mathematical Statistics **26**(4), 641–647 (1955)
2. Basten, T., van Benthum, E., Geilen, M., Hendriks, M., Houben, F., Igna, G., Reckers, F., de Smet, S., et al.: Model-driven design-space exploration for embedded systems: the octopus toolset. In: Margaria, T., Steffen, B. (eds.) ISoLA 2010, Part I. LNCS, vol. 6415, pp. 90–105. Springer, Heidelberg (2010)
3. van den Berg, F., Hooman, J., Hartmanns, A., Haverkort, B., Remke, A.: Computing response time distributions using iterative probablistic model checking. In: Computer Performance Engineering, LNCS, vol. 9272. Springer (2015) (to appear)
4. van den Berg, F., Remke, A., Haverkort, B.R.: A domain specific language for performance evaluation of medical imaging systems. In: 5th Workshop on Medical Cyber-Physical Systems, pp. 80–93. Schloss Dagstuhl (2014)
5. van den Berg, F., Remke, A., Mooij, A., Haverkort, B.: Performance evaluation for collision prevention based on a domain specific language. In: Balsamo, M.S., Knottenbelt, W.J., Marin, A. (eds.) EPEW 2013. LNCS, vol. 8168, pp. 276–287. Springer, Heidelberg (2013)
6. Censor, Y.: Pareto optimality in multiobjective problems. Applied Mathematics and Optimization **4**(1), 41–59 (1977)
7. Conway, R.: Some tactical problems in digital simulation. Management Science **10**(1), 47–61 (1963)

8. Devroye, L.: Sample-based non-uniform random variate generation. In: Proceedings of the 18th Winter simulation conference, pp. 260–265. ACM (1986)
9. Ghodsi, R., Skandari, M., Allahverdiloo, M., Iranmanesh, S.: A new practical model to trade-off time, cost, and quality of a project. Australian Journal of Basic and Applied Sciences 3(4), 3741–3756 (2009)
10. Groote, J., Osaiweran, A., Wesselius, J.: Analyzing the effects of formal methods on the development of industrial control software. In: 27th IEEE International Conference on Software Maintenance, pp. 467–472. IEEE (2011)
11. Hartmanns, A., Hermanns, H.: The modest toolset: an integrated environment for quantitative modelling and verification. In: Ábrahám, E., Havelund, K. (eds.) TACAS 2014 (ETAPS). LNCS, vol. 8413, pp. 593–598. Springer, Heidelberg (2014)
12. Haveman, S., Bonnema, G., van den Berg, F.: Early insight in systems design through modeling and simulation. Procedia Computer Science 28, 171–178 (2014)
13. Igna, G., Vaandrager, F.: Verification of printer datapaths using timed automata. In: Margaria, T., Steffen, B. (eds.) ISoLA 2010, Part II. LNCS, vol. 6416, pp. 412–423. Springer, Heidelberg (2010)
14. Johnson, J.: Designing with the Mind in Mind: Simple Guide to Understanding User Interface Design Rules. Morgan Kaufmann (2010)
15. Wilson, J.: Gantt charts: A centenary appreciation. European Journal of Operational Research 149(2), 430–437 (2003)

Stream Processing on Demand
for Lambda Architectures

Johannes Kroß[1]($\boxtimes$), Andreas Brunnert[1], Christian Prehofer[1],
Thomas A. Runkler[2], and Helmut Krcmar[3]

[1] fortiss GmbH, Guerickestr. 25, 80805 Munich, Germany
{kross,brunnert,prehofer}@fortiss.org
[2] Siemens AG, Corporate Technology, Otto-Hahn-Ring 6, 81739 Munich, Germany
thomas.runkler@siemens.com
[3] Technische Universität München, Boltzmannstr. 3, 85748 Garching, Germany
krcmar@in.tum.de

Abstract. Growing amounts of data and the demand to process them
within time constraints have led to the development of big data systems.
A generic principle to design such systems that allows for low latency
results is called the lambda architecture. It defines that data is analyzed
twice by combining batch and stream processing techniques in order to
provide a real time view. This redundant processing of data makes this
architecture very expensive. In cases where process results are not continuously required to be low latency or time constraints lie within several
minutes, a clear decision whether both processing layers are inevitable
is not possible yet. Therefore, we propose stream processing on demand
within the lambda architecture in order to efficiently use resources and
reduce hardware investments. We use performance models as an analytical decision-making solution to predict response times of batch processes
and to decide when to additionally deploy stream processes. By the example of a smart energy use case we implement and evaluate the accuracy
of our proposed solution.

Keywords: Lambda architecture · Big data · Performance · Model ·
Evaluation

1 Introduction

With the increasing ubiquity of information and communication technology
(ICT) and the emergence of the Internet of things (IoT) the available data
amount is growing exponentially. Simultaneously, technologies have been developed to store, manage and analyze these diverse and high volumes of data, also
known as *big data* [30]. These circumstances allow for applying analytics in order
to gain knowledge and support decision-making. For more and more usage scenarios, these analytical capabilities must also meet specific time requirements
such as real-time [17]. One common approach to design big data systems that
can cover many use cases is the lambda architecture [26]. It mainly consists of a

© Springer International Publishing Switzerland 2015
M. Beltrán et al. (Eds.): EPEW 2015, LNCS 9272, pp. 243–257, 2015.
DOI: 10.1007/978-3-319-23267-6_16

batch layer and a speed layer. The former iteratively processes a set of historical data in batches while the latter processes the arriving data stream in parallel to incrementally analyze latest data. By joining the output of both layers query results always reflect current data.

Nowadays, various complementary technologies with different characteristics exist to build a big data system and there is hardly one technology solution that fits most use cases of an organization. Although the lambda architecture simply is a generic design framework which offers a solution for many use cases, nonetheless, a variety of technologies can be applied for the batch or speed layer. Examples for the batch layer are Hadoop MapReduce [5], Apache Pig [7], and Apache Spark [9] and for the speed layer Apache Storm [10], Apache Spark Streaming [9], Apache Samza [8], or Amazon Kinesis [2]. This multitude leads to the development of complex system of systems, which often results in performance issues and high resource requirements [14]. Furthermore, the lambda architecture intends to process all data twice in both layers. Batch processes also analyze data from the ground up in each iteration to ensure fault tolerance in case of hardware failures or human mistakes [26]. These fundamental ideas require costly resources. For use cases where time constraints are not continuously needed or lie between several minutes, it can be often an important question whether a speed layer is really required or not. However, this question can usually not be answered during system development nor in test systems under realistic workload. As stream processing heavily utilizes main memory, the speed layer can also become an expensive investment [24].

Therefore, we propose a speed layer or stream processing, respectively, on demand. The idea is to exclusively use batch processes as often as possible and switch on stream processing only when batch processes are likely to exceed response time constraints. In this way, computing power is utilized more efficiently and resources can be saved as well as be available for other processes. In case of virtualized environments, investments can be directly decreased by reducing cloud service resources. In order to switch on stream processing at the right time, it is inevitable to predict the response time of succeeding batch iterations. For this purpose, we use performance models. They allow to describe performance influencing factors of software systems and to predict performance metrics such as response time, throughput and utilization by means of analytical solvers or simulation engines [13]. Therefore, we integrate estimated resource demands into the model based on measurements from batch processes to simulate an accurate system behavior. This enables us to efficiently schedule stream processes.

In this paper, we first give a detailed description of our proposed approach in Section 2 and how we use performance models to support decision-making. In Section 3, we validate our approach in an experiment. We describe the selected use case, the setup and sample algorithm for the batch layer, and the prototype performance model to predict batch processes. Afterwards, we discuss the experimental results we derived for different workload scenarios. In Section 4, we reflect related work in the area of the lambda architecture and, finally, conclude our paper with providing an outlook for future work in Section 5.

2   Stream Processing On Demand

In order to make decisions about when to switch on stream processing, we use performance models as an analytical solution. As illustrated in Figure 1, the iterative process is divided into two main steps in which the following Sections 2.1 and 2.2 are structured. First, one batch iteration and, potentially, a concurrent stream process are started within the lambda architecture. Second, after the batch process has ended, a decision-making model is used to decide whether stream processing is required in the next batch process iteration or not. Basis of decision-making is a performance model which is used to predict the response time of a batch process. Afterwards, the procedure is repeated.

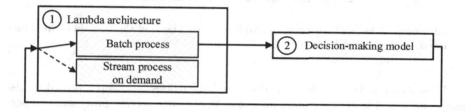

Fig. 1. Stream Processing On Demand Process

2.1   Data Processing in the Lambda Architecture

As already mentioned, our focus is on data processing, namely batch and stream processing, within lambda architecture and not storing data sets or results. Figure 2 illustrates the data flow and structure of batch and speed layer that differ from each other. Starting point is a shared data source which either streams the same data into each processing layer or gets accessed by each layer to retrieve data. Within the batch layer, all data are stored in a data set. A special characteristic of the data set is that it is append-only and data are not updated or removed [26]. Batch processes use the data set to operate on. In doing so, each batch process usually analyzes a huge set of historical data which leads to response times of minutes or hours for one batch job. The results are written to separate views, which is also considered as serving layer by Marz and Warren [26] for batch results. Batch processes constantly run iteratively and start from the beginning once a batch job has finished. If a batch process starts, only data that have been created before are included. Consequently, data that arrive during the current batch process are only included in the next new batch process.

Since all data are analyzed in each cycle, each new result view can replace its predecessor. As the batch layer does not rely on incremental processing, it has the advantage of being a robust system where everything can be recomputed and reconstructed in case of hardware or software failures or human mistakes [26].

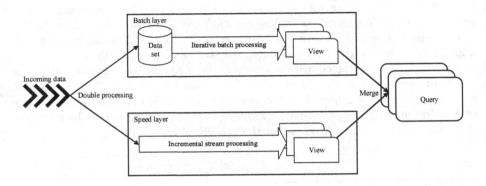

Fig. 2. Composition and data flow of batch and speed layer of the lambda architecture (adapted from Marz and Warren [26])

In contrast to the batch layer, the speed layer does not keep a record of historical data and solely uses main memory. As of today, stream processes run permanently and analyze each incoming message. They incrementally calculate and immediately update their result views. Thus, both layers include separated views and, in practice, usually different technologies are used as underlying databases because of their distinct requirements regarding read and write operations. In order to receive a holistic result, the view of both layers have to be merged in a query.

Although both layers process the same data, the results of queries that merge views only reflect data that are processed once at the time of the query. The purpose of the speed layer is to analyze the data prior to the batch layer and enable low latency by incremental updated result views. As a result, a past view of the speed layer can be discarded as soon as a subsequent batch job has finished.

A typical implementation of the lambda architecture as illustrated in Figure 2 would be to use Apache Kafka [6] - a publish-subscribe messaging system - as shared source for incoming data. For the batch layer, HDFS can be used as data set and Hadoop MapReduce for batch processing. For storing batch results, which Marz and Warren [26] also describe as serving layer, ElephantDB[1] represents a specialized database for this purpose. For the speed layer Apache Storm [10] is an example of an appropriate technology and Apache Cassandra [4] of a database.

---

[1] https://github.com/nathanmarz/elephantdb

2.2   Decision-Making Model

To decide when to switch on stream processing, we predict the response time of
succeeding batch processes and build a decision-making model. To comprehend
why it is necessary to predict the succeeding batch processes, the chronological
sequence of batch and stream processes as intended by the lambda architecture
is illustrated in Figure 3. As already mentioned, results of batch processes are
not available until they finish, while results of stream processes are incremental
and can be queried at any time. Supposing one *batch process i* has ended and
a decision must be made at time y on whether additional stream processes are
needed afterwards or not, the earliest point in time where results of stream processes can be reasonably used is at time z. *Stream process j* considers only data
newer than time y. Therefore, a batch process is required that has analyzed data
before time y. However, the corresponding *batch process j* will only start after
time y and end at a given time z. Thus, a decision must already be made at time
y, if *batch process k* violates time-constraints so stream processes are switched
on at time y. Consequently, query results after time z will have consistently
incorporated all data.

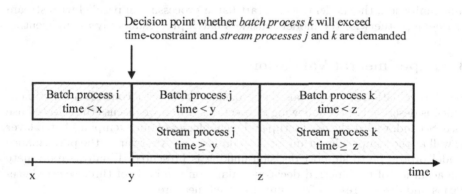

Fig. 3. Chronological sequence of batch and stream processes

The above mentioned response time prediction is part of our decision-making
model. Its procedure is depicted in Figure 4. Starting point is a finished batch
process iteration. The response time of the second next batch iteration is predicted by using a performance model, which takes two inputs - the time constraint for the duration of a batch process and the load intensity. The latter
means information about the incoming data of the batch layer. For instance,
this can be in the form of a variable distribution as modeled by the LIMBO
tool [22]. The prediction can be accomplished by means of simulation or analytical solving. If the predicted response time does not lie within the specified

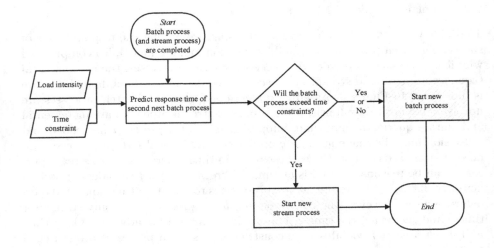

Fig. 4. Decision-making model

time limitation, the model tries to start batch processing in parallel with stream processing, otherwise the model considers batch processing only as sufficient.

3   Experimental Validation

For the evaluation of our proposed approach, we conduct a controlled experiment which is described in the following Subsections. First, we discuss the selected use case. Second, we list the used setup and technologies of our exemplary batch layer as well as the sample algorithm for data processing. Afterwards, the performance model prototype to support decision-making is presented. Finally, we evaluate the accuracy of the inferred decision-making on the basis of three selected scenarios and discuss results from our observed measures.

3.1   Use Case and Design Options

To represent incoming data and their distribution, we pick the example of a common smart energy use case as illustrated in Figure 5.

Here, several hundred wind turbines are positioned in several wind farms in different geographic locations with long distances onshore or offshore. In order to operate efficiently, they measure several thousand parameters per turbine such as pressure, temperature or vibrations of rotor blades. As they are subject to various influences, wind turbines are not always in operation and do not measure data, for instance, if they are defect or are maintained. While onshore wind turbines and wind farms, respectively, tend to have a time-based availability between 95-99%, the values for offshore wind farms with distance less than 12km range from

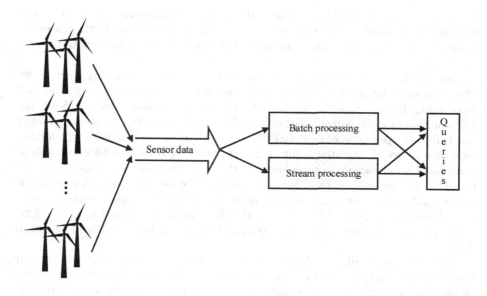

Fig. 5. Data processing of wind power facilities

67.4% to 90.4% [19]. However, wind turbines include also downtimes, if wind is too strong or too weak which is described by the metric energy-based availability. Faulstich et al. [20] compared time-based and energy-based availability of wind turbines. In an extreme case where the downtime due to defects and the downtime due to wind speed does not overlap, the energy-based availability lies within 90.4-95,2%.

Dependent on a wind turbine's availability, we assume it either produces a set of measurement data with constant volume or does not produce any output data. As a result, wind turbines generate not only immense amount of heterogeneous data, but also variable load which makes it difficult to predict the production rate of data. As soon as data are generated, they flow into a central data center where they are processed. Dependent on the use case, data are handled in different ways. They can be gathered and stored in a central repository where batch processing can be used to extract, transform, and load (ETL) data and to apply complex analytics. This procedure usually lies in the range of minutes or hours and is not suitable for real-time requirements. For this purpose, stream processing can be used to directly process data as they stream in. Here, analytical algorithms may be designed in a simpler and less complex way than at batch processing as well as implemented in slightly different way as they produce incremental results.

In scenarios where low latency results are required and normally stream processing is chosen, but also analysis of historical data by batch processing need to be incorporated for conclusive results, the lambda architecture is an appropriate solution that allows for serving such use cases. Therefore, on both processing layers, stream and batch, the same kind of algorithm is implemented and results are joined.

Sensor data can be used for a variety of analytical scenarios such as for condition monitoring, diagnostics, predictive analytics or maintenance, and load forecasting. For our experiment, we concentrate on the latter example. Since the introduction of energy exchange such as the continuous intraday spot market of the European power exchange (EPEX), power can be bargained in 15-minute intervals up to 45 minutes before delivery which enables providers as well as consumers to efficiently act on short notice. In this case, the time-constraint is within 15 minutes. Typical forecast methods for short-term load forecasting include different exponential smoothing methods such as an autoregressive integrated moving average (ARIMA) model [33] or recurrent neural networks [29]. Furthermore, these algorithms are often applied on a sliding window of historical data.

Therefore, we will use this smart energy scenario as an example for our proposed approach and generate sensor data that are processed by one central system in similarly way as we have modeled it in a previous work [23]. The generator produces comma-separated values (CSV) files that represent measurements from wind turbines of one wind farm. Listing 1 shows the file structure and syntax.

Listing 1. Example of generated monitoring data from wind turbines

```
id,    timestamp,                   power,   param1,  ...   paramN
12,    2015-04-01 08:23:04.125,     12.67,   value1,  ...   value1
15,    2015-04-01 08:23:03.973,     13.49,   value2,  ...   value2
13,    2015-04-01 08:23:04.096,     12.59,   value3,  ...   value3
...
```

Each line represents a measurement of one wind turbine consisting of a *id*, *timestamp*, a *power* value and several hundred more parameters which we generated randomly and do not include in our succeeding analytic algorithms.

3.2   Implementation of the Batch Layer

To examine the accuracy of response time prediction for batch processes, we setup the batch layer using HDFS to store data sets and Hadoop MapReduce for batch processing. For simplicity, we installed a single node cluster in pseudo-distributed mode so Apache Hadoop runs only on one machine, but their daemons have their own Java processes. In order to do load forecasting and apply the data generator as mentioned in Section 3.1, we implemented a simple moving average algorithm in a Hadoop MapReduce job. It is based on an example algorithm[2].

---

[2] https://github.com/jpatanooga/Caduceus/

The MapReduce programming model intends to implement one map and one reduce function. The former takes a key/value pair as input and produces a set of key/value pairs, whereas the latter takes a key and set of associated values and combines the values to another smaller set [18]. In our case the map function is implemented as

Listing 2. Map function pseudo code

```
map(Object key1, String value1):
    // key1:   file name
    // value1: measurements of wind turbines of one farm
    for each line l in value:
        kv = parse(l)
        emit({kv.id, kv.timestamp}, {kv.timestamp, kv.power})
```

The function is called for each file within a given folder. It receives one CSV file and its value, which are multiple rows of measurement data of wind turbines. The algorithm reads every line and parses it in order to filter the *id* of a wind turbine, the *timestamp* of the measurement and the *power* value that describes the generated power to that time. Afterwards it releases a composite key containing the *id* and *timestamp*, and the values *timestamp* and *power*. By using a composite key Hadoop sorts the ids of wind turbines and, in a secondary sort, the timestamp for each id. Subsequently, the reduce method results in a simpler design as displayed in Listing 3.

Listing 3. Reduce function pseudo code

```
reduce(Object key, Iterator<object> values):
    // key:    an object containing id and timestamp
    // values: power values ordered by timestamp
    result = simpleMovingAverage(values)
    emit(id, result)
```

The reduce function is called for each different wind turbine and calculates the actual simple moving average. It receives the key object and a list of values as input which contains timestamps and power values sorted by the former. The function itself calculates the *result* and emits it with the corresponding wind turbine *id*.

3.3 Performance Model Prototype

We use the Palladio component model (PCM) [12] for our performance model. PCM is an annotated software architecture model that allows for describing performance relevant factors of software architecture, execution environment and usage profile [13]. Such performance models enable software architects and performance engineers to predict performance metrics such as response time, utilization or throughput by means of simulation or analytical solving.

PCM is divided into several sub-models. In the repository model, we specify a batch process as a software component with its service effect specification (SEFF) to describe the resource demands of the provided service. In the resource environment model, we describe the hardware resources and processing rates on which a batch process will be executed. The concrete assignment of modeled batch processes to resources is determined in the allocation model. Finally, we specify the load intensity from wind turbine measurements in the usage model.

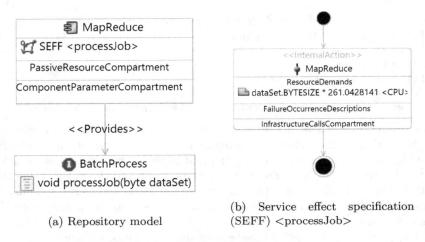

(a) Repository model

(b) Service effect specification (SEFF) <processJob>

Fig. 6. Modeling a batch process with the Palladio component model

Figure 6 shows the substantials of modeling the batch process in our performance model. As shown in Figure 6a, we specify one interface *BatchProcess* with the method *processJob* to analyze an input data set. The implementation of the interface and its method is modeled by the component *MapReduce* with the corresponding SEFF. As illustrated in Figure 6b the SEFF itself solely consists of a CPU resource demand in dependence on an incoming data set size. The data set size is specified in the usage model, in our case, in gigabyte.

In order to define the CPU resource demand and simulate a realistic system behavior we integrated measurements into our performance model. Therefore, we measured response times of the MapReduce job described in Section 3.2 while running it. Afterwards, we used an approximation with response times, which is also implemented by the LibReDe library [32], to estimate the required CPU time each process takes per transaction. One transaction means exactly one batch process that analyzes a set of messages. In our case, the resulting resource demand we estimated is 261 as represented in Figure 6b.

In order to predict results, PCM instances must be first transferred to be either simulated or solved analytically. Available model transformations are a model-to-text transformation like SimuCom [12], queuing Petri nets (QPN) transformations as well as a transformation to layered queuing networks (LQN). Brosig et al. [13] evaluated these model transformations with regards to their efficiency and accuracy. In our application scenario, time is critical and the model need to be solved as efficiently as possible so resulting predictions are available at an early opportunity and the next batch process can be initiated. Therefore, we recommend the use of a model transformation to LQNs. It showed to be the most efficient solution as it is an analytical solver [13].

The performance model prototype has the limitation that is does not reflect the scheduling of processes itself within a cluster, for instance, as accomplished by Apache Hadoop YARN. Therefore, we assume sufficient available resources so batch and stream processes always run without interference.

3.4 Controlled Experiment

To conduct our experiments we run the mentioned data generator to produce CSV files for 10 wind farms with 100 wind turbines each, whereas one wind turbine approximately produces one measurement every second. Afterwards, we run the implemented Hadoop MapReduce job which reads only data measured within a sliding window of 24 hours. While the batch process is running, meanwhile we determine the incoming data volume. After the batch process is finished, we predict the response time of the second next batch process using our performance model. For the immediate succeeding batch process, we exactly know the data volume it will process as we know the historical data distribution and tracked new arrived data. For the batch process to be predicted, the data volume must be estimated. Therefore, a variety of specialized tools and algorithms exist to classify and forecast workload such as the approach by Herbst et al. [21]. As we target an efficient solution and a short-term forecast is required, namely, only the next point, we only use a naïve forecast in this study. It does not involve any computational overhead and simply takes the value of the latest observation as next forecast point in contrast to other methods such as cubic smoothing splines or ARIMA 101 that are more appropriate for scenarios with strong trends or noises [21]. In our case, the next forecast point equals the arrived data volume which has not been absorbed by the last batch process yet. Afterwards, we trigger the performance models with the predicted load intensity as input, and compare the predicted response time with the eventual measured response time.

As already mentioned, the aim is to minimize the usage of the speed layer. The level of potential resource reductions and costs savings that can be achieved depends on the characteristics of the underlying workload and variations in data distributions. The effectiveness of our solution itself, however, depends on how well the data volume is predicted and, especially, how accurate batch processes are predicted. Therefore, we concentrate on the latter in this controlled experiment and perform three selected scenarios with different load intensities by assuming different availabilities of wind turbines based on Faulstich et al. [19,20] to evaluate the accuracy of our solution.

In the first scenario, we assume the wind turbine availability (WTA) is constant during two following batch iterations. Consequently, the measurement data wind turbines produce do also not fluctuate so the predicted load intensity using a naïve forecast is very close to the actual measured load intensity. In the second scenario, we assume an increase of the WTA of 5 % for the subsequent batch process and, vice versa, we assume a decrease in a final third scenario. For each

Table 1. Measured and predicted results of batch processes

Scenario	WTA	Fluctuation	PRT	MRT	RE
	85 %	± 0 %	12.78 minutes	12.17 minutes	5.01 %
1	90 %	± 0 %	13.53 minutes	13.60 minutes	0.51 %
	95 %	± 0 %	14.28 minutes	15.47 minutes	7.69 %
	85 %	+ 5 %	12.78 minutes	13.82 minutes	7.53 %
2	90 %	+ 5 %	13.53 minutes	15.03 minutes	9.98 %
	90 %	− 5 %	13.53 minutes	12.58 minutes	7.55 %
3	95 %	− 5 %	14.28 minutes	13.17 minutes	8.43 %

scenario, we conduct several experiments with different WTA to also validate the prediction accuracy under different load intensities. Afterwards we compare predicted response times (PRT) with eventual measured response times (MRT) of the batch process and calculate the relative error (RE) of the PRT. The results are listed in Table 1.

For a WTA of 85% and no fluctuation during the following batch process, we predict the response time for the batch process to be 12.78 minutes. We measured a MRT of 12.17 minutes which leads to a RE of 5.01%. For a WTA of 90%, the RE of the predicted response time is only 0.51 % and 7.69% for a WTA of 95%.

In the second scenario, for a 85% WTA and a 5% increase of available wind turbines during the following batch iteration, the PRT is 12.78 minutes and the MRT 13.82 minutes with a 7.53% RE. Here, the PRT equals the same PRT as in the experiment for first scenario with a 85% WTA since the naïve forecast, as already mentioned, uses the last observation point, namely 85%, as next prediction point. The same occurrence also applies for the following experiments. The highest RE with 9.98% appeared for a WTA of 90% with +5% fluctuation at which the PRT is 13.53 minutes and the MRT 15.03 minutes.

For a decrease of the 5% WTA in the last scenario, we measured REs in the range similar to the former scenario. With a starting point of 90% WTA, the PRT is 13.53 minutes and the MRT 12.58 minutes. For 95% WTA, the PRT equals 14.28 minutes and MRT 13.17 minutes.

In our experiments, we showed that we are able to predict the response times of a batch process or MapReduce job, respectively, with RE between 0.51% and 9.98%. With regards to our exemplary use case, power can be traded every quarter of an hour in the intraday spot market. Assuming a fluctuating workload and a maximum acceptable response time of 14 minutes remaining one minute buffer, we would be able to accurately schedule stream processing in the second scenario, namely, not to switch on in the first experiment and to switch on stream processing in the second experiment as the MRT exceeds the time-constraint with 15.03 minutes. For a decreasing fluctuation, we would proper schedule stream processing for a starting WTA of 90%. However, for the last experiment in Table 1, we would have left the speed layer switched on as the PRT lies over 14 minutes in contrast to the MRT which is mainly caused by the naïve forecast.

4   Related Work

Similar to our use case, Sequeira et al. [31] propose a system based on the lambda architecture to analyze energy consumption. Martnez-Prieto et al. [25] adapted the architecture for semantic data and Casado and Younas [15] give an extensive review about technologies for the lambda architecture. Regarding optimization or efficient resource usage of the architecture, however, related research mainly focuses on the processing layers itself. For instance, Aniello et al. [3] and Rychl et al. [28] specify on scheduling stream processes, while Alrokayan et al. [1] concentrate on scheduling batch processes.

Regarding predicting batch processes, there is comprehensive research available, for instance, specialized for MapReduce jobs [11], [34], [35] as well as for big data applications in cloud infrastructures [16].

To overcome redundancy regarding software development and infrastructure complexity, approaches such as storm-yarn[3] or by Nabi et al. [27] exist to integrate stream processing in the Apache Hadoop environment. Summingbird[4] is an open source library that allows to write algorithms that can be used for batch as well as stream processing.

5   Conclusion and Future Work

This paper introduced a novel approach to use resources more efficiently when implementing the lambda architecture. It is applicable for usage scenarios where time constraints of queries are not permanently required to be low or lie within several minutes. To reduce processing power, we propose to switch on stream processing on demand in cases where batch processes are likely to exceed time requirements. By using historical information of incoming data and naïve forecasting to classify workload, we predicted the response time of succeeding batch iterations. Therefore, we used performance models in which we integrated estimated resource demands based on measurements. The results allow us to make decisions when additional stream processes are required or, vice versa, can be saved to reduce resource usage. If hardware provision is used in a as-a-service manner, it allows for reducing costs directly.

For future work we plan to automate the process illustrated in Figure 1. This involves to automatically measure incoming data during each batch iteration, apply workload forecasting techniques and trigger solving the performance model. Another challenge is to also integrate the speed layer into our test environment. This will enable us to examine our approach and its efficiency for successive batch iterations for a lengthy period of time. Furthermore, we will integrate other workload forecasting techniques besides the naïve forecast to evaluate possible prediction enhancements and scheduling optimizations.

---

[3] https://github.com/yahoo/storm-yarn
[4] https://github.com/twitter/summingbird

References

1. Alrokayan, M., Vahid Dastjerdi, A., Buyya, R.: Sla-aware provisioning and scheduling of cloud resources for big data analytics. In: Proceedings of the 2014 IEEE International Conference on Cloud Computing in Emerging Markets, pp. 1–8. IEEE (2014)
2. Amazon Web Services: Amazon Kinesis (2015). http://aws.amazon.com/kinesis/ (accessed: April 28, 2015)
3. Aniello, L., Baldoni, R., Querzoni, L.: Adaptive online scheduling in storm. In: Proceedings of the 7th ACM International Conference on Distributed Event-based Systems, pp. 207–218. ACM, New York (2013)
4. Apache Cassandra: The Apache Cassandra project (2015). http://cassandra.apache.org/ (accessed April 28, 2015)
5. Apache Hadoop: Welcome to Apache Hadoop! (2015). http://hadoop.apache.org/ (accessed April 28, 2015)
6. Kafka, A.: A high-throughput distributed messaging system (2015). http://kafka.apache.org/ (accessed April 28, 2015)
7. Apache Pig: Welcomt to Apache Pig! (2014). https://pig.apache.org/ (accessed April 28, 2015)
8. Apache Samza: Samza (2015). http://samza.apache.org/ (accessed April 28, 2015)
9. Apache Spark: Lightning-fast cluster computing (2015). https://spark.apache.org/ (accessed April 28, 2015)
10. Apache Storm: Storm, distributed and fault-tolerant realtime computation (2015). http://storm.apache.org/ (accessed April 28, 2015)
11. Barbierato, E., Gribaudo, M., Iacono, M.: Performance evaluation of nosql big-data applications using multi-formalism models. Future Generation Computer Systems **37**, 345–353 (2014)
12. Becker, S., Koziolek, H., Reussner, R.: The palladio component model for model-driven performance prediction. The Journal of Systems and Software **82**(1), 3–22 (2009)
13. Brosig, F., Meier, P., Becker, S., Koziolek, A., Koziolek, H., Kounev, S.: Quantitative evaluation of model-driven performance analysis and simulation of component-based architectures. IEEE Transactions on Software Engineering **41**(2), 157–175 (2015)
14. Brunnert, A., Vögele, C., Danciu, A., Pfaff, M., Mayer, M., Krcmar, H.: Performance management work. Business & Information Systems Engineering **6**(3), 177–179 (2014)
15. Casado, R., Younas, M.: Emerging trends and technologies in big data processing. Concurrency and Computation: Practice and Experience **27**(8), 2078–2091 (2015)
16. Castiglione, A., Gribaudo, M., Iacono, M., Palmieri, F.: Modeling performances of concurrent big data applications. Practice and Experience, Software (2014)
17. Chen, C.L.P., Zhang, C.Y.: Data-intensive applications, challenges, techniques and technologies: a survey on big data. Information Sciences **275**, 314–347 (2014)
18. Dean, J., Ghemawat, S.: Mapreduce: Simplified data processing on large clusters. Communications of the ACM **51**(1), 107–113 (2008)
19. Faulstich, S., Hahn, B., Tavner, P.J.: Wind turbine downtime and its importance for offshore deployment. Wind Energy **14**(3), 327–337 (2011)
20. Faulstich, S., Lyding, P., Tavner, P.: Effects of wind speed on wind turbine availability (2011)

21. Herbst, N.R., Huber, N., Kounev, S., Amrehn, E.: Self-adaptive workload classification and forecasting for proactive resource provisioning. Concurrency and Computation: Practice and Experience **26**(12), 2053–2078 (2014)
22. von Kistowski, J., Herbst, N.R., Kounev, S.: LIMBO: A tool for modeling variable load intensities. In: Proceedings of the 5th ACM/SPEC International Conference on Performance Engineering, pp. 225–226. ACM, New York (2014)
23. Kroß, J., Brunnert, A., Prehofer, C., Runkler, T.A., Krcmar, H.: Model-based performance evaluation of large-scale smart metering architectures. In: Proceedings of the 4th International Workshop on Large-Scale Testing, pp. 9–12. ACM, New York (2015)
24. Liu, X., Iftikhar, N., Xie, X.: Survey of real-time processing systems for big data. In: Proceedings of the 18th International Database Engineering & Applications Symposium, pp. 356–361. ACM, New York (2014)
25. Martnez-Prieto, M.A., Cuesta, C.E., Arias, M., Fernnde, J.D.: The solid architecture for real-time management of big semantic data. Future Generation Computer Systems 47, 62–79 (2015), special Section: Advanced Architectures for the Future Generation of Software-Intensive Systems
26. Marz, N., Warren, J.: Big data: principles and best practices of scalable real-time data systems. Manning Publications Co. (2015)
27. Nabi, Z., Wagle, R., Bouillet, E.: The best of two worlds: integrating ibm infosphere streams with apache yarn. In: Proceedings of the 2014 IEEE International Conference on Big Data, pp. 47–51. IEEE (2014)
28. Rychlý, M., Škoda, P., Smrž, P.: Heterogeneity-aware scheduler for stream processing frameworks. International Journal of Big Data Intelligence **2**(2), 70–80 (2015)
29. Schäfer, A.M., Zimmermann, H.-G.: Recurrent Neural Networks Are Universal Approximators. In: Kollias, S.D., Stafylopatis, A., Duch, W., Oja, E. (eds.) ICANN 2006. LNCS, vol. 4131, pp. 632–640. Springer, Heidelberg (2006)
30. Schermann, M., Hemsen, H.: Buchmller, C., Bitter, T., Krcmar, H., Markl, V., Hoeren, T.: Big data - an interdisciplinary opportunity for information systems research. Business & Information. Systems Engineering **6**(5), 261–266 (2014)
31. Sequeira, H., Carreira, P., Goldschmidt, T., Vorst, P.: Energy cloud: Real-time cloud-native energy management system to monitor and analyze energy consumption in multiple industrial sites. In: Proceedings of the 2014 IEEE/ACM 7th International Conference on Utility and Cloud Computing, pp. 529–534. IEEE (2014)
32. Spinner, S., Casale, G., Zhu, X., Kounev, S.: LibReDE: a library for resource demand estimation. In: Proceedings of the 5th ACM/SPEC International Conference on Performance Engineering (ICPE 2014), pp. 227–228. ACM, New York (2014)
33. Taylor, J.W.: An evaluation of methods for very short-term load forecasting using minute-by-minute british data. International Journal of Forecasting **24**(4), 645–658 (2008)
34. Verma, A., Cherkasova, L., Campbell, R.H.: Aria: automatic resource inference and allocation for mapreduce environments. In: Proceedings of the 8th ACM International Conference on Autonomic Computing, pp. 235–244. ACM, New York (2011)
35. Vianna, E., Comarela, G., Pontes, T., Almeida, J., Almeida, V., Wilkinson, K., Kuno, H., Dayal, U.: Analytical performance models for mapreduce workloads. International Journal of Parallel Programming **41**(4), 495–525 (2013)

An AnyLogic Simulation Model for Power and Performance Analysis of Data Centres

Björn F. Postema$^{(\boxtimes)}$ and Boudewijn R. Haverkort

Centre for Telematics and Information Technology, University of Twente,
Enschede, The Netherlands
{b.f.postema,b.r.h.m.haverkort}@utwente.nl
http://www.utwente.nl/ewi/dacs/

Abstract. In this paper we propose a simulation framework that allows for the analysis of power and performance trade-offs for data centres that save energy via power management. The models are cooperating discrete-event and agent-based models, which enable a variety of data centre configurations, including various infrastructural choices, workload models, (heterogeneous) servers and power management strategies. The capabilities of our modelling and simulation approach is shown with an example of a 200-server cluster. A validation that compares our results, for a restricted model with a previously published numerical model is also provided.

Keywords: Data centres · Simulation · Discrete-event models · Agent-based models · Power management · Performance analysis · Power-performance trade-off · Cascading effect · Transient analysis · Steady-state analysis

1 Introduction

In 2012-2013, the global power consumption of *data centres* (DCs) was approximately 40 GW; this number is still increasing [7]. Hence, being able to evaluate the effect of energy-savings measures is valuable. One such energy-savings measure is *power management* (PM), which tries to lower the power state of servers, while performance is kept intact. Moreover, the so-called cascade effect (to be discussed later; cf. [8]) on energy consumption in infrastructure, strengthens the effects of PM strategies.

This paper aims to obtain insight in power usage and system performance (measured in terms of throughput and response times) in early DC design phases. It presents high-level models to estimate DC power consumption and performance. We will present and simulate cooperating models for (a) IT

B.F. Postema—The work in this paper has been supported by the Dutch national STW project Cooperative Networked Systems (CNS), as part of the program "Robust Design of Cyber- Physical Systems" (CPS).

B.R. Haverkort—The work in this paper has been supported by the EU FP7 project Self Energy-supporting Autonomous Computions (SENSATION; grant no. 318490).

© Springer International Publishing Switzerland 2015
M. Beltrán et al. (Eds.): EPEW 2015, LNCS 9272, pp. 258–272, 2015.
DOI: 10.1007/978-3-319-23267-6_17

equipment, (b) the cascade effect, (c) the system workload, and (d) power management. The value of our models is shown through the analysis and simulation of an example DC. Our models combine *discrete-event* models and *agent-based* models. Simulating these models sheds light on the above-mentioned power-performance trade-off. For the construction of our models, the multi-method simulation tool ANYLOGIC [1] is used. ANYLOGIC supports a mixture of three common methodologies to build simulation models: (a) *system dynamics*, (b) *process-centric/discrete-event* modelling, and (c) *agent-based* modelling. In this paper, we do not use system dynamics. Discrete-event modelling is a suitable approach for the analysis of systems that encompass a continuous process, that can be divided into discrete parts. Each part is characterised by triggering an event. As [15, p.6] states about discrete-event simulation:

> Discrete-event simulation concerns the modeling of a system as it evolves over time by a representation in which the state variables change instantaneously at separate points in time. These points in time are the ones at which an event occurs, where an event is defined as an instantaneous occurence that may change the state of the system.

Agent-based modelling allows to model individual behaviour to obtain global behaviour with so-called communicating agents. It allows to easily specify heterogeneous populations. As [15, p.694] states about agent-based simulation:

> We define an agent-based simulation to be a DES where entities (agents) do, in fact, interact with other entities and their environment in a major way.

This paper contributes by taking the first steps towards accurate insight in both power and performance by presenting simple queueing models of IT equipment that are easy to extend and allow heterogeneity. Also, a model for the cascading effect is taken into account, and workloads can be based on general probability distributions or on measurement data. Moreover, the insight in power and performance has strong visual support for transient and steady-state analysis. Next steps that follow from this research involve refining and validation of models for more realistic case studies based on measurements and knowledge obtained from cooperation with the project partner Target Holding that allocated their IT equipment in the Centrum voor Informatie Technologie (CIT) data centre in Groningen, the Netherlands.

Over the last few years, various authors have proposed models for the analysis of the power-performance trade-off in data centres. Numerical solutions to compute power and performance for DCs based on Markov models have been proposed in [14], [9], [11], fluid analysis has been proposed in [17] and stochastic Petri nets in [16], [5], [12]. All these numerical approaches allow for the rapid computation of trade-offs, but are often limited in their modelling capabilities, thus leaving them useful for only few metrics under limiting assumptions. Simulation using ANYLOGIC, as we propose here, might be slower, however, it can handle a wider variety of DCs than numerical analysis and scales well to larger systems (as we will see).

The paper is further organised as follows. First, the DC and its context are described in Section 2. Section 3 continues from this system description by introducing all models, metrics and visualisation. A case study with a 200-server example and model validation are presented in Section 4, followed by Section 5 with the conclusions and future work.

2   System Description

In [2], important customer demands for DCs are distinguished, that direct choices on the system architecture, namely: *availability, scalability, flexibility, security* and *performance*. The minimum requirements for a server are *location, space, power supply, network accessibility* and *healthy environment conditions*. The demands from the customer and server requirements drive the choice of the most relevant components in a typical DC. Therefore, a data centre consist of various components, as described in [3], which are typically: Automatic Transfer Switches (ATSs), Uninterruptible Power Supplies (UPSs), Power Distribution Units (PDUs), servers, chillers, coolers, network equipment and devices for monitoring and control.

Through the network the DC becomes accessible from the outside world. The **workload** of a DC is the amount of work that is expected to be done by the DC. The workload of a DC is an important indication for functionality and efficiency. An indication of the workload in a DC is the number of *jobs* per time unit that arrive via the network, together with the length (distribution) of the jobs. Jobs sent through the network arrive in a buffer of a load balancer, that schedules the jobs. We assume that storage and network equipment guarantee negligible job losses in this buffer.

Energy consumption can be reduced in DCs in several ways [8]. One way is **power management** (PM), that aims to switch servers into a lower power state to reduce power consumption, while performance is kept intact. The challenge is to minimise the number of idle servers but prevent unacceptable performance degradation. Sometimes energy consumption reduces at the cost of performance, resulting in a trade-off. We will illustrate such trade-offs later in the paper.

3   Data Center Models

Section 3.1 presents an overview of all implemented agent-based models based on Section 2. These agent-based models are built from underlying queueing models, state-chart models and functions for analysis, which are detailed in Sections 3.2-3.5. Finally, power and performance metrics are presented in Section 3.6.

3.1   Model Overview

All relevant entities are modelled as agents, which enables easy extension towards heterogeneous entities. An overview of all agents is given in the UML diagram in Figure 1.

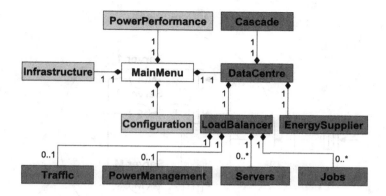

Fig. 1. All implemented agents in one UML diagram.

The `MainMenu` agent links to the agents `PowerPerformance`, `Infrastructure` and `Configuration` with visual representation of the results (light grey). The other agents, i.e., `DataCentre`, `Cascade`, `LoadBalancer`, `EnergySupplier`, `Traffic`, `Power Management`, `Servers` and `Jobs` are the DC models, including a visual representation (dark grey). In the upcoming subsections, the models inside these agents are discussed. The models inside the agent-based models are queueing models, state-chart models and functions for analysis.

3.2 IT Equipment Model

Jobs arrive in a queue in a load balancer. The load balancer decides to which server the jobs should be dispatched depending on the state information.

Figure 2 shows an $G|G|1|\infty|\infty$ queue of the load balancer. Jobs arrive in a FIFO buffer in the load balancer according to a general arrival process (left-most queue) and are served (big circle) in one of the M servers after injection of the job in one of the server queues and waiting for service there.

In order to compute response times, the `LoadBalancer` agent flags a job with a time stamp before it enters the load balancer queue. When a job is finished it compares the time stamp with its current time stamp to compute a response time sample.

Each `Server` agent comprises a $G|G|1|\infty|\infty$ queue with FIFO buffer. The jobs from the load balancer are injected and arrive at the server queue. At most one job at a time is served with a generally distributed service time (with mean value $1/\mu$). If a server has been switched off, then no jobs are routed to it.

The main reason for this modelling approach, instead of directly using an $G|G|M|\infty|\infty$ queue, is that any scheduling algorithm based on the state information of the server can be implemented in this framework, and it also allows for heterogeneous servers.

The *power state* of a server indicates how the server is used and how much power is consumed for that use. The server state can be described with

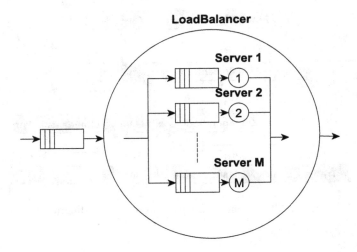

Fig. 2. Load balancer and servers queueing models.

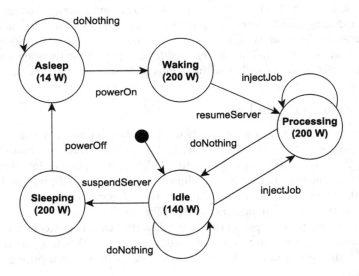

Fig. 3. State-chart model of server with sleep power states.

a state-chart model that switches between the low power consuming inactive `Asleep` state and the high power consuming active states `Idle` and `Processing`, that is controlled by external agents via messages; as depicted in Figure 3. Initially, the server is idle, i.e., the initial state is `Idle`. When the server is active, it can switch between the power state `Processing` (200 W) and `Idle` (140 W). When a server receives a sleep message, it first needs time to suspend the system in power state Sleeping (200 W). After a generally distributed time with mean $1/\alpha_{sl}$, the server is in power state Asleep (14 W). Power state Waking (200 W), which takes extra time before the server starts processing the first job, i.e., after a generally distributed time with mean $1/\alpha_{wk}$ the server is back on. The cycle to shut down and boot a server follows the following sequence of power states: Idle (140 W) → Shutting Down (200 W) → Off (0 W) → Booting (200 W)→ Processing (200 W). The servers leave the power state Booting after a generally distributed time with mean $1/\alpha_{bt}$ and the power state Shutting Down after a generally distributed time with mean $1/\alpha_{sd}$. The power consumption values as used here are taken from [10].

The used power state model is highly abstract and could be refined, e.g., based on recent results for CPU-intensive workloads [13].

The currently implemented *job scheduling* depends on the power state of servers. Initially, a random idle server is selected. If no idle server is present, an off server is selected. In case only active servers are available, a random server is selected. Another variant of a scheduling mechanism is to inject a job in the server with the shortest queue. In case there are multiple shortest queues, a random server is chosen; such (and other) variants can all be easily implemented in our framework.

3.3 Cascade Model

The cascade effect, as elaborated on before, occurs in many DC infrastructure components that consume power based on server power consumption.

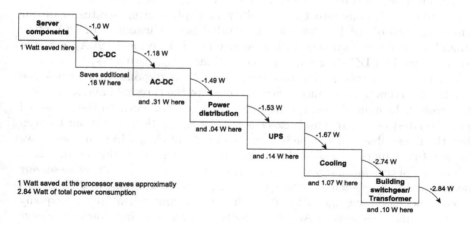

Fig. 4. EnergyLogic's cascade effect model.

The model for the cascade effect in DCs from [8], as depicted in Figure 4, is used in the Cascade agent. For each unit of power used by the servers, other DC infrastructure components, e.g., DC-DC, AC-DC, Power distribution, UPS, cooling, building switchgear/transformer "waste" power in a linear relation.

Hence, energy savings at the level of the server has great impact on the overall energy usage. The `Cascade` agent computes the power consumption metrics via simple linear functions.

3.4 Workload

Based on the description from Section 3.2, jobs enter the load balancer in a $G|G|1|\infty|\infty$ queue following a generally distributed inter-arrival time. In ANY-LOGIC, the most common probability distributions are pre-implemented functions, e.g., exponential, normal, uniform and Erlang. The agent `Job` is added to the buffer after an inter-arrival time based on a function call that generates a random variable for the specified probability distribution. Additionally, in combination with the `Traffic` agent, custom discrete and continuous probability distributions can be defined using, e.g., frequency tables or observed samples. In this paper, we only discuss generally distributed times with time-constant means and jobs with fixed mean lengths, yet our simulation does allow time-varying means in order to support realistic time-varying workload with heterogeneous jobs obtained from measurements in data centres.

3.5 Power Management Strategies

Without application of PM, all servers in the DC are either processing or idle. PM, however, aims to switch servers into lower power states to reduce power consumption when the workload is low, while performance is kept intact. The `PowerManagement` agent has functions to decide when servers need to be put to sleep or even switched off, and when servers need to be switched on.

In order to demonstrate the capability of implementing strategies in our framework, two of the functions are illustrated here. Customers of DCs often demand a certain performance with a Service Level Agreement (SLA), e.g., the response time in a DC should never exceed 25 ms ($R_{\mathrm{thres}} = 0.025\,\mathrm{s}$).

The *threshold strategy* tries to stay as close to this response time as possible by putting servers to sleep until it gets too close to the threshold and servers are again woken. In more detail, the response time gets too close to the threshold when the latest observed sample exceeds 80 % of R_{thres}. Servers are put to sleep when the latest observed sample is lower than 60 % of R_{thres}. In future work, we will investigate more advanced threshold strategies, e.g., including hysteresis.

The aim of the *shut-down strategy* is to achieve a workload of all active servers that is equal to a pre-defined percentage, e.g., a server workload of 20 % means a server spends on average 20 % of the time processing, when jobs are equally scheduled among all servers. As a consequence, servers are shut down to achieve that goal. The only exception to this rule is when there are not enough servers in the DC.

3.6 Power-Performance Metrics

Quantitative metrics are used to provide insight into power and performance in DCs.

Power Consumption. An infrastructure component c has power consumption $P_c(t)$ (in Watt) at time t (in seconds). Power consumption $P_{\text{server}_i}(t)$ of server i depends on the server's power state. The total power consumption of K servers $P_{\text{servers}}(t)$ at time t:

$$P_{\text{servers}}(t) = \sum_{i=1}^{K} P_{\text{server}_i}(t). \tag{1}$$

The power consumption of other system components (like infrastructure), $P_{\text{other}}(t) = \sum_j P_j(t)$, where $j \neq \text{server}_i$ from all other components is computed through the cascade model. The total power consumption then equals the sum of power consumption by all components, i.e., $P_{\text{total}}(t) = P_{\text{other}}(t) + P_{\text{servers}}(t)$. The mean power consumption up to time t is computed as:

$$E[P_{\text{total}}(t)] = \frac{1}{t} \int_{x=0}^{t} P_{\text{total}}(x)dx. \tag{2}$$

Note that this integral is not explicitly computed, but that an efficient discretisation takes place. This discretisation takes full advantage of the fact that events trigger changes in the power consumption, i.e., there is a piecewise linear function for the power consumption over time. The mean power consumption up to time t, where k events occur at time $e_0, e_1, \ldots, e_k$ within the interval $[0, t]$ with a fixed first event $e_0 = 0$ and a fixed last event $e_k = t$, is computed as:

$$E[P_{\text{total}}(t)] = \frac{1}{e_k - e_0} \sum_{i=0}^{k} \int_{x=e_i}^{e_{i+1}} P_{\text{total}}(x)dx \tag{3}$$

$$= \frac{1}{e_k - e_0} \sum_{i=0}^{k} (e_i - e_{i-1}) P_{\text{total}}(e_i) \tag{4}$$

Response Time. This is the delay R_i (in ms) from the moment a job i enters until the moment it leaves the DC. So, each job will report its response time R_i. Given m observations, the mean response time is computed as:

$$E[R] = \frac{1}{m} \sum_{i=1}^{m} R_i. \tag{5}$$

Power State Utilisation. The power state utilisation $\rho_i(t)$ is the percentage of servers in a particular power state i at time t, with $\rho_i(t) \in [0, 1]$. The sum of all power state utilisations at time t is exactly 100%, i.e., $\sum_i \rho_i(t) = 1$.

The mean power state utilisation up to time t is computed as:

$$E[\rho_i(t)] = \frac{1}{t} \int_{x=0}^{t} \rho_i(x)dx. \tag{6}$$

In practice, the integral is not explicitly computed, but an efficient discretisation takes place, similar as done for the mean power consumption. The mean power state utilisation up to time t, where k events occur at time $e_0, e_1, \ldots, e_k$ within the interval $[0, t]$ with a fixed first event $e_0 = 0$ and a fixed last event $e_k = t$, is computed as:

$$E[\rho_i(t)] = \frac{1}{e_k - e_0} \sum_{i=0}^{k} \int_{x=e_i}^{e_{i+1}} \rho_i(x) dx \tag{7}$$

$$= \frac{1}{e_k - e_0} \sum_{i=0}^{k} (e_i - e_{i-1}) \rho_i(e_i) \tag{8}$$

3.7   Visualisation

The `PowerPerformance` and `Infrastructure` agents are implemented to show visuals and "live" values obtained from the simulation runs.

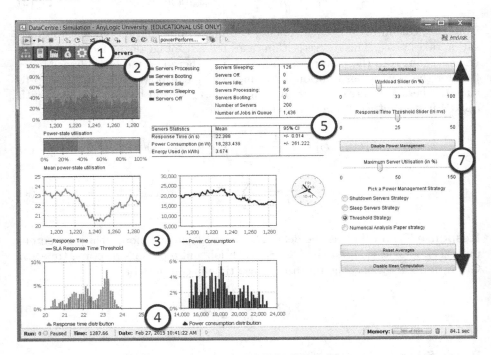

Fig. 5. The dashboard for the IT equipment.

Figure 5 shows an intuitive dashboard with results and configuration parameters of the DC model. The top line shows a menu bar with **(1)** links to the model, visuals and configuration. A cumulative utilisation plot **(2)** shows "live" how many servers are in each power state. A stack chart below this plot shows the mean cumulative utilisation, i.e., how many servers are in each power state. Furthermore, two time plots **(3)** show "live" power consumption (left) and live response time (right) of the simulation. Two histogram plots (4) show the distribution of samples used to compute the means of power consumption (left) and response time (right). The values of the means are displayed in a small table including confidence intervals (5); the exact way how these confidence intervals are computed is not clear (to us) from the documentation, hence, these should be handled with care. Table (6) shows the exact number of servers in each power state, the total number of servers in the DC and the total number of jobs in the queue(s). Configuration options (7) can be used to change the behaviour of the simulation on the fly: adjusting the server workload, the PM strategies, reset the averages and disable averages are the main configuration options. Additional configuration options are available in the `Configuration` agent, like changing the arrival, service, and booting time distributions.

4   Results

First, an example of a data centre with a 200-server computational cluster is elaborated to illustrate the capabilities of the simulation models in Section 4.1. Next, steps are taken for model validation by comparison of the results obtained from simulation to results obtained from models that are solved numerically in Section 4.2.

4.1   Case Study: Computational Cluster

We address a DC that needs to be installed with 200 servers. A *Service Level Agreement* (SLA) permits a response time of at most 25 s. Jobs are served, and, require on average 1 s service time. Furthermore, we require that at most 33 % of all servers are processing, which is not unusual [4]. Booting and shutting down of servers require exactly 100 s and going to sleep and waking up need only 10 s. The *Power Usage Efficiency* (PUE) of the DC is 1.5, i.e., 1 W saved at server level corresponds to 1.5 W saved in total; this is in line with the cascade effect model of Section 3.3. Furthermore, all the other IT equipment (that is, the non-servers) consume 1000 W, in total.

Table 1 shows an overview of workload (λ), service time distribution (μ), IT equipment specifications (mean booting time α_{bt}, mean shutting down time α_{sd}, mean sleeping time α_{sl} and mean waking time α_{wk} of servers), number of servers (n), PUE and power consumption by other IT equipment (P_{otherIT}). Figure 6 shows the power consumption in each power state, combined with a legend for time-cumulative utilisation plot for the shut-down strategy.

First assume that the exact workload is known at all times, and the shut-down strategy (as described in Section 3.5) is applied. Figure 7 shows transient behaviour in a time-cumulative utilisation plot. The x-axis represents the model time t (in s) and the y-axis shows the percentage of servers in each of the power states. The workload without PM is around 33 %. With PM switched on, 50 % of all servers is shut down, such that 66 % of all active servers are processing jobs.

Table 1. DC configuration and workload.

λ	$\exp(33.0)$	μ	$\exp(1.0)$
α_{bt}	$\det(100)$	α_{sd}	$\det(100)$
α_{sl}	$\det(10)$	α_{wk}	$\det(10)$
n	200 servers	PUE	1.5
$P_{otherIT}$	1000 W		

■ Servers Processing (200 W) ■ Servers Booting (200 W)

■ Servers Sleeping (14 W) ■ Servers Off (0 W)

■ Servers Idle (140 W)

Fig. 6. Legend and power consumption in power-states.

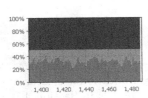

Fig. 7. Time-cumulative utilisation plot with shutdown strategy.

Fig. 8. Time-response time plot with threshold strategy.

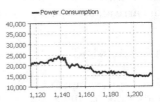

Fig. 9. Time-power consumption plot with threshold strategy.

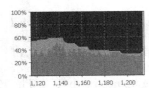

Fig. 10. Time-cumulative utilisation plot with threshold strategy.

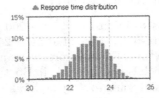

Fig. 11. Response time samples distribution with threshold strategy.

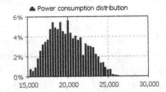

Fig. 12. Power consumption samples distribution with threshold strategy.

Furthermore, the mean power consumption is $\approx 18\,\mathrm{kW}$ and the mean response time is $\approx 1\,\mathrm{s}$.

In practice, the future workload is not exactly known. If workload prediction is inaccurate, late response of the PM strategy can dramatically increase the number of jobs in the system. Such situations have lead to worse performance, either by dropped jobs or large queues.

The threshold strategy (as described in Section 3.5) is based on response times rather than on the workload to control the power state of servers. For this strategy, the mean values are computed and time plots are generated (as can be seen from Figure 8–10). The mean response time $E[R] \approx 23\,\mathrm{s}$ and mean power consumption $E[P_{\mathrm{servers}}] \approx 20\,\mathrm{kW}$.

Figure 8 shows a time-response time plot with again on the x-axis the model time t and on the y-axis a green line interpolating between the response time samples. A horizontal red line is drawn to indicate the response time threshold $R_{\text{thres}} = 25$ s. Moreover, Figure 9 depicts a time-power consumption plot with model time t on the x-axis and a blue line that interpolates between power consumption $P_{\text{servers}}(t)$ samples on the y-axis. Furthermore, Figure 10 shows a time-cumulative utilisation plot. The x-axis represents the model time t (in s) and the y-axis shows the percentage of servers in each of the power states.

As seen in Figure 8–10, servers wake (for $t \in [1120, 1140]$), because the observed response times are approaching the threshold. Therefore, power consumption increases from ≈ 20 kW to ≈ 25 kW and the response time decreases from ≈ 24 s to ≈ 21 s. The next step is to put servers to sleep again (for $t \in [1140, 1220]$), because the perceived response time is fine. As a consequence, response times increase again from ≈ 21 s to ≈ 23 s, but power consumption decreases from ≈ 25 kW to ≈ 15 kW.

4.2   Model Validation

For a simpler but very similar model, numerical solutions using stochastic Petri net (SPN) models have been presented in [16], also to compute mean response time and mean power consumption, again to analyse the power-performance trade-offs caused by PM (but no response time and power consumption distributions).

Table 2. DC configuration and workload.

λ	exp(1.0)	μ	exp(1.0)
α_{bt}	exp(0.01)	α_{sd}	n.a.
n	2-10 servers	β	exp(0.005)

In this paper, we compare the power-performance metrics obtained from our simulation DC models to similar metrics found in the numerical approach, that was presented in [16]. Therefore, the DC model is configured to exactly the same rates, power management strategy, number of servers and job scheduling as with the numerical solution. While this validation covers only a few scenarios, this comparison does show the feasibility of expressing models with the exact same data centre scenario that approach the same power and performance values.

Table 2 shows the configuration and workload. The Poissonian arrival rate $\lambda = 1.0$ jobs/s, $\alpha_{\text{bt}} = 0.01$ servers/s, and $\mu = 1.0$ jobs/s. A special PM strategy is implemented with an exponentially distributed release time with rate $\beta = 0.005$ servers/s that determines the number of servers shutting down per second when idle; note that deterministic time-outs are not allowed in stochastic Petri nets, which explains why the time-out has been chosen like this with the numerical approach. The number of servers is scaled from 2 to 10. Time spend on shutting down a server is ignored.

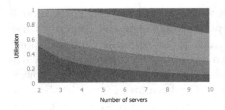

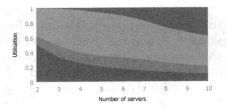

Fig. 13. Cumulative utilisation plot when scaling the number of servers for numerical analysis.

Fig. 14. Servers-cumulative utilisation plot when scaling the number of servers for simulation.

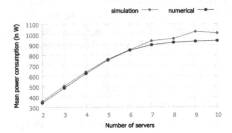

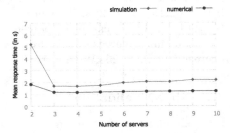

Fig. 15. Mean power consumption for various number of servers for simulation and numerical analysis.

Fig. 16. Mean response time for various number of servers for simulation and numerical analysis.

Figure 13 and Figure 14 show cumulative power state utilisation plots for the servers with the PM strategy, for respectively the SPN-based numerical analysis and our simulation. The x-axis represents the number of servers n and the y-axis shows the percentage of servers in each of the power states (from top to bottom: red = off, orange = idle, green = booting and blue = processing). The plots confirm each other as the plots approach similar shape, but different values; the plots are not completely the same, which is partly the case due to the fact that we run a stochastic simulation, which, in essence, is a statistical experiment. Another reason for the observed difference lies in the implementation of the job scheduling: the SPN-based models use only one general buffer, whereas our simulation models use a separate buffer per server.

Figure 15 shows the mean power consumption for various servers in a DC. The x-axis represents the number of servers n and the y-axis shows mean power consumption (in W). The curves for simulation and numerical analysis have similar shape, but with different values, which range, respectively, from 353 W and 342 W for 2 servers to 982 W and 939 W for 10 servers.

Figure 16 shows the mean response times for various servers in a DC with the PM strategy. The shape of both curves are again very similar, but with different values, which are for simulation and numerical analysis respectively from 5.2 s and 1.84 s for 2 servers to 2.19 s and 1.25 s for 10 servers.

Another reason, for the different values in the curves, is that numerical analysis has no load balancer, but an implicit way for scheduling jobs. First, jobs are scheduled to a random idle server with both numerical analysis and simulation. Otherwise, the jobs are scheduled to a random off server. If all servers are booting or processing, numerical analysis keeps the jobs in the buffer and simulation inject the job in a random server.

5 Conclusions and Future Work

The contribution of this paper is the presentation of a new ANYLOGIC-based tool with an intuitive dashboard, effective for obtaining quick insights in transient and steady-state behaviour of heterogeneous DC with any possible workload and PM strategies. Furthermore, the ANYLOGIC environment enables to easily extend, refine and adapt DC models to many other scenarios.

Insight is obtained in the power and performance in DCs with varying number of servers, PM strategies and workloads. Relevant metrics are derived from the qualitative DC demands, including power consumption, response time and power state utilisation. These metrics are estimated by gathering samples from a mixture of discrete-event and agent-based models for IT equipment, PM and workload, implemented in ANYLOGIC. Furthermore, a cascade model enables the computation of total power consumption. Our approach is illustrated with a 200-server case study.

A well known open-source toolkit CLOUDSIM [6] allows to simulate cloud computing scenarios and allows to specify (textually) DC models with virtual machines, applications, users, scheduling and provisioning. This tool obtains utilisation, response times, execution times and energy consumption metrics from simulation runs. It is future work to to investigate the capabilities of CLOUDSIM in comparison to our ANYLOGIC-based simulation models.

Foreseen future extensions to the presented models are, among others, (i) the analyses of other PM strategies, e.g., based on number of jobs in the system and on hysteresis based strategies; (ii) energy-efficiency measures based on dynamic voltage and frequency scaling; (iii) power consumption of scaling workload; (iv) large-scale DC setting with heterogeneous servers, and a mixture of job sizes and inter-arrival times; (v) virtualisation; and (vi) thermal-aware DCs. Other future work includes the validation of the models with actual measurements from a DC.

References

1. AnyLogic: AnyLogic: Multimethod Simulation Software (2000). http://www.anylogic.com/
2. Arregoces, M., Portolani, M.: Data Center Fundamentals. Cisco Press, Indianapolis (2003)
3. Barroso, L.A., Hölzle, U.: The Datacenter as a Computer: An Introduction to the Design of Warehouse-Scale Machines. Synthesis Lectures on Computer Architecture 8(3), 1–154 (2009). http://www.valleytalk.org/wp-content/uploads/2013/10/WSC_2.4_Final-Draft.pdf

4. Birke, R., Chen, L.Y., Smirni, E.: Data Centers in the Wild: A Large Performance Study. Tech. rep., IBM Research (2012). http://domino.research.ibm.com/library/cyberdig.nsf/papers/0C306B31CF0D3861852579E40045F17F/File/rz3820.pdf

5. Bruneo, D., Lhoas, A., Longo, F., Puliafito, A.: Analytical Evaluation of Resource Allocation Policies in Green IaaS Clouds. In: Proc. of 3rd Int. Conf. on Cloud and Green Computing, pp. 84–91. IEEE Computer Society Washington, DC (2013)

6. Calheiros, R.N., Ranjan, R., Beloglazov, A., De Rose, C.A.F., Buyya, R.: CloudSim: A Toolkit for Modeling and Simulation of Cloud Computing Environments and Evaluation of Resource Provisioning Algorithms. Software: Practice and Experience **41**(1), 23–50 (2011)

7. Datacenter Dynamics Intelligence: Global Data Center Power 2013. Tech. rep., Datacenter Dynamics (2013). http://www.dcd-intelligence.com/Products-Services/Open-Research/Global-Data-Center-Power-2013

8. Emerson Network Power: Energy Logic: Reducing Data Center Energy Consumption by Creating Savings that Cascade Across Systems. White Paper of Emerson Electric Co (2009)

9. Gandhi, A.: Dynamic Server Provisioning for Data Center Power Management. Phd thesis, Carnegie Mellon University (2013). http://citeseerx.ist.psu.edu/viewdoc/summary?doi=10.1.1.376.4361

10. Gandhi, A., Doroudi, S., Harchol-Balter, M., Scheller-Wolf, A.: Exact Analysis of the M/M/k/setup Class of Markov Chains via Recursive Renewal Reward. In: Proc. of the ACM SIGMETRICS Int. Conf. on Measurement and Modeling of Computer Systems, vol. 41, pp. 153–166. ACM, New York (2013)

11. Ghosh, R., Naik, V.K., Trivedi, K.S.: Power-Performance Trade-offs in IaaS Cloud: A Scalable Analytic Approach. In: Proc. of 41st Int. Conf. on Dependable Systems and Networks Workshops, pp. 152–157. IEEE Computer Society Washington, DC (2011). http://ieeexplore.ieee.org/lpdocs/epic03/wrapper.htm?arnumber=5958802

12. Katoen, J.P., Noll, T., Santen, T., Seifert, D., Wu, H.: Performance Analysis of Computing Servers using Stochastic Petri Nets and Markov Automata. Technical report, RWTH Aachen University (2013)

13. von Kistowski, J., Block, H., Beckett, J., Lange, K.D., Arnold, J.A., Kounev, S.: Analysis of the Influences on Server Power Consumption and Energy Efficiency for CPU-Intensive Workloads. In: Proc. of the 6th ACM/SPEC International Conference on Performance Engineering, pp. 223–234. ACM Press, New York (January 2015). http://dl.acm.org/citation.cfm?id=2668930.2688057

14. Kuhn, P.J., Mashaly, M.: Performance of Self-Adapting Power-Saving Algorithms for ICT Systems. In: Int. Symposium IFIP/IEEE on Integrated Network Management, pp. 720–723. IEEE (2013)

15. Law, A.M.: Simulation Modeling and Analysis, 5th edn. McGraw-Hill (2015)

16. Postema, B.F., Haverkort, B.R.: Stochastic Petri Net Models for the Analysis of Trade-Offs in Data Centres with Power Management. In: Klingert, S., Chinnici, M., Rey Porto, M. (eds.) E2DC 2014. LNCS, vol. 8945, pp. 52–67. Springer, Heidelberg (2015)

17. Stefanek, A., Hayden, R.A., Bradley, J.T.: Fluid Analysis of Energy Consumption using Rewards in Massively Parallel Markov Models. In: Proc. of the 2nd ACM/SPEC International Conference on Performance Engineering (ICPE), pp. 121–132 (2011)

Simulation Techniques

Rare Event Simulation with Fully Automated Importance Splitting

Carlos E. Budde[1(✉)], Pedro R. D'Argenio[1], and Holger Hermanns[2]

[1] FaMAF, Universidad Nacional de Córdoba – CONICET, Córdoba, Argentina
{cbudde,dargenio}@famaf.unc.edu.ar
[2] Fakultät für Mathematik und Informatik, Universität des Saarlandes,
Saarbrücken, Germany
hermanns@cs.uni-saarland.de

Abstract. Probabilistic model checking is a powerful tool for analysing probabilistic systems but it can only be efficiently applied to Markov models. Monte Carlo simulation provides an alternative for the generality of stochastic processes, but becomes infeasible if the value to estimate depends on the occurrence of rare events. To combat this problem, intelligent simulation strategies exist to lower the estimation variance and hence reduce the simulation time. Importance splitting is one such technique, but requires a guiding function typically defined in an *ad hoc* fashion by an expert in the field. We present an automatic derivation of the importance function from the model description. A prototypical tool was developed and tested on several Markov models, compared to analytically and numerically calculated results and to results of typical *ad hoc* importance functions, showing the feasibility and efficiency of this approach. The technique is easily adapted to general models like GSMPs.

1 Introduction

Nowadays, systems are required to have a high degree of resilience and dependability. Determining properties that fail with extremely small probability in complex models can be computationally very demanding. Though these types of properties can be efficiently calculated using numerical tools, such as the model checker PRISM [8], this is limited to finite Markov models, and, moreover, the representation through an adequate data structure needs to fit in the computer memory. Beyond this class of models calculations are limited to Monte Carlo simulation methods. However, standard Monte Carlo simulation may easily need an enormous amount of sampling to obtain the desired confidence level of the estimated probability, in order to compensate for the high variance induced by the rare occurrences of the objective property.

Supported by ANPCyT project PICT-2012-1823, SeCyT-UNC program 05/BP12 and their related projects, EU 7FP grant agreements 295261 (MEALS) and 318490 (SENSATION), by the DFG as part of SFB/TR 14 AVACS, by the CAS/SAFEA International Partnership Program for Creative Research Teams, and by the CDZ project CAP (GZ 1023).

© Springer International Publishing Switzerland 2015
M. Beltrán et al. (Eds.): EPEW 2015, LNCS 9272, pp. 275–290, 2015.
DOI: 10.1007/978-3-319-23267-6_18

To reduce this considerable need for simulation runs, efficient Monte Carlo simulation techniques have been tailored to deal with rare events. These can be largely divided into two conceptually different techniques: *importance sampling* and *importance splitting* methods. *Importance sampling* (see [12] and references therein) modifies the sampling distribution in a way that increases the chance to visit the set of rare states. This introduces a bias in the resulting point estimate which needs to be corrected by weighing it with the corresponding *likelihood ratio* [7]. The change of measure requires some understanding of the system under study. A bad choice of measure may have a negative impact on the simulation.

Instead we focus on *importance splitting* techniques, see e.g. [11,18,19]. Importance splitting works by decomposing the state space in multiple levels where, ideally, the rare event is at the top and a level is higher as the probability of reaching the rare event grows. Thus the estimation of the rare probability is obtained as the product of the estimates of the (not so rare) conditional probabilities of moving one level up. As a consequence, the effectiveness of this technique crucially depends on an adequate grouping of states into levels. *Importance functions* are the means to assign a value to each state so that, if perfect, such value is directly related to the likelihood of reaching the rare event. So, a state in the rare set should receive the highest importance and the importance of a state decreases according to the probability of reaching a rare state from it.

Usually, an expert in the area of the system provides the importance function in an *ad hoc* manner. Again, a badly chosen importance function can deteriorate the effectiveness of the technique. With some notable exceptions [3,5,14], automatic derivation of importance functions has received scarce attention.

In this article we provide a simple but effective technique to derive automatically an importance function. It leads the definition of the different levels for importance splitting techniques. The algorithm works by applying inverse breadth first search on the underlying graph of the stochastic process, labelling each state with the shortest distance to a rare state. The importance of each state is then defined as the difference between the maximum distance and its actual distance. Obviously this technique still requires a finite system which fits in the computer memory, but it is not limited to Markov models.

In particular, we focus on the RESTART method [18,19], though the approach presented here can be applied to other importance splitting techniques. We show correctness and effectiveness by performing some significant experimentation in several known case studies. We limit experiments to Markov models to compare the simulated results against numerically obtained values (using PRISM) in order to show correctness. The effectiveness is shown by comparing the performance of the simulation under the automatically calculated importance function against the performance under *ad hoc* importance functions.

The paper is organised as follows. Sec. 2 introduces the models and the type of properties we deal with. Sec. 3 presents the criteria to decide when to stop the simulation. Importance splitting is described in Sec. 4. The algorithm to derive the importance function and the tool that supports it are described in Sec. 5. Experimental results are presented in Sec. 6. The paper concludes in Sec. 7.

2   Formal Models and Properties

Although the technique presented in this paper can be applied to generalised semi-Markov process, we only focus here on discrete-time and continuous-time Markov chains since it is our interest to validate, among other things, the correctness of the technique against values obtained analytically or numerically.

Definition 1 (DTMC). *A* discrete-time Markov chain *or* DTMC *is a tuple* $\mathcal{M} = (S, \mathbf{P}, AP, L)$ *where* $S \neq \varnothing$ *is a countable set of states, the* transition probability function $\mathbf{P} : S \times S \to [0, 1]$ *satisfies* $\forall s \in S \,.\, \sum_{s' \in S} \mathbf{P}(s, s') = 1$, *AP is a set of* atomic propositions *and* $L : S \to 2^{AP}$ *is the* labelling function.

The transition probability function $\mathbf{P}$ specifies for each state s the probability $\mathbf{P}(s, s')$ of jumping to another state s' by means of a single transition. Notice this depends solely on s and s', there is no information about the path which led into s. This is called the *memoryless property* of markovian systems. The states s' for which $\mathbf{P}(s, s') > 0$ are denoted the *successors* of s. The imposed constraint ensures $\mathbf{P}$ is a distribution.

DTMCs remain in the current state for a single time unit before jumping to a successor state. In contraposition, state jumps in continuous-time Markov chains are described with probabilistic timing information. This means that in order to perform a transition both the probability of the successor state and the probability of sojourn time in the current state need to be defined.

Definition 2 (CTMC). *A* continuous-time Markov chain *or* CTMC *is a tuple* $\mathcal{M} = (S, \mathbf{R}, AP, L)$ *where* $\mathbf{R} : S \times S \to \mathbb{R}_{\geq 0}$ *is the* transition rate function *and all the other elements are like in Definition 1.*

The non-negative real value $\lambda = \mathbf{R}(s, s')$ states the speed rate, sampled from an exponential distribution, at which the transition $s \to s'$ would be taken. A null value indicates there is no such transition, and a positive value indicates the probability of jumping to state s' within t time units is $1 - e^{-\lambda t}$.

We focus both on transient and steady state properties. The transient properties we consider aim to calculate the probability of reaching a set of *goal states* G before visiting any *reset state* in the set R. This is characterised by the PCTL formula

$$\mathsf{P}(\neg R \,\mathsf{U}\, G) \tag{1}$$

where U denotes the unbounded until operator from LTL and $\mathsf{P}(\Phi)$ denotes the probability of observing any state that satisfies formula Φ. For simulation purposes we need the probability of reaching a state either in G or in R to be 1. This type of property is recurring in the literature of rare event simulation, see e.g. [1,2,4,19]. Though not in this paper, bounded until properties of the form $\mathsf{P}(\neg R \,\mathsf{U}^{\leq t}\, G)$ can be addressed by our tool in the same way as (1).

While transient properties focus on probabilities of traversing a system from a state to a class of states, steady state analysis focuses on the quantification of a property once the system has reached an equilibrium. In particular, the steady state probability of a set of *goal states* G is the portion of time in which a state in G is visited in the long run and is characterised by the CSL property

$$\mathsf{S}\left(G\right) \tag{2}$$

Though less frequently, this type of property has also appeared in the literature of importance splitting [16,19]. More generally, properties of these type quantify the ratio of goal events G w.r.t. reference events S in the long run such as, e.g., the ratio between lost and sent packets or measures like throughput.

3   Stopping Criteria

The efficacy of Monte Carlo simulation depends on the precision of the estimated parameter. Confidence intervals are use to convey a notion of how far the estimated value may be from the actual value. Confidence intervals are bounds surrounding the computed point estimate and they are characterised by two numbers: the *confidence level* and the *precision*. The first gives information on, roughly speaking, how likely it is for the real population parameter (e.g. the probability of visiting some goal state) to be located within these bounds. The second defines the length of the interval. It is precisely from these two values that the number of required simulation runs is dynamically determined.

These intervals can be constructed in different ways depending on the nature of the sampled population and the parameter to estimate. Following the Central Limit Theorem consider a "big enough" sample $\{x_i\}_{i=0}^{N-1}$ of independently simulated runs, where x_i is the outcome of the i-th run. If $\overline{X}$ is the mean value of the sample, then

$$\left[\; \overline{X} - z_{1-\frac{\alpha}{2}}\frac{\hat{\sigma}}{\sqrt{N}} \;\; , \;\; \overline{X} + z_{1-\frac{\alpha}{2}}\frac{\hat{\sigma}}{\sqrt{N}} \;\right] \tag{3}$$

is a confidence interval for the estimated rare event probability, with confidence level $100(1 - \alpha)\%$ and semi-precision (or error margin) $\hat{\sigma}/\sqrt{N} \cdot z_{1-\frac{\alpha}{2}}$ [9]. Here the constant value $z_{1-\frac{\alpha}{2}}$ represents the $1 - \frac{\alpha}{2}$ quantile of a unit normal variate, uniquely determined by the confidence level chosen by the user, and $\hat{\sigma}$ is the observed sample variance. This is the method of choice to build confidence intervals around estimates of steady state properties like (2). Since nothing is known about the distribution of the x_i samples, namely the long run paths generated on the model, then no tighter bound can be inferred.

The situation is different for transient properties like (1) because every path will almost surely end as soon as it reaches either some goal or reset state. This defines a Bernoulli experiment. Hence, each x_i in the sample $\{x_i\}_{i=0}^{N-1}$ takes either value 1 if run i reaches a goal state in G or 0 otherwise (i.e., it reaches a state in R). Thus, if m is the number of runs that reached a goal state, $\overline{X} \doteq \sum_{i=0}^{N-1}\frac{x_i}{N} = \frac{m}{N} = \hat{p}$, is the estimate of (1).

The previous analysis also shows that $\hat{\sigma} = \hat{p}(1 - \hat{p})$, which is used by the *normal approximation interval* to narrow the length of the interval with respect to eq. (3). Since the precision of the interval has been fixed by the user, this translates into smaller values for N and hence shorter simulation times. There exist however better fitted confidence intervals, specially tailored for situations when the proportion parameter p takes extreme values (viz. $p \approx 0$ or $p \approx 1$). The *Wilson score interval* is one such method [20], and the technique of choice to build confidence intervals whenever dealing with property (1).

4   Rare Event Simulation Through Importance Splitting

The use of Monte Carlo simulation for the estimation of parameters that depend on the occurrence of rare events (i.e. events that occur with very low probability) may easily become an extremely time demanding process due to the high variance induced by these rare occurrences. Since the confidence level and precision are requirements for the estimation, eq. (3) shows that high values of $\hat{\sigma}$ can only be countered by increasing N, which could grow exponentially on the model size [7].

Importance splitting (IS for short) attempts to speed up the occurrence of a rare event, i.e. visiting goal states by generating a drift of the simulations towards them. The first known reference is due to Kahn and Harris for splitting particles in a physical context [6]. The work by José and Manuel Villén-Altamirano stands amongst the most relevant modern contributions. They introduced RESTART, a version of IS with multiple thresholds, fixed splitting and deterministic discards of unpromising simulations [15–19]. Garvels provides a thorough analysis of splitting techniques for rare event simulation in his PhD thesis [2]. For a broad survey of importance splitting see [11] and references therein.

The general idea in IS is to favour the "promising runs" that approach the rare event by saving the states they visit at certain predefined checkpoints. Replicas of these runs are created from those checkpoint states, which continue evolving independently from then on. Contrarily, simulation runs deemed to steer away from the rare event are identified and killed, avoiding the use of computational power in fruitless calculi. The likelihood of visiting a goal state from any other state s is called the *importance* of s. The variation in such importance is what determines when should a simulation be split or killed, as the importance value crosses some given *thresholds* up or down, respectively.

From a statistical point of view, IS decomposes the probability of reaching a goal state into several conditional probabilities, each one of them representing the probability of crossing a threshold given that the lower thresholds have already being crossed. The general idea is that sampling each conditional probabilities is easier, i.e. incurs in less variance per estimation, than attempting to sample the rare event at once. Take for instance a buffer where the measure of interest is the probability of exceeding a capacity $C \in \mathbb{N}$, starting from a non-empty state. Denote by c the buffer occupancy and by $\{c > C\}$ the set of states where such capacity is exceeded. The sought value is in consequence

$$p = \mathsf{P}(\{c > C \mid c > 0\}) = \mathsf{P}\left(\{c > C \mid c \geqslant \tfrac{C}{2}\}\right)\mathsf{P}\left(\{c \geqslant \tfrac{C}{2} \mid c > 0\}\right) = p_1\,p_2\ .$$

The state space has been divided into the disjoint regions $\{c \geqslant \frac{C}{2}\}$ and $\{c < \frac{C}{2}\}$, covered by p_1 and p_2 respectively. We say $C/2$ is a *simulation threshold* and this can be easily generalised to $n < C$ thresholds:

$$p = \prod_{i=0}^{n-1} \mathsf{P}(\{c \geqslant c_{i+1} \mid c > c_i\}) = \prod_{i=0}^{n-1} p_i \tag{4}$$

In this equation the i-th threshold is $T_i = c_i$, namely a buffer with c_i elements, $c_0 \doteq 0$ and $c_n \doteq C$. The probability of reaching c_{i+1} elements in the buffer once the "simulation is above T_i" is denoted p_i.

Using this partition of the state space, IS generates the desired estimate $\hat{p} \approx p$ by approximating each of the conditional probabilities p_i. The resulting $\hat{p}_i$ estimates are then multiplied to compose $\hat{p} = \hat{p}_1 \hat{p}_2 \cdots \hat{p}_n$ [2,19]. Notice IS will perform efficiently as long as all thresholds are chosen such that $p_i \gg p$. Only then will the step-wise estimation present a lower variance than traditional Monte Carlo runs. Thresholds are intimately related to the importance of states and could be though of as key importance values.

We focus on the IS technique RESTART [19]. A RESTART run can be represented graphically as in Fig. 1 where the horizontal axis represents the simulation progress and the vertical axis the importance value of the current state. The run starts from an initial state of the model and evolves until the first threshold T_1 is crossed *upwards*. This takes the path from zone Z_0 below threshold T_1 into zone Z_1 between T_1 and T_2. As this happens the state is saved and $r_1 - 1$ replicas or *offsprings* of the path are created. See A in Fig. 1, where the *number of splittings* for T_1 is $r_1 = 3$. This follows the idea of rewarding promising simulations: up-crossing a threshold suggests the path heads towards a goal state.

From then on the r_1 simulations will evolve independently. As they continue, one of them may hit the upper threshold T_2, activating the same procedure as before: $r_2 - 1$ offsprings are generated from it and set to evolve independently. See B on T_2; here, the splitting is $r_2 = 2$.

However, it could also happen that some simulation hits T_1 again, meaning this path is leading *downwards*. That is an ill-willed simulation steering away from the goal set, and RESTART deals with it discarding the run right away (see C in Fig. 1). In each zone Z_i there

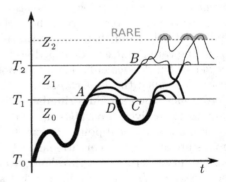

Fig. 1. RESTART importance splitting

exists nonetheless an *original simulation*, which crossed threshold T_i upwards generating the $r_i - 1$ offsprings. This run is allowed to survive a down-crossing of threshold T_i (see D in Fig. 1).

In this setting all simulations reaching a goal state went through the replication procedure, which stacked up on every threshold crossed. Simply counting these hits would introduce a bias, because the *relative weight* of the runs in upper zones decreases by an amount equal to the number of splittings of the threshold.

In consequence, each rare event observed is pondered by the relative weight of the simulation from which it stemmed. If all the goal states exist beyond the uppermost threshold like in Fig. 1, then this is a matter of dividing the observed quantity of rare events by the constant $\text{SPLIT}_{\text{MAX}} \doteq \prod_{i=1}^{n} r_i$. Otherwise more involved labelling mechanism must be implemented.

5   Fully Automated Importance Splitting

Importance splitting simulations are entirely guided by the *importance function* which defines the importance of each state. This function conveys the locations where the simulation effort should be intensified, and it is from its definition that many other settings of IS are usually derived. Importance functions are defined in most situations in an *ad hoc* fashion by an expert in the field of the particular system model under study. With a few exceptions in some specific areas [3,5,14], automatic derivation of importance functions is still a novel field for general systems. Here we present an efficient mechanism to automatically derive this function from the model description.

Importance Function Derivation. The importance of a state s is formally defined as the probability of observing a rare event after visiting s. Therefore if one could track or at least conjecture a path leading from s to a goal state, some notion of the distance between them may be determined and used to choose an appropriate importance for s, where shortest paths should be favoured.

Input: system model $\mathcal{M}$
Input: goal state set $G \neq \varnothing$
$g(G) \leftarrow 0$
queue.push(G)
repeat
    $s \leftarrow$ queue.pop()
    for all $s' \in \mathcal{M}$.predecessors(s) **do**
        if s' not visited **then**
            $g(s') \leftarrow g(s) + 1$
            queue.push(s')
        end if
    end for
until queue empties **or** s_0 visited
$g(s) \leftarrow g(s_0)$ for every non visited state s
$f(s) \leftarrow g(s_0) - g(s)$ for every state s
return f

Fig. 2. Importance function derivation

The core idea is simple enough: starting from the rare event itself, i.e. from the subset G of goal states perform a simultaneous backwards-reachability analysis and label each visited states layer with decreasing importance. This way the shortest path leading from each state into G is computed by means of a Breadth-First Search routine of complexity $\mathcal{O}(k \cdot n)$ where n is the size of the state space and k is the branching degree. Albeit $k = n$ in the worst case, k is normally several orders of magnitude smaller than n. The pseudo-code is described in Fig. 2, where s_0 stands for the initial state of the model $\mathcal{M}$.

From the description of RESTART it is implied that s_0 has minimum importance. Therefore its distance to the subset G is the largest one our algorithm will consider, allowing for an incomplete traversal of the state space on the average case. More precisely, on a first run the states are labelled with their distance to the subset G. This solely affects states at smaller or equal distance from G than s_0, revealing at the same time the maximum distance $\text{DIST}_{\text{MAX}} \doteq f(G) = g(s_0)$.

Bluemoon Tool. To verify the feasibility of the proposed algorithm the proto-typical tool *Bluemoon*[1] was developed and run on several sample models. The software is currently implemented as a module for the probabilistic model checker PRISM [8]. All functionality related to the Markov chains was borrowed from PRISM, namely the models description syntax and its internal ADT representation. We also took advantage of its model checking algorithms to tag the special states of relevance for the importance labelling and the simulations.

The rare event probability estimation is carried out in five distinctive steps:

1. first the model is composed from its high-level description, using an internal column-major sparse matrix representation to favour the later construction of the importance function;
2. then the special states are identified, which includes the goal states and either the reset or the reference states, depending on whether property (1) or (2) was queried respectively. Both this step and the previous model construction are done by means of the mechanisms already provided by PRISM;
3. afterwards the states are labelled with their importance. This can either be done with the algorithm of Fig. 2, or with an *ad hoc* user expression. The importance function is currently represented with a vector of integers;
4. before simulating the proper environment is constructed, which comprises choosing the number of splitting and initial effort per simulation, determining the number of thresholds and defining them, and transforming the model ADT into a row-major sparse matrix, better fitted to forward references;
5. finally several independent RESTART simulations are run, "dynamically" checking at the end of each run whether the stopping criteria was met.

The importance function can also be specified *ad hoc* on invocation of the tool. The user can request any importance assignment of his choice, as long as it comes expressed as an integer expression which PRISM can evaluate on every state. This includes the extreme case of *no importance*: the user can ask for simulations to be run in a pure Monte Carlo style with no splitting involved.

All these options can be determined by the user in the command line, along the confidence criteria (confidence level, precision, method to use) and the *initial effort* to spend per simulation. "Effort" may have one of two different meanings: for transient properties like (1), it stands for the number of independent simulations launched per main iteration, and for steady state properties like (2) it means the maximum number of reference states to visit per simulation.

Regarding the fourth step, the user is allowed to choose the number of splittings, which else defaults to the minimal value 2. To minimise the variance incurred per partial estimation p_i (see eq. (4)) this value will be the same for all thresholds [13,17]. In addition, the number of splittings should ideally be the inverse of the conditional probability p_i [2,19]. The selection of the thresholds is performed once the importance function is built with those two conditions in

---

[1] Named after the kindred English expression «*once in a blue moon*».

mind. Thus, we use the *adaptive multilevel splitting* technique [1, 2] which aims to locate the thresholds so that all conditional probabilities p_i are approximately the same, and moreover, they are the inverse of the selected number of splittings.

Notice the importance function could have been derived using the algorithm in Fig. 2, or evaluated on each state if it was defined *ad hoc* by the user. Given the nature of the algorithm presented, the former option ensures all states will be located above the uppermost threshold. If on the other hand the importance was decided arbitrarily by the user, some goal states might not be given the maximum value and could, during the later selection of the thresholds, end up below the uppermost threshold. This anomaly was discussed in [15] and it is detected and hence countered by our tool.

6 Experimental Validation

With the aim to validate our approach, we selected four case studies from the literature and analysed them using the Bluemoon tool. To validate correctness, the results estimated with simulation were compared against the analytic solution whenever this was available, and also against a numerical solution of the corresponding logical property as computed by PRISM.

In all cases several independent experiments were launched. In each case we compute interval estimates $\hat{p} \pm e_m$ for the probability p of observing the rare event, where e_m is the error margin of the confidence interval. The precision and confidence level were fixed a priori, and each simulation continued until either the specified confidence criteria was met or a wall time limit was reached.

For each experiment we varied some model parameter, testing the performance of the simulation methods for decreasing values of p, overall ranging from magnitudes of $p \approx 1.63 \cdot 10^{-2}$ to $p \approx 2.02 \cdot 10^{-15}$. From now on IFUN will denote "importance function". To validate the performance of our approach, for each model and parameter value, we tested three simulation strategies: RESTART using the automatically built IFUN, RESTART using a few *ad hoc* importance functions, and standard Monte Carlo. Different split values where tested on the importance splitting runs.

The obtained estimates and total simulation times in seconds are presented in tables comparing the performance of the different simulations. We have repeated each experiment a given number of times and each table entry contains the average of the estimated probabilities $\hat{p}$ in each repetition and the average of the total execution time of each repetition. Moreover, a '*' next to an entry indicates that at least one of the repetitions of the experiment did not finish before the wall time limit. All experimentation was carried out in 8-cores 2.7 GHz Intel Xeon E5-2680 processors, each with 32 GiB 1333MHz of available DDR3 RAM.

CTMC Tandem Queue. Consider a tandem network consisting of two connected queues. Customers arrive at the first queue following a Poisson process with parameter λ. After being served by server 1 at rate μ_1 they enter the second queue, where they are attended by server 2 at rate μ_2. The event of interest is an overflow in the second queue for maximum capacity C. This model has received

Table 1. Transient analysis of CTMC tandem queue

Split	IFUN	$C = 8$		$C = 10$		$C = 12$		$C = 14$	
		$\hat{p}$ avg	time	$\hat{p}$ avg	time	$\hat{p}$ avg	time	$\hat{p}$ avg	time
2	auto	5.61e-06	12.2	3.13e-07	51.7	1.90e-08	214.4	1.11e-09	2995.8
	q_2	5.55e-06	11.4	3.18e-07	101.1	1.91e-08	425.9	1.20e-09	2775.4
	q_1+2q_2	5.66e-06	23.8	3.09e-07	403.3	1.91e-08	1631.8	1.17e-09	6712.9 *
	q_1+q_2	5.51e-06	24.9	3.07e-07	422.0	1.87e-08	6821.0	–	–
6	auto	6.08e-06	4.0	3.14e-07	40.2	1.88e-08	180.2	1.15e-09	2819.1
	q_2	5.52e-06	4.2	3.19e-07	66.5	1.86e-08	192.1	1.14e-09	2836.8
	q_1+2q_2	5.57e-06	23.3	3.12e-07	400.4	1.85e-08	6766.5	–	–
	q_1+q_2	5.59e-06	23.3	3.17e-07	397.5	1.90e-08	6575.3	–	–
15	auto	5.75e-06	2.1	3.47e-07	26.5	1.99e-08	167.8	1.13e-09	2858.2
	q_2	5.88e-06	1.9	3.10e-07	28.3	1.91e-08	444.2	1.17e-09	2794.2
	q_1+2q_2	5.79e-06	22.0	3.19e-07	396.2	1.85e-08	6764.4	–	–
	q_1+q_2	5.29e-06	24.1	3.21e-07	392.1	1.93e-08	6432.1	–	–
M.C.		5.67e-06	21.3	3.04e-07	394.6	1.88e-08	6522.8	–	–
Prism		5.59e-06		3.15e-07		1.86e-08		1.14e-09	

considerable attention in the literature [2–4, 16, 19]. We follow the setting from [2] which has an exact analytic solution. The first queue is initially empty and the second has a single customer. We measure the probability of full occupancy in the second queue before it empties, i.e. an instance of Property (1). As in [2, p. 84] the model was tested for the values $\lambda = 3$, $\mu_1 = 2$, $\mu_2 = 6$. The maximum capacities tested for the second queue were $C \in \{8, 10, 12, 14\}$. Simulations had to reach a 95% confidence level with precision equal to 20% of the estimated parameter within 2 hours of wall time. Results are reported in Table 1 with the full estimation process time expressed in seconds. Standard Monte Carlo usually took the longest, and it failed to meet the stopping criteria for the biggest queue size. With a few meaningless exceptions the automatically derived IFUN was the fastest option, beating in performance all *ad hoc* versions. As side remark we notice the splitting value affected the performance, but mostly for the small queue sizes regarding the automatic function.

We also study long run behaviour for the same setting. In this case the rare event was the saturation of the second queue and hence, following Property (2), we estimated the steady state probability of such saturated state. This time simulations were requested to reach a confidence level of 95% with precision equal to 10% of the estimated parameter within 2 hours of wall time. The obtained estimates are shown in Table 2 for capacities $C \in \{10, 15, 20, 25\}$. Standard Monte Carlo met the criteria only for the smallest queue size. Importance splitting simulations did much better but only when good importance functions were employed. The automatically built IFUN and the best *ad hoc* IFUN "q_2" were the only ones to finish in the majority of the cases. These only failed (in all) experiments for $C = 25$ and Split $= 2$, and in particular q_2 also failed to finish

Table 2. Steady state analysis of CTMC tandem queue

Split	IFUN	$C = 10$		$C = 15$		$C = 20$		$C = 25$	
		$\hat{p}$ avg	time	$\hat{p}$ avg	time	$\hat{p}$ avg	time	$\hat{p}$ avg	time
2	auto	3.38e-06	26.1	1.61e-08	257.6	7.33e-11	3866.0	–	–
	q_2	3.38e-06	50.0	1.62e-08	873.9	7.16e-11	6073.4 *	–	–
	q_1+2q_2	3.29e-06	62.5	1.63e-08	502.5	7.27e-11	3568.2	–	–
	q_1+q_2	3.36e-06	175.8	–	–	–	–	–	–
6	auto	3.42e-06	21.8	1.59e-08	30.0	7.42e-11	137.6	3.30e-13	1955.4
	q_2	3.36e-06	29.3	1.61e-08	45.7	7.54e-11	51.0	3.36e-13	181.7
	q_1+2q_2	3.47e-06	89.1	1.60e-08	676.2	7.39e-11	3819.7 *	–	–
	q_1+q_2	3.33e-06	125.2	–	–	–	–	–	–
15	auto	3.37e-06	30.1	1.61e-08	99.8	7.52e-11	184.8	3.26e-13	424.0
	q_2	3.36e-06	14.7	1.63e-08	114.9	7.41e-11	120.0	3.30e-13	341.9
	q_1+2q_2	3.31e-06	143.4	1.61e-08	1608.6	–	–	–	–
	q_1+q_2	3.47e-06	148.7	–	–	–	–	–	–
M.C.		3.33e-06	201.2	–		–		–	
Prism		3.36e-06		1.62e-08		7.42e-11		3.29e-13	

4 out of 5 repetitions for $C = 20$ and Split $= 2$. We observe that in some few cases the automatic IFUN performed worse than q_2 (cf. $C = 25$, Split $= 6$).

DTMC Tandem Queue. We model the same tandem system as a DTMC. Here, each of the three possible events may happen in a single time unit. For each event, we set the following probabilities per time unit: $a_0 = 0.1$ for arrivals on the first queue, $a_1 = 0.14$ for packet transition between queues, and $a_2 = 0.19$ for departures from the second queue. We simulated the system for the overflow levels $C \in \{15, 20, 25, 30, 35\}$ of the second queue. The CTMC and DTMC version have the same state space but the underlying graph structures are slightly different. Hence the automatically derived IFUN are different but all ad hoc IFUNs are the same in both types of models.

We set the confidence level at 95%, the precision at 10% and the wall time limit at 4 h. The results are reported in Table 3. For $C \geqslant 25$ all standard Monte Carlo simulations failed and just a fraction of the RESTART ones finished. Notice that the ad hoc "$q_1 + 2q_2$" IFUN lead on performance for some configurations. Notwithstanding, and with the sole exception of $C = 25$ for splittings 6 and 15, the automatically derived importance function outperformed all tested ad hoc versions. This was true also for $C = 35$, where only some of the simulations using q_2 and the automatic version of the IFUN finished in time.

Mixed Open/Closed Queue Network [4, Sec. 4.1]. This model consists of two parallel queues handled by one server: an *open queue* q_o, that receives packets from an external source, and a *prioritised closed queue* q_c, that receives (sends) packets from (to) some internal system buffer. Elements in q_o are served at rate μ_{11} unless q_c has packets which are handled first at rate μ_{12}. Packets in *internal circulation* are served at rate μ_2 and sent back to q_c. If there is

Table 3. Steady state analysis of DTMC tandem queue

Split	IFUN	C = 15		C = 20		C = 25		C = 30	
		$\hat{p}$ avg	time	$\hat{p}$ avg	time	$\hat{p}$ avg	time	$\hat{p}$ avg	time
2	auto	4.81e-07	20.4	1.28e-08	85.1	3.24e-10	275.8	7.91e-12	699.5 *
	q_2	4.97e-07	24.6	1.27e-08	83.3	3.15e-10	281.0	8.17e-12	652.8 *
	q_1+2q_2	5.02e-07	15.2	1.31e-08	27.6	3.24e-10	112.2	7.75e-12	3274.7 *
	q_1+q_2	5.05e-07	164.4	1.30e-08	312.7	3.28e-10	2136.3	9.37e-12	2486.3 *
6	auto	4.92e-07	25.9	1.26e-08	81.1	3.18e-10	10674.6 *	7.23e-12	699.9 *
	q_2	4.97e-07	29.8	1.28e-08	213.6	3.31e-10	5404.3 *	–	–
	q_1+2q_2	4.98e-07	2494.6 *	1.29e-08	157.9	3.24e-10	1917.1	8.12e-12	11364.6 *
	q_1+q_2	4.90e-07	165.3	1.28e-08	1247.0	3.15e-10	2215.1 *	7.88e-12	4394.2 *
15	auto	4.96e-07	45.1	1.28e-08	155.4	3.37e-10	1298.6 *	8.61e-12	572.3 *
	q_2	4.98e-07	134.7	1.26e-08	142.4	3.36e-10	839.3 *	7.63e-12	13763.1 *
	q_1+2q_2	4.93e-07	253.9	1.28e-08	4175.4	3.12e-10	635.7 *	–	–
	q_1+q_2	4.97e-07	240.3	1.22e-08	1108.0	3.18e-10	2370.0 *	8.97e-12	4424.2 *
M.C.		4.87e-07	596.4	1.23e-08	–	–		–	
Prism		4.94e-07		1.28e-08		3.22e-10		7.96e-12	

only one circulating internal packet, the system is an $M/M/1$ queue with server breakdowns.

Starting from an empty system, we estimate the probability that q_o reaches maximum capacity b before both queues are emptied again. The setting is as in [4]: one packet in internal circulation, $\mu_{11} = 4$, $\mu_{12} = 2$, $\mu_2 \in \{0.5, 1.0\}$ and capacities $b \in \{20, 40\}$. We set the confidence at 95%, the precision at 10% and the wall time limit at 8 h. Results are reported in Table 4. For the cases in which $b = 40$ none of the simulations met the desired confidence within the time limit. Thus, in the respective columns on the table, we show instead the minimum and maximum estimations of the repetitions. Note that these estimations are nonetheless very close to the value reported by PRISM. Experiments for $b = 20$ favour the automatic IFUN overwhelmingly for both failure rates and all splitting. A speedup of at least 148x was gained in comparison to both *ad hoc* importance assignments. This is particularly surprising regarding "q_o" which seem to be a sensible choice when comparing to the previous tandem queue systems.

Queueing System with Breakdowns [7, Sec. 4.4]. Consider a system where sources of type $i \in \{1, 2\}$ have exponential on/off times with parameters α_i and β_i respectively. These sources, whenever active, send packets at rate λ_i to the only system buffer. Queued packets are handled by a server which breaks down at rate γ and gets fixed at rate δ, processing at rate μ when functional. We estimate the probability of the buffer reaching maximum capacity K before emptying.

As in [7] we start with a single packet in the queue and a broken server. There are five sources of each type and, initially, all are down except for one of type 2. The sources parameters are $(\alpha_1, \beta_1, \lambda_1) = (3, 2, 3)$ and $(\alpha_2, \beta_2, \lambda_2) = (1, 4, 6)$. The server parameters are $(\gamma, \delta, \mu) = (3, 4, 100)$ and the queue capacities tested were $K \in \{20, 40, 80, 160\}$. We set the confidence level at 95%, the precision at 10% and the wall time limit at 2.5 h. Results are shown in Table 5, where s_d refers

Table 4. Mixed Open and Closed Queueing Network

Split	IFUN	$\mu_2 = 1.0$				$\mu_2 = 0.5$			
		$b = 20$		$b = 40$		$b = 20$		$b = 40$	
		$\hat{p}$ avg	time	$\hat{p}$ min	$\hat{p}$ max	$\hat{p}$ avg	time	$\hat{p}$ min	$\hat{p}$ max
	auto	5.79e-07	11.1	5.68e-13	5.69e-13	3.91e-08	131.2	2.02e-15	2.03e-15
2	q_o	5.97e-07	1485.1	5.67e-13	5.69e-13	3.92e-08	19690.9	1.99e-15	2.02e-15
	q_c+q_o	5.95e-07	1493.6	5.68e-13	5.70e-13	3.91e-08	19733.3	2.01e-15	2.03e-15
	auto	5.83e-07	11.2	5.67e-13	5.69e-13	3.90e-08	132.7	1.95e-15	2.04e-15
5	q_o	5.97e-07	1490.2	5.68e-13	5.70e-13	3.91e-08	20118.6	2.01e-15	2.05e-15
	q_c+q_o	5.94e-07	1491.3	5.68e-13	5.68e-13	3.92e-08	19753.9	2.01e-15	2.02e-15
	auto	6.04e-07	16.8	5.68e-13	5.69e-13	3.86e-08	133.0	2.01e-15	2.03e-15
9	q_o	5.96e-07	1481.2	5.68e-13	5.69e-13	3.91e-08	19816.7	2.02e-15	2.03e-15
	q_c+q_o	5.96e-07	1481.0	5.65e-13	5.69e-13	3.92e-08	19763.1	2.02e-15	2.03e-15
M.C.		6.04e-07	1400.5	–	–	4.02e-08	18417.3	–	–
Prism		5.96e-07		5.68e-13		3.91e-08		2.02e-15	

Table 5. Multiple-source queue with breakdowns

Split	IFUN	$K = 20$		$K = 40$		$K = 80$		$K = 160$	
		$\hat{p}$ avg	time	$\hat{p}$ avg	time	$\hat{p}$ avg	time	$\hat{p}$ avg	time
	auto	1.63e-02	5.4	4.54e-04	9.8	3.72e-07	478.5	2.43e-13	2464.1 *
2	q	1.65e-02	5.5	4.62e-04	19.0	3.71e-07	162.9	2.45e-13	3691.3 *
	$q+s_d$	1.62e-02	20.2	4.63e-04	448.6	3.75e-07	880.6 *	2.42e-13	9034.9 *
	$q+s_u$	1.63e-02	181.4	4.48e-04	537.2	–	–	–	–
	auto	1.64e-02	5.8	4.60e-04	9.1	3.66e-07	84.4	2.47e-13	1809.8 *
3	q	1.67e-02	5.9	4.54e-04	17.1	3.73e-07	87.5	2.41e-13	4105.4 *
	$q+s_d$	1.63e-02	16.4	4.62e-04	53.3	3.73e-07	242.4 *	2.46e-13	4709.4 *
	$q+s_u$	1.61e-02	115.3	4.61e-04	824.0	3.68e-07	3537.7	2.45e-13	5145.1 *
	auto	1.64e-02	6.2	4.72e-04	8.1	3.71e-07	91.5	2.45e-13	2836.0
5	q	1.64e-02	6.3	4.62e-04	17.3	3.70e-07	103.2	2.47e-13	1154.5
	$q+s_d$	1.66e-02	7.3	4.60e-04	59.5	3.73e-07	856.4	2.47e-13	1823.5
	$q+s_u$	1.65e-02	49.5	4.62e-04	159.7	3.74e-07	367.9	2.50e-13	1251.1 *
	auto	1.60e-02	6.3	4.80e-04	7.8	3.75e-07	109.7	2.46e-13	886.4
9	q	1.62e-02	6.6	4.54e-04	18.5	3.67e-07	136.6	2.44e-13	591.4
	$q+s_d$	1.60e-02	5.5	4.65e-04	26.7	3.72e-07	153.9	2.47e-13	4446.8 *
	$q+s_u$	1.61e-02	18.0	4.57e-04	67.6	3.72e-07	348.1	2.44e-13	1885.6
	auto	1.66e-02	6.5	4.98e-04	9.0	3.74e-07	134.9	2.45e-13	1251.0
15	q	1.61e-02	6.6	4.66e-04	18.9	3.75e-07	367.5	2.43e-13	2812.3 *
	$q+s_d$	1.63e-02	5.6	4.68e-04	23.5	3.72e-07	321.7	2.47e-13	1879.9 *
	$q+s_u$	1.65e-02	11.3	4.56e-04	36.5	3.70e-07	285.1	2.42e-13	1427.6
M.C.		1.65e-02	0.4	4.58e-04	11.8	–	–	–	–
Prism		1.63e-02		4.59e-04		3.72e-07		2.45e-13	

to the number of *sources down* and $s_u = 10 - s_d$ refers to the number of *sources up*. Standard Monte Carlo failed for $K \geq 80$, and from all *ad hoc* importance functions only one, "q", showed a relatively stable good behaviour. With very few exceptions (cf. $(K, \text{Split}) \in \{(80, 2), (160, 5), (160, 9)\}$) the automatic IFUN was the best importance assignment observed. Furthermore this came at very low cost, since the function derivation times were 0.1 s, 0.5 s, 2.1 s and 8.3 s for capacities $K = 20, 40, 80, 160$ respectively.

7   Concluding Remarks

Related Work. There have been some few incursions in automatic derivation of importance functions. Sewards et. al construct their function based on the logical property to be checked [5], which must support some "layered restatement" of its syntax or resource to approximate heuristics. In [14] the approach from Booth & Hendriks as reported in [10] is applied to stochastic Petri nets. However no simulation times are reported in this case and the technique proposed requires solving several instances of ILP, known to be an NP-complete problem. These works, like ours, are based on static analysis of the model or property. Instead, in [3] importance is assigned to states applying reversed simulation sequentially on each of them. This requires some knowledge on the stationary distribution of the system, and the applicability of the approach is shown for finite DTMCs.

Further Discussions. Overall the presented algorithm obtained, with very little computational overhead, an IFUN which rivalled the best *ad hoc* alternatives. For transient properties like (1), the derived function performed even better than the quasi-optimal versions from the literature. This was particularly noticeable in the queueing system with breakdowns from [7], where very complex internal behaviours make it hard to distil a good *ad hoc* importance assignment. In some cases however the best *ad hoc* IFUN met the stopping criteria faster for steady state properties like (2), but in all scenarios either both automatic and the best *ad hoc* IFUN finished before the wall time limit or none did, see Table 2 and 3. There were also situations where one or two experiment repetitions failed to finish in time, but those who did took much less than the time limit, as e.g. Table 3 for $C = 30$, Split = 6. This could be due to peculiarities of RESTART discussed in [2,11]. In this direction, it would be good to study the performance of our technique under other importance splitting algorithms such as [2]. Though we have only reported an average on the point estimators, we remark all experiments behave according to the confidence parameters when compared to the numerically calculated values reported by PRISM.

Our algorithm works nicely as long as the number of transitions outgoing each state is significantly lower than the number of states. If instead, the underlying graph of the Markov Chain is highly connected, two problems arise. On the one hand, the BFS algorithm approaches quadratic complexity where the large majority of the computation is unproductive, spent on visiting already visited nodes. On the other hand, the eventually derived IFUN will most likely run on a very small domain as a consequence of a short minimal distance between the initial state and a rare sate. This actually happens in a case study taken from [19] (and not reported in this paper) were a huge amount of computation was spent on the derivation of the IFUN, spending a total amount of time that largely surpassed standard Monte Carlo simulation. In spite of this, the automatic IFUN performed better than the IFUN proposed in [19].

To conclude we would like to highlight the generality of our approach, here limited to Markov chains exclusively with numerical validation purposes. To show this however the current tool should be exported out of PRISM into a wider framework with a more expressive model description syntax.

Acknowledgments. We thank Raúl E. Monti who helped on early developments of the tool. The experiments were performed on the Mendieta Cluster from CCAD at UNC (http://ccad.unc.edu.ar).

References

1. Cérou, F., Guyader, A.: Adaptive multilevel splitting for rare event analysis. Stochastic Analysis and Applications **25**(2), 417–443 (2007)
2. Garvels, M.J.J.: The splitting method in rare event simulation. PhD thesis, University of Twente (2000)
3. Garvels, M.J.J., Van Ommeren, J.-K.C.W., Kroese, D.P.: On the importance function in splitting simulation. Eur. Trans. Telecommun. **13**(4), 363–371 (2002)
4. Glasserman, P., Heidelberger, P., Shahabuddin, P., Zajic, T.: Multilevel splitting for estimating rare event probabilities. Operations Research **47**(4), 585–600 (1999)
5. Jegourel, C., Legay, A., Sedwards, S.: Importance Splitting for Statistical Model Checking Rare Properties. In: Sharygina, N., Veith, H. (eds.) CAV 2013. LNCS, vol. 8044, pp. 576–591. Springer, Heidelberg (2013)
6. Kahn, H., Harris, T.E.: Estimation of particle transmission by random sampling. National Bureau of Standards Applied Mathematics Series **12**, 27–30 (1951)
7. Kroese, D.P., Nicola, V.F.: Efficient estimation of overflow probabilities in queues with breakdowns. Performance Evaluation **36**, 471–484 (1999)
8. Kwiatkowska, M., Norman, G., Parker, D.: PRISM 4.0: Verification of Probabilistic Real-Time Systems. In: Gopalakrishnan, G., Qadeer, S. (eds.) CAV 2011. LNCS, vol. 6806, pp. 585–591. Springer, Heidelberg (2011)
9. Law, A.M., Kelton, W.D., Kelton, W.D.: Simulation modeling and analysis, vol. 2. McGraw-Hill, New York (1991)
10. L'Ecuyer, P., Demers, V., Tuffin, B.: Rare events, splitting, and quasi-Monte Carlo. ACM Trans. Model. Comput. Simul. 17(2) (April 2007)
11. L'Ecuyer, P., Le Gland, F., Lezaud, P., Tuffin, B.: Splitting techniques. In: Rare Event Simulation using Monte Carlo Methods, pp. 39–61. J. Wiley & Sons (2009)
12. L'Ecuyer, P., Mandjes, M., Tuffin, B.: Importance sampling in rare event simulation. In: Rare Event Simulation using Monte Carlo Methods, pp. 17–38. J. Wiley & Sons (2009)
13. L'Ecuyer, P., Tuffin, B.: Approximating zero-variance importance sampling in a reliability setting. Annals of Operations Research **189**(1), 277–297 (2011)
14. Reijsbergen, D., de Boer, P.-T., Scheinhardt, W., Haverkort, B.: Automated Rare Event Simulation for Stochastic Petri Nets. In: Joshi, K., Siegle, M., Stoelinga, M., D'Argenio, P.R. (eds.) QEST 2013. LNCS, vol. 8054, pp. 372–388. Springer, Heidelberg (2013)
15. Villén-Altamirano, J.: RESTART method for the case where rare events can occur in retrials from any threshold. Int. J. Electron. Commun. (AEÜ) **52**, 183–189 (1998)

16. Villén-Altamirano, J.: Rare event RESTART simulation of two-stage networks. European Journal of Operational Research **179**(1), 148–159 (2007)
17. Villén-Altamirano, M., Martínez-Marrón, A., Gamo, J., Fernández-Cuesta, F.: Enhancement of the accelerated simulation method restart by considering multiple thresholds. In: Proc. 14th Int. Teletraffic Congress, pp. 797–810 (1994)
18. Villén-Altamirano, M., Villén-Altamirano, J.: RESTART: A method for accelerating rare event simulations. Analysis **3**, 3 (1991)
19. Villén-Altamirano, M., Villén-Altamirano, J.: The Rare Event Simulation Method RESTART: Efficiency Analysis and Guidelines for Its Application. In: Kouvatsos, D.D. (ed.) Next Generation Internet: Performance Evaluation and Applications. LNCS, vol. 5233, pp. 509–547. Springer, Heidelberg (2011)
20. Wilson, E.B.: Probable inference, the law of succession, and statistical inference. Journal of the American Statistical Association **22**(158), 209–212 (1927)

Speed-Up of Stochastic Simulation of PCTMC Models by Statistical Model Reduction

Cheng Feng[⊠] and Jane Hillston

LFCS, School of Informatics, University of Edinburgh, Edinburgh, Scotland, UK
s1109873@sms.ed.ac.uk, jane.hillston@ed.ac.uk
http://www.quanticol.eu

Abstract. We present a novel statistical model reduction method which can significantly boost the speed of stochastic simulation of a population continuous-time Markov chain (PCTMC) model. This is achieved by identifying and removing agent types and transitions from the simulation which have only minor impact on the evolution of population dynamics of target agent types specified by the modeller. The error induced on the target agent types can be measured by a normalized coupling coefficient, which is calculated by an error propagation method over a directed relation graph for the PCTMC, using a limited number of simulation runs of the full model. Those agent types and transitions with minor impact are safely removed without incurring a significant error on the simulation result. To demonstrate the approach, we show the usefulness of our statistical reduction method by applying it to 50 randomly generated PCTMC models corresponding to different city bike-sharing scenarios.

1 Introduction

Continuous time Markov chains (CTMC) have been widely used to study population dynamics in many domains such as ecology [1], system biology [2] and computer networking [3]. Recently, with the widespread adoption of wireless communication techniques, large-scale collective ICT systems comprised of many communicating entities and without centralised control, have become feasible and their pervasive, transparent nature makes it crucial that their dynamic behaviour is predicted prior to deployment. Population CTMCs (PCTMC) have been proposed as a suitable tool to model such systems.

Currently, there are two typical approaches to analyse PCTMC models. One is through stochastic simulation, which is usually computationally expensive as performance metrics can only be derived after many simulation runs. The other is to build an analytical model in the form of initial value problems by fluid-limit theory [4] or moment closure techniques [5]. However, due to the intrinsic spatial-heterogeneity in collective systems (the same agent can exhibit different behaviour in different positions), the analytical model can be unresolvable due to the number of coupled ODEs in the model. Moreover, the question also arises of whether the fractured population is large enough to justify fluid/moment closure techniques [6]. As a result, in many circumstances, stochastic simulation is the

© Springer International Publishing Switzerland 2015
M. Beltrán et al. (Eds.): EPEW 2015, LNCS 9272, pp. 291–305, 2015.
DOI: 10.1007/978-3-319-23267-6_19

only option to analyse such models. Nevertheless, although the spatial property of collective systems increases the complexity of the underlying PCTMC models, on the other hand, we also find that it gives us the possibility to decouple parts of the model. For instance, two agents which are located far away from each other are generally less likely to influence each other than another two agents located in close proximity. In this paper, we propose a statistical model reduction method which can significantly boost the speed of stochastic simulation of PCTMC models by identifying and removing those unimportant agents and transitions with respect to some target agents which we are interested in, after a few simulation runs. The error caused by removing these agents and transitions can be controlled by the modeller by setting an acceptable error threshold.

Specifically, in order to evaluate the coupling coefficient between two agent types (which tell us how much error will be caused to the population dynamics of one agent type if we discard agents of the other type and their associated transitions), we build a directed relation graph (DRG) for the PCTMC model. We are inspired by the DRG first introduced by Lu and Law [7] for species and reaction reduction in the numerical simulation of chemical reaction mechanisms for hydrocarbon oxidation. This approach has since been improved by many researchers in the combustion research domain. Examples include DRG with error propagation [8], DRG with sensitivity analysis [9], etc.

In our DRG for PCTMC models, the vertices are the agent types and the directed edges are coupling coefficients between agent types. In the DRG for species in chemical reaction mechanisms for hydrocarbon oxidation, the technique is used to reduce a deterministic model (a set of coupled ODEs), and the coupling coefficients between species can be computed using experimental data. In contrast, in our work we reduce a stochastic model when there is no experimental data available to establish the value of the coupling coefficients. Instead, we designed two statistical reduction algorithms in which we evaluate the coupling coefficients between agent types based on a limited number of simulation runs of the full PCTMC model. After that, an error propagation method is applied to identify those agent types and transitions which can be discarded without leading to significant error. These agents are removed from the simulation, and the remaining model contains only necessary agent types and transitions, which are tightly coupled to the identified target agents. The whole process is fast, and has low computational cost compared to the total simulation cost.

To demonstrate the efficiency of this method, we will apply it to a family of 50 models for bike-sharing systems based on random topology of stations and random values of parameters. The model is specified in PALOMA [10], a process algebra recently designed for the modelling of collective adaptive systems, which makes it easy to generate many variations of the same model. The underlying PCTMC model can be automatically generated according to the population semantics of this modelling language. Note that although we illustrate the application of our method on PCTMC models derived from PALOMA, our method is not limited to PALOMA models. Moreover, to our knowledge, this is the first work which applies the DRG method outside the combustion simulation domain.

2   A Brief Introduction of PCTMC

A CTMC is a Markovian stochastic process defined on a finite state space and evolving in continuous time. In this paper, we specifically consider PCTMC models of interacting agents [11], in which we assume that there are a number of distinct agent types, each of which has a potential population. Agents interact via a set of transitions. Transitions will change one or more agents from one type to another. In general, a PCTMC model can be expressed as a tuple $\mathcal{P} = (\mathbf{X}, E, \mathbf{x_0})$:

- $\mathbf{X} = (x_1, ..., x_n) \in \mathbb{Z}_{\geq 0}^n$ is an integer vector with the ith ($1 \leq i \leq n$) component representing the current population level of an agent type i.
- $E = \{e_1, ..., e_m\}$ is the set of transitions, of the form $e = (r_e(\mathbf{X}), \mathbf{d}_e)$, where:
    1. $r_e(\mathbf{X}) \in \mathbb{R} \geq 0$ is the rate function, associating with each transition the rate of an exponential distribution, depending on the global state of the model.
    2. $\mathbf{d}_e \in \mathbb{Z}^n$ is the update vector which gives the net change for each element of $\mathbf{X}$ caused by transition e.
- $\mathbf{x_0} \in \mathbb{Z}_{\geq 0}^n$ is the initial state of the model.

Transition rules can be easily visualised in the chemical reaction style, as

$$x_1 + ... + x_n \to (x_1 + d_e^1) + ... + (x_n + d_e^n) \quad \text{at } r_e(\mathbf{X})$$

where d_e^i ($1 \leq i \leq n$) denotes the net change on the population of agents of type i caused by transition e. The tuple $\mathcal{P}$ contains all the information that is needed for the discrete event simulation of a PCTMC model using standard simulation algorithms, like SSA [12]. Clearly, the speed of stochastic simulation is dependent on the number of agent types, the populations of agents and the number of transitions in the model.

3   Directed Relation Graph with Error Propagation

The DRG method with error propagation was proposed in [8] to efficiently recognise removable species and reactions in the numerical simulation of large scale chemical kinetic mechanisms. In this section, we will focus on the modification of this method for application to PCTMC models.

Specifically, in the DRG for a PCTMC model, each vertex represents an agent type in the PCTMC. There exists an edge from vertex i to vertex j if and only if the removal of agents of type j and their associated transitions would directly induce an error in the evolution of the population dynamics of agents of type i. This effect can be quantified by a normalized coupling coefficients c_{ij}, defined as

$$c_{ij} = \frac{|\sum_{e \in E} r_e\ d_e^i\ \delta_e^j|}{\max(Prod_i, Cons_i)} \tag{1}$$

where r_e is the rate of transition e, d_e^i is the net change to the population of agent type i caused by transition e, δ_e^j equals 1 if agent type j is involved in transition e otherwise it is 0 (we say agent type j is involved in transition e if and only if the net change on the population of agents of type j caused by transition e is non-zero, or the rate of transition e depends on the population of agents of type j), and

$$Prod_i = \sum_{e \in E} \mathbf{1}_{d_e^i > 0} \, r_e \, d_e^i \tag{2}$$

$$Cons_i = -\sum_{e \in E} \mathbf{1}_{d_e^i < 0} \, r_e \, d_e^i \tag{3}$$

which are the total production and consumption of agents of type i respectively. c_{ij} is bounded between 0 and 1 since

$$\left| \sum_{e \in E} r_e \, d_e^i \, \delta_e^j \right| = \left| \sum_{e \in E} \mathbf{1}_{d_e^i > 0} \, r_e \, d_e^i \, \delta_e^j + \sum_{e \in E} \mathbf{1}_{d_e^i < 0} \, r_e \, d_e^i \, \delta_e^j \right| = |Prod_{ij} - Cons_{ij}| \tag{4}$$

where $Prod_{ij}$ ($Cons_{ij}$) is the total production (consumption) of agents of type i from transitions in which agent type j is involved. Then, as $0 \le Prod_{ij} \le Prod_i$ and $0 \le Cons_{ij} \le Cons_i$, it can be inferred that $-Cons_i \le Prod_{ij} - Cons_{ij} \le Prod_i$, which is equivalent to $|Prod_{ij} - Cons_{ij}| \le \max(Prod_i, Cons_i)$.

Furthermore, we say agent type j and its associated transitions are *removable* if $c_{ij} \le \theta$, where i is the target agent type which we are interested in and θ is an *acceptable error threshold* given by the modeller. Moreover, note that coupling coefficients are not symmetric, since it is not necessarily the case that $c_{ij} = c_{ji}$.

3.1   Group-Based Direct Coupling Coefficient

We will remove agents (and their transitions) one by one until the cumulative induced error reaches the acceptable error threshold. Since a transition typically involves more than one agent type, when we consider removing an agent type, we cannot assume that it is independent of the agent types that have already been removed. The following equation gives the coupling coefficient of an agent type which also takes into account those agent types which have already been removed:

$$c_{ij,\{S\}} = \frac{\left| \sum_{e \in E} r_e \, d_e^i \, \delta_e^{j,\{S\}} \right|}{\max(Prod_i, Cons_i)} \tag{5}$$

where S is the set of agent types which have already been removed, $\delta_e^{j,\{S\}}$ equals 1 as long as agent type j or an agent type in S is involved in transition e, otherwise it is 0.

3.2   Indirect Coupling Coefficient

For those agent types which are not directly connected in the DRG, by using an error propagation method, we can evaluate the indirect coupling coefficient between two agent types. Specifically, indirect coupling is quantified by path dependent coefficient $c_{ij,\sigma}$, which is the product of the direct coupling coefficients along the path σ between agent types i and j. The influence of removing agent type j on agent type i is characterized by coefficient C_{ij}, which is the maximum of the path dependent coefficients:

$$c_{ij,\sigma} = \prod_{xy \in \sigma} c_{xy} \tag{6}$$

$$C_{ij} = \max_{all\ paths\ \sigma} c_{ij,\sigma} \tag{7}$$

Figure 1 shows part of a DRG, in which the indirect coupling coefficient between i and j is $C_{ij} = c_{im} \times c_{mj} = 0.24$, since $c_{im} \times c_{mj} > c_{ik} \times c_{kl} \times c_{lj}$.

Similarly, taking into consideration agent types that have already been removed, we define:

$$c_{ij,\{S\},\sigma} = \prod_{xy \in \sigma} c_{xy,\{S\}} \tag{8}$$

$$C_{ij,\{S\}} = \max_{all\ paths\ \sigma} c_{ij,\{S\},\sigma} \tag{9}$$

where $C_{ij,\{S\}}$ is the indirect coupling coefficient from agent type i to j given a set of agent types $\{S\}$ which have already been removed.

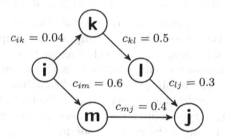

Fig. 1. Part of a directed relation graph with five agent types

4   Statistical Model Reduction

In this section, we give the details of our statistical model reduction method. Given the target agent types which we are going to investigate, the dynamics in a PCTMC and the acceptable error threshold for those agent types, after a few simulation runs of the full model our method can accurately identify those agent types and transitions which can be removed from the simulation without leading to an error beyond the acceptable threshold, based on the DRG with error propagation. For simplicity, we only deal with a single target agent type in the introduction of the two algorithms, but it can be readily seen that they can both be extended to cope with multiple target agent types.

4.1 Statistical Transition Rate Evaluation

First of all, since the rate of a transition, $r_e(\mathbf{X})$ depends on the current state of the PCTMC, we cannot evaluate c_{ij} in Equation (1) solely from the system description since r_e may evolve during the simulation period. However, by learning from past simulation runs, we can give a statistical rate for each transition. Specifically, we let $r_e^k = \frac{N_e^k}{T}$, where N_e^k is the firing count of transition e during the kth simulation run, T is the time length of a simulation run, r_e^k is the statistical rate of transition class e in the kth simulation. We use the statistical transition rate to estimate the coupling coefficient, which gives the overall contribution of an agent type to the evolution of the population dynamics of the target agent type during a simulation.

4.2 Model Reduction Algorithms

We introduce two automatic model reduction algorithms both based on DRG with error propagation, but with different sampling methods for deducing the coupling coefficients and different reduction criteria.

Algorithm with Fixed Length Sampling. We first introduce a straightforward reduction algorithm. Specifically, we let the user define a number K, which is the number of simulation runs before the reduction process is actually carried out. At the end of the Kth simulation run, we approximate the rate of transition e by $r_e = \frac{\sum_{k=1,\ldots,K} N_e^k}{K \times T}$, i.e. the average of the statistical transition rate over the K runs. Then, we initiate the DRG and compute the coupling coefficients for the target agent type with respect to all other agent types. We delete an agent type which has the minimal coupling coefficient with the target agent type each time, and then update the coupling coefficients of remaining agent types, taking into account the newly removed agent type. The reduction process is terminated when removing an agent type will exceed the acceptable error threshold given by the user. Algorithm 1 gives the pseudo code for this algorithm.

Algorithm with Flexible Length Sampling. The second algorithm is more strict in the sense that we only remove agent types after we reach $\Pr(C_{ti,\{S\}} < \theta) \geq p$, where p is the confidence level. That is to say, that given a set of already removed agents $\{S\}$, we only remove agent type i when we achieve confidence level p that the coupling coefficient $C_{ti,\{S\}}$ is less than the acceptable error threshold θ.

In order to test whether $\Pr(C_{ti,\{S\}} < \theta) \geq p$ holds, we raise two hypotheses, which are: H_0: $\Pr(C_{ti,\{S\}} < \theta) \geq p_0$ and H_1: $\Pr(C_{ti,\{S\}} < \theta) \leq p_1$ where (p_1, p_0) is an indifference region where we say that the probability is sufficiently close to p that we are indifferent with respect to which of the two hypotheses H_0 or H_1 is accepted. Moreover, we require the probability of accepting H_1 when actually H_0 holds to be less than α, and the probability of accepting H_0 when actually H_1 holds to be less than β.

Algorithm 1.. Reduction with fixed length sampling

User specifies the target agent type t, number of sampling simulation runs K and the acceptable error threshold θ.

Simulate the full model K runs

Compute statistical rates for all transitions.

Initiate the DRG, and compute the coupling coefficients from the target agent type to any other agent types.

repeat

    find agent type i which has the minimal coupling coefficient $C_{ti,\{S\}}$ among the remaining agent types, where $\{S\}$ is set of removed agent types.

    if $C_{ti,\{S\}} < \theta$ **then**

        remove agent type i and the transitions it is involved.

        add agent type i to $\{S\}$, and update the coupling coefficient of remaining agent types.

    end if

until $C_{ti,\{S\}} \geq \theta$

Use the reduced model in the following simulation runs.

Furthermore, we let

$$\frac{p_{1m}}{p_{0m}} = \frac{p_1^{d_m}(1-p_1)^{m-d_m}}{p_0^{d_m}(1-p_0)^{m-d_m}}$$

where m is the current number of simulation runs, d_m is the number of simulation runs for which $C_{ti,\{S\}} < \theta$ holds. Applying the idea behind the sequential probability ratio test (SPRT) [13], we can accept H_0 when $\frac{p_{1m}}{p_{0m}} \leq \frac{\beta}{1-\alpha}$ and accept H_1 when $\frac{p_{1m}}{p_{0m}} \geq \frac{1-\beta}{\alpha}$. Accepting H_0 means that with confidence level p removing agent type i and the associated transitions will cause error in the target agent type of less than θ, thus the agent type and transitions are removable. Accepting H_1 means that we are highly confident that we cannot remove agent type i, thus the reduction process should be terminated.

It can be seen that we can only reach the removal criterion of an agent type, or the termination criterion of the reduction process, after at least $K = \min(\log\frac{p_1}{p_0} \frac{\beta}{1-\alpha}, \log\frac{1-p_1}{1-p_0} \frac{1-\beta}{\alpha})$ simulation runs, when $d_m = m$ or $d_m = 0$. Thus, we will start the reduction process after K simulation runs. Specifically, at the end of the Kth simulation run, we compute the average statistical rate for each transition as $r_e = \frac{\sum_{k=1,\ldots,K} N_e^k}{K \times T}$, and then build the DRG and compute the coupling coefficients using the average statistical transition rates. Again, we find the agent type which has the minimal coupling coefficient with the target agent type. Then, we remove an agent type each time repeatedly if hypothesis H_0 holds. We simulate the model with the reduced version. Compared with the first algorithm, the difference is that we will continue to check whether there are more agent types that are removable until we reach hypothesis H_1. Note that when some agent types and transitions are removed, we lose information about the rate of those removed transitions in the following runs. Thus, we will use

---

Algorithm 2.. Reduction with flexible length sampling

User specifies t, θ, $p0$, $p1$, α and β

Simulate the full model $K = \min(\log_{\frac{p_1}{p_0}} \frac{\beta}{1-\alpha}, \log_{\frac{1-p_1}{1-p_0}} \frac{1-\beta}{\alpha})$ runs

Compute statistical transition rates for each simulation run, and the average statistical transition rates for the K runs.

Initiate the DRG and compute the coupling coefficients from the target agent type t to any other agent types using the average statistical transition rates.

Point A:

repeat

    find agent type i which has the minimal coupling coefficient $\overline{C}_{ti,\{S\}}$ using the average statistical transition rates among the remaining agent types, where $\{S\}$ is set of removed agent types.

    Compute $C_{ti,\{S\}}$ for each simulation run

    if $\frac{p_{1m}}{p_{0m}} \leq \frac{\beta}{1-\alpha}$ **then**

        remove agent type i and the transitions it is involved.

        add agent type i to $\{S\}$, and update the coupling coefficient of remaining agent types using the average statistical transition rates.

    else if $\frac{p_{1m}}{p_{0m}} \geq \frac{1-\beta}{\alpha}$ **then**

        stop the reduction process.

    end if

until $\frac{p_{1m}}{p_{0m}} > \frac{\beta}{1-\alpha}$

Simulate the reduced model, after each run go to Point A unless the reduction process has stopped.

---

the average rate of those removed transitions in the previous simulation runs in order to calculate the coupling coefficients if they are needed. Algorithm 2 gives the pseudo code for this algorithm.

4.3 Comparison of the Two Algorithms

It is obvious that the algorithm with fixed length sampling is easier to implement and has less computational cost for the reduction process. However, the algorithm with flexible length sampling has the advantage of more stringent error control on the simulation result. We will report on a comprehensive test of these two reduction algorithms in Section 6, using 50 randomly generated city bike-sharing models. But before that, we first briefly introduce the modelling language we use to generate our models.

5   Modelling Language and Model Definition

In this section, we give a brief introduction of the modelling language, PALOMA, and the model definition of the bike-sharing scenario using this language.

5.1   PALOMA

PALOMA is a stochastic process algebra, specifically designed to support the construction of formal models of large collective adaptive systems in which agents are distributed over a discrete set of named locations, $\mathcal{L}$. Agents are parame-terised by a *location*, denoted by ℓ, $\ell \in \mathcal{L}$. There is a finite set of action types $\mathcal{A}$, and actions may be undertaken spontaneously or may be induced by a message of the same type, sent by another agent in the system. All spontaneous actions are assumed to have a duration governed by an exponential distribution and characterised by a rate r. A model P consists of a number of agents composed in parallel. The language has the following grammar:

$$\pi ::= !(\alpha, r)@\mathbf{IR}\{\overrightarrow{\ell}\} \mid ?(\alpha, p)@\mathbf{Pr}\{v\} \mid !!(\alpha, r)@\mathbf{IR}\{\overrightarrow{\ell}\} \mid ??(\alpha, p)@\mathbf{Wt}\{v\} \mid (\alpha, r)$$
$$S(\ell) ::= \pi.S'(\ell') \mid S_1(\ell) + S_2(\ell) \mid M$$
$$P ::= S(\ell) \mid P \parallel P$$

Agents can change their states and locations by different actions:

Spontaneous action with broadcast message emission: $!(\alpha, r)@\mathbf{IR}\{\overrightarrow{\ell}\}$ describes that the agent performs an action α, $\alpha \in \mathcal{A}$, *spontaneously* with rate r. During the occurrence of the action, a broadcast message, also typed α, is emitted. The *influence range* of the broadcast is defined by the location vector $\overrightarrow{\ell}$, which gives a list of locations where agents can *potentially* be influenced by this message. For example, $\overrightarrow{\ell} = range(d)$ denotes that the influence range is a set of locations whose distance from the location of the sender agent is less than a specific threshold d. Another frequently used definitions of influence range is $\overrightarrow{\ell} = local$, which represent that the influence range of the broadcast message is restricted to the location of the sender agent.

Spontaneous action with unicast message emission: $!!(\alpha, r)@\mathbf{IR}\{\overrightarrow{\ell}\}$ also describes a spontaneous action of type α, rate r and influence range $\overrightarrow{\ell}$. The difference is that here the message is a unicast, meaning that at most one agent can receive the message.

Action induced by a broadcast message: $?(\alpha, p)@\mathbf{Pr}\{v\}$ describes that the agent performs an action α *immediately* after receiving and accepting a broadcast message of type α. Whether the agent receives the broadcast message is decided by two factors. Firstly, the agent must be located within the influence range of the message; otherwise, the message will be ignored. Secondly, the value $v \in [0, 1]$ gives the probability that the message is received by the agent given that it is within the influence range of the broadcast. v can be defined dynamically. For instance, $v = 1/|S(\ell)|$ denotes that the message reception probability is dependent on the number of agents in state S in location ℓ, where $| \cdot |$ is an operator which gives the number of agents in a particular state and location. Formally, the definition of v follows this grammar:

$$v ::= c \quad \mid \quad dist(\ell_1, \ell_2) \quad \mid \quad |S(\ell)| \quad \mid \quad v \, (op) \, v$$

where c is a constant real number, $dist(\ell_1, \ell_2)$ is the distance between locations ℓ_1 and ℓ_2, (op) is a basic arithmetic operator. Once the message has been received, the agent decides whether to accept it. Here, a constant value $p \in [0, 1]$ encodes the probability that the agent will accept the message. This can be thought of as the agent choosing to respond to a spontaneous action of the given type with probability p. The definition of v and p supports a rich set of possible interaction patterns between agents.

Action induced by a unicast message: $??(\alpha, p)@\mathbf{Wt}\{v\}$ describes that the agent performs an action α immediately after receiving and accepting a unicast message of type α. Here, $v \in \mathbb{R}^+$ gives the *weight* of the agent to be the receiver of this unicast message. The definition of v follows the same grammar as previously, but with a different value domain. The weights are used to resolve between several potential receiver agents: suppose there are n agents denoted by $S_1(\ell_1), S_2(\ell_2), ..., S_n(\ell_n)$, which can potentially receive the unicast message, with weights $v_1, v_2, ..., v_n$. Then, the probability that agent $S_1(\ell_1)$ receives the message is v_1/Σ, where Σ denotes $\sum_{i=1}^{n} v_i$, the sum of the associated weights of all potential receivers. If there is no potential receiver, the message is simply discarded. The value $p \in [0,1]$ is a distinct probability deciding whether a received message is accepted or not. Note that if the selected agent does not accept the unicast message, the message is discarded; it cannot be passed to any other potential receiver agent.

Spontaneous action without message emission: (α, r) denotes that the agent performs a spontaneous action named α with a rate r governed by a negative exponential distribution. No message is sent out during the firing of this action.

Alternative behaviours are represented by the standard choice operator, $+$. A choice between spontaneous actions is resolved via the race policy, based on their corresponding rates. A choice between two induced actions of the same type within a single component is not allowed. M denotes a constant name for an agent. Compositionality is proved by the parallel operator.

5.2   Model Definition

Here we present the PALOMA model for the template city bike-sharing scenario which can be automatically parsed to a PCTMC model via the population semantics introduced in [10]. Suppose that there are n bike stations in the city, and each one has a number of available bikes and slots. Therefore, we represent the available bikes and slots in Station i (for $i = 1, ..., n$) by agents as follows:

$$Slot(\ell_i) = ??(return, 1)@\mathbf{Wt}\{1\}.Bike(\ell_i) \quad Bike(\ell_i) = ??(borrow, 1)@\mathbf{Wt}\{1\}.Slot(\ell_i)$$

Both $Slot(\ell_i)$ and $Bike(\ell_i)$ are passive. They can only be induced to make a *return* (returning a bike to this station) or *borrow* (borrowing a bike from this station) action by a unicast message, and when this happens they switch role.

The agents to represent the bike stations are defined as:

$$Station(\ell_i) = !(SlotAvailable_i, \gamma)@\mathbf{IR}\{range(1)\}.Station(\ell_i) +$$
$$!(BikeAvailable_i, \gamma)@\mathbf{IR}\{range(1)\}.Station(\ell_i)$$

A bike station performs both $BikeAvailable_i$ and $SlotAvailable_i$ self-jump spontaneous actions with broadcast message emission at the rate of γ. The influence range of the broadcast messages is defined by the function $range(d)$, which means that only agents in locations whose distance to the location of the sender station is less than d can potentially be influenced by this message.

The agents representing bike users are defined as follows:

$$Pedestrian(\ell_i) = (seekb_i, b_i).SeekBike(\ell_i) + \sum_{j \neq i}(walk_{ij}, w_{ij}).Pedestrian(\ell_j)$$

$$SeekBike(\ell_i) = \sum_{j=1}^{m}?(BikeAvailable_j, 1)@\mathbf{Pr}\{v_1\}.Walk2Station_j(\ell_i)$$

$$Walk2Station_j(\ell_i) = (W2S_{ij}, w2s_{ij}).CheckBikeNum(\ell_j)$$

$$CheckBikeNum(\ell_i) = ?(BikeAvailable_i, 1)@\mathbf{Pr}\{v_2\}.BorrowBike(\ell_i)$$

$$BorrowBike(\ell_i) = !!(borrow, o)@\mathbf{IR}\{local\}.Biker(\ell_i)$$

$$Biker(\ell_i) = (seeks_i, s_i).SeekSlot(\ell_i) + \sum_{j \neq i}(ride_{ij}, r_{ij}).Biker(\ell_j)$$

$$\cdots$$

$$ReturnBike(\ell_i) = !!(return, o)@\mathbf{IR}\{local\}.Pedestrian(\ell_i)$$

where

$$v_1 = \theta_0 + \theta_1 \frac{d - dist(\ell_i, \ell_j)}{d} + \theta_2 \frac{|Bike(\ell_j)|}{|Bike(\ell_j)| + |Slot(\ell_j)|} \tag{1}$$

$$v_2 = \frac{|Bike(\ell_i)|}{|Bike(\ell_i)| + \sigma} \tag{2}$$

As can be seen from the definition, when the user agent is in the *Pedestrian* state, it travels from location ℓ_i to location ℓ_j at the rate of w_{ij} by performing a spontaneous action $walk_{ij}$ without message emission. It may also seek a bike at the rate of b_i, and enter into the *SeekBike* state.

The user agent in the *SeekBike*(ℓ_i) state can do a *BikeAvailable$_j$* action induced by a broadcast message sent by a station agent in location ℓ_j and goes to the *Walk2Station$_j$*(ℓ_i) state, which represents that the user is walking from location ℓ_i to the bike station in location ℓ_j. The probability of receiving a bike available message from the station in location ℓ_j is defined in Equation (1). It can be interpreted as follows: the users tend to borrow a bike from a closer bike station with more available bikes, and θ_1, θ_2 are associated weights of those factors, θ_0 is the noise term (imagine that the user checks the bike numbers in nearby stations using a smart phone application). The user in the *Walk2Station$_j$*(ℓ_i) state can do a spontaneous action $W2S_{ij}$ at the rate of $w2s_{ij}$, where $1/w2s_{ij}$ is the expected time to walk from ℓ_j to the bike station in ℓ_i.

The user in the *CheckBikeNum*(ℓ_i) state can only do a *BikeAvailable$_i$* action induced by a broadcast message sent by the station in ℓ_i. The probability of receiving the message is defined in Equation (2), where σ is a very small real number to avoid a zero denominator. This ensures that the user can only go to the *BorrowBike*(ℓ_i) state if the bike station is not empty. The borrow bike action *borrow* is fired at rate o. Meanwhile, a unicast message *borrow* is sent out, and the user becomes a *Biker*.

A user agent in the *Biker* state can perform actions and become a *Pedestrian* again in a similar fashion, thus we do not give the details due to lack of space.

Finally, the initial population of agents are given in the following definition:

$$\ldots \parallel Pedestrian(\ell_i)[n_p^i] \parallel Slot(\ell_i)[n_s^i] \parallel Bike(\ell_i)[n_b^i] \parallel Station(\ell_i) \parallel \ldots$$

where $Pedestrian(\ell_i)[n_p^i]$ is syntactic sugar which represents n_p^i copies of $Pedestrian(\ell_i)$ in parallel.

6   Experiments

The usefulness of a reduction algorithm can be evaluated by the size of the reduced model (the proportion of removed agent types and transitions), the decrease of simulation time, and the error caused by the reduction. Thus, in order to do the evaluation, we simulate the bike-sharing models with and without applying the reduction algorithms. To make our experiments more thorough, we generated 50 bike-sharing models each with 30 locations. There are 50 pedestrians and a bike station which is equipped with 25 available bikes and 5 available slots initially in each location in the simulation. The topology of the locations and the value of parameters in each model are generated randomly.

We simulate each model without reduction for 500 runs. Next, we randomly pick the bike agents in 2 bike stations as our target agent types, denoted as t_1 and t_2. The two reduction algorithms are applied in the simulation of these models with different acceptable error thresholds with respect to t_1 and t_2. For each value of the acceptable error threshold, we also simulate each model for 500 runs (including the sampling runs), and compare the size of the reduced model and the decrease of simulation time with the full model without reduction. In the simulation with the reduction algorithm with fixed length sampling, we set the sampling length to 50 simulation runs. In the simulation with the reduction algorithm with flexible length sampling, we set $p0 = 0.95$, $p1 = 0.9$, $\alpha = \beta = 0.1$.

Figure 2 gives the proportional reduction of simulation time, agent types and transitions with different reduction algorithms and varying acceptable error thresholds. It can be seen that both reduction algorithms can significantly reduce the size of the model and simulation time. Observe that the larger the value of the error threshold, the more agent types, transitions and simulation time can be reduced within the model. This reflects the soundness of our reduction algorithms from another perspective. Moreover, it can be seen that the algorithm with fixed length sampling tends to remove more agent types and transitions from the simulation. The reduction in simulation time cost when applying the flexible sampling method is smaller compared with fixed length sampling method both due to a proportionally smaller reduction of agent types and transitions, and the larger computational cost of the reduction process.

Furthermore, in order to measure the error caused by reduction, we evenly sample the population of target agents at 200 time points along the simulation. The error in the population of a target agent type t at a time point i can be quantified by: $Error_{t,i} = \frac{|\overline{x_{t,i}^f} - \overline{x_{t,i}^r}|}{\overline{x_{t,i}^f}}$ where $\overline{x_{t,i}^r}$ and $\overline{x_{t,i}^f}$ are the average population of agents of type t at time point i in the 500 simulation runs with and

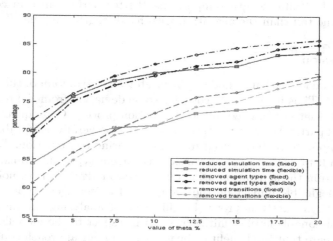

Fig. 2. The proportional reduction of simulation time, agent types and transitions (y-axis) with different reduction algorithms and acceptable error thresholds (x-axis).

Table 1. The average error caused by reduction with different algorithms and acceptable error thresholds.

Value of θ	2.5%	5%	7.5%	10%	12.5%	15%	17.5%	20%
Fixed Length Sampling	1.5%	2.8%	3.1%	3.4%	3.5%	3.9%	4.1%	4.6%
Flexible Length Sampling	1.1%	1.8%	2.9%	3.0%	3.3%	3.8%	4.1%	4.4%

without reduction. If we treat each $Error_{t,i}$ where $t \in \{t_1, t_2\}$, $i \in (1, 2, \ldots, 200)$ as an error sample, then the average error caused by a reduction algorithm in our experiments can be measured by:

$$\overline{Error_{t1,t2}} = \frac{\sum_{i=1}^{200}(Error_{t1,i} + Error_{t2,i})}{2 \times 200}$$

where $t1$, $t2$ are the target agent types in our experiments. Table 1 gives the average error caused by reduction with different algorithms and acceptable error thresholds. Table 2 shows the 95th percentiles of the error samples (95% of the error samples are below this value). We find that the average error caused by both reduction algorithms is significantly smaller than the acceptable error threshold we assign to the target agents. We can also observe that the reduction

Table 2. The 95th percentiles of error caused by reduction with different algorithms and acceptable error thresholds.

Value of θ	2.5%	5%	7.5%	10%	12.5%	15%	17.5%	20%
Fixed Length Sampling	2.8%	4.7%	6.8%	8.4%	11.9%	15.8%	16.2%	18.2%
Flexible Length Sampling	1.6%	4.3%	6.3%	7.1%	9.9%	11.6%	13.6%	16.7%

algorithm with flexible length sampling has better performance in controlling the error in the tail than the algorithm with fixed length sampling.

6.1 Discussion

As modellers we know that a model is inevitably an abstraction of the system in the real world. Thus it inevitably contains some deviation from the real system, due to details that are omitted in the abstraction process. Consequently, except for the case of particular safety critical systems, it is generally acceptable to allow some minor noise to be introduced to a model during construction. Taking this perspective a little further, we can consider the agent types and transitions that we removed from the simulation using our reduction algorithms as noise factors which have negligible impact on the evolution of target agent types. Thus our method can significantly improve the efficiency of analysing the model whilst retaining a reasonable level of faithfulness with respect to the modelled system.

We anticipate that the benefit to be gained from our approach could be particularly valuable in statistical model checking since it usually requires thousands of simulation runs in order to check whether a hypothesis holds. For example, for the bike-sharing system, suppose we want to check whether the following hypothesis holds: $\Pr(\mathbf{G}_{[0,100]} 0 < x_b < C) \geq 95\%$ where x_b is the number of bike agents in a station, C is the capacity of that station. This means we require that in the first 100 time points, the probability of the station being empty or full should be less than 5%. Thus, if we set the bike agents in that station as our target agent type, the simulation speed can be significantly boosted by removing those agent types and transitions that are loosely-coupled to the target agent type, as illustrated by the sample simulation runs presented in the previous section.

Moreover, when applying the reduction algorithm with flexible length sampling in the bike-sharing model, we find that in general more than 90% removable agent types are identified at the end of $\log_{\frac{p_1}{p_0}} \frac{\beta}{1-\alpha}$ simulation runs which is the minimal number of runs to reach the removal criterion. Thus, identifying removable agent types and transitions should in general be much quicker than reaching the criterion to accept or reject a hypothesis in statistical model checking. We plan to explore and exploit this promising application of our approach in future work.

7 Conclusion

We have presented two statistical model reduction algorithms for the stochastic simulation of PCTMC models. Both algorithms are based on investigating the coupling coefficients between agent types in the model by building a directed relation graph and applying an error propagation method to measure agent types which are not directly related. We have shown that our reduction algorithms can significantly reduce the computational cost of the simulation. Moreover, the error caused by the reduction is well-controlled by the acceptable error threshold set by the modeller. We have proposed that our reduction method could be very useful in statistical model checking for PCTMC models. We are going to investigate this idea further in the near future.

Acknowledgement. The authors would like to thank Daniël Reijsbergen, Vashti Galpin and Stephen Gilmore for their helpful comments on an earlier draft of this work. This work is supported by the EU project QUANTICOL, 600708.

References

1. Allen, L.J., Allen, E.J.: A comparison of three different stochastic population models with regard to persistence time. Theoretical Population Biology **64**(4), 439–449 (2003)
2. Calder, M., Vyshemirsky, V., Gilbert, D., Orton, R.: Analysis of Signalling Pathways Using Continuous Time Markov Chains. In: Priami, C., Plotkin, G. (eds.) Transactions on Computational Systems Biology VI. LNCS (LNBI), vol. 4220, pp. 44–67. Springer, Heidelberg (2006)
3. Cerotti, D., Gribaudo, M., Bobbio, A.: Markovian agents models for wireless sensor networks deployed in environmental protection. Rel. Eng. & Sys. Safety **130**, 149–158 (2014)
4. Tribastone, M., Gilmore, S., Hillston, J.: Scalable differential analysis of process algebra models. IEEE Transactions on Software Engineering **38**(1), 205–219 (2012)
5. Guenther, M.C., Stefanek, A., Bradley, J.T.: Moment Closures for Performance Models with Highly Non-linear Rates. In: Tribastone, M., Gilmore, S. (eds.) UKPEW 2012 and EPEW 2012. LNCS, vol. 7587, pp. 32–47. Springer, Heidelberg (2013)
6. Hillston, J.: Challenges for Quantitative Analysis of Collective Adaptive Systems. In: Abadi, M., Lluch Lafuente, A. (eds.) TGC 2013. LNCS, vol. 8358, pp. 14–21. Springer, Heidelberg (2014)
7. Lu, T., Law, C.K.: A directed relation graph method for mechanism reduction. Proceedings of the Combustion Institute **30**(1), 1333–1341 (2005)
8. Pepiot-Desjardins, P., Pitsch, H.: An efficient error-propagation-based reduction method for large chemical kinetic mechanisms. Combustion and Flame **154**(1), 67–81 (2008)
9. Niemeyer, K.E., Sung, C.J., Raju, M.P.: Skeletal mechanism generation for surrogate fuels using directed relation graph with error propagation and sensitivity analysis. Combustion and Flame **157**(9), 1760–1770 (2010)
10. Feng, C., Hillston, J.: PALOMA: A Process Algebra for Located Markovian Agents. In: Norman, G., Sanders, W. (eds.) QEST 2014. LNCS, vol. 8657, pp. 265–280. Springer, Heidelberg (2014)
11. Bortolussi, L., Hillston, J., Latella, D., Massink, M.: Continuous approximation of collective system behaviour: A tutorial. Performance Evaluation **70**(5), 317–349 (2013)
12. Gillespie, D.T.: Exact stochastic simulation of coupled chemical reactions. The Journal of Physical Chemistry **81**(25), 2340–2361 (1977)
13. Younes, H.L.S., Simmons, R.G.: Probabilistic Verification of Discrete Event Systems Using Acceptance Sampling. In: Brinksma, E., Larsen, K.G. (eds.) CAV 2002. LNCS, vol. 2404, pp. 223–235. Springer, Heidelberg (2002)

Author Index

Printed in the United States
By Bookmasters